Rochester Symposium on Developmental Psychopathology

Volume 7

ADOLESCENCE: OPPORTUNITIES AND CHALLENGES

Edited by:

Dante Cicchetti & Sheree L. Toth

Mt. Hope Family Center
University of Rochester

UNIVERSITY OF ROCHESTER PRESS

First published 1996

University of Rochester Press
34-36 Administration Building, University of Rochester
Rochester, New York, 14627, USA
and at PO Box 9, Woodbridge, Suffolk IP12 3DF, UK

ISBN 1–878822–67–5

Library of Congress Cataloging-in-Publication Data

Adolescence: Opportunities and Challenges / edited by Dante Cicchetti
& Sheree L. Toth.
    p.   cm.—(Rochester Symposium on Developmental
Psychopathology, ISSN 1065–6511 ; v. 7)
  Includes bibliographical references and indexes.
  ISBN 1–878822–67–5 (alk. paper)
  1. Adolescent psychopathology—Congresses.   2. Adolescent
psychology—Congresses.   I. Cicchetti, Dante.   II. Toth, Sheree L.
III. Series : Rochester Symposium on Developmental Psychopathology
(Series) : v. 7.
RJ503.A312   1996
816.89′022—dc20   96–28192
                                                                   CIP

British Library Cataloguing-in-Publication Data

A catalogue record for this book is
available from the British Library

This publication is printed on acid-free paper
Printed in the United States of America

Table of Contents

List of Contributors

JAY BELSKY, PH.D.
    The Pennsylvania State University, University Park, PA 16802
JACK BLOCK, PH.D.
    University of California at Berkeley, Berkeley, CA 94720
DAVID A. BRENT, M.D.
    Western Psychiatric Institute and Clinic, University of Pittsburgh,
    Pittsburgh,
RONALD E. DAHL, M.D.
    Western Psychiatric Institute and Clinic, University of Pittsburgh School of
    Medicine, Pittsburgh, PA 15213
DAVIDO DUPREE, PH.D.
    Graduate School of Education, University of Pennsylvania, Philadelphia, PA
    19104
JACQUELYNNE S. ECCLES, PH.D.
    Institute for Social Research, University of Michigan, Ann Arbor, MI 48106
CYNTHIA T. GARCÍA COLL, PH.D.
    Department of Education, Brown University, Providence, RI 02912
PER F. GJERDE, PH.D.
    University of California at Santa Cruz, Santa Cruz, CA 95064
DANIEL P. KEATING, PH.D.
    Ontario Institute for Studies in Education and Canadian Institute for
    Advanced Research, Toronto, Ontario, CANADA M5S 1V6
ROBERT KEGAN, PH.D.
    The Massachusetts School of Professional Psychology and The Clinical
    Development Institute, Harvard University, Cambridge, MA 02138
CATHERINE LORD, PH.D.
    The University of Chicago, Chicago, IL 60637
SARAH E. LORD, PH.D.
    Institute for Social Research, University of Michigan, Ann Arbor, MI 48106
DARLA J. MACLEAN, PH.D.
    Brock University, St. Catherines, Ontario, CANADA L2S 3A1
GRACE MORITZ, A.C.S.W.
    Western Psychiatric Institute and Clinic, University of Pittsburgh, Pitts-
    burgh, PA 15213

ROBERT W. ROESER, PH.D.
    Institute for Social Research, University of Michigan, Ann Arbor, MI 48106
NEAL D. RYAN, M.D.
    Western Psychiatric Institute and Clinic, University of Pittsburgh School of
    Medicine, Pittsburgh, PA 15213
JUDITH G. SMETANA, PH.D.
    Warner Graduate School of Education and Human Development, University
    of Rochester, Rochester, NY 14627
MARGARET BEALE SPENCER, PH.D.
    Graduate School of Education, University of Pennsylvania, Philadelphia, PA
    19104
LAURENCE STEINBERG, PH.D.
    Temple University, Philadelphia, PA 19122
HEIDIE A. VÁZQUEZ GARCÍA, B.A.
    Department of Education, Brown University, Providence, RI 02912

Preface

Adolescence, defined as the period generally encompassing the second decade of life, has not been accorded the attention that other periods in the life span have received (Takanishi, 1995). Interestingly, the first article in the *Annual Review of Psychology* devoted exclusively to adolescence did not appear until the 1988 issue (Petersen, 1988), almost 40 years after the first issue was published! Moreover, in the history of this important compendium, only two chapters have been devoted exclusively to adolescence, the second appearing in 1995 (Compas, Hinden, & Gerhardt, 1995).

The reasons for the relative paucity of research on adolescence as compared to other developmental periods can be attributed to a number of factors, including: the belief that adolescence is a transitional period and therefore does not warrant significant attention, the conceptualization that the perturbations in adjustment problems during adolescence are age- and stage-specific and therefore not likely to persist, and concerns related to the methodological challenges posed by the period of adolescence that stem from the rapid changes in physical, biological, cognitive, social-cognitive, and interpersonal development (see, for example, Brooks-Gunn & Reiter, 1990; Furstenberg, 1990; Harter, 1990; Keating, 1990; Steinberg, 1990). The pervasive and long-standing belief in the importance of early experience, as well as the generally negative view of adolescents as difficult individuals who do not require research attention because even lay people understand this period of development (Brooks-Gunn & Petersen, 1984; Petersen, 1988), have further contributed to the reduced attention accorded adolescence. Thus, scientists interested in adolescent development needed to develop sound research paradigms, as well as to convince their colleagues and the field of the importance of investigating this period of development before progress in understanding adolescence could occur.

Historically, adolescence has most typically been described as a period of life characterized by turmoil, including emotional upheaval, conflict with parents, alienation, and identity crisis (Erikson, 1950, 1968; Weiner, 1992). In fact, since its introduction into the literature at the turn of the century, the turmoil hypothesis (Hall, 1904) has permeated theoretical accounts of adolescent development and has been used as an explanatory construct for what was considered to be an adolescent propensity for deviant and psychopathological behavior (Strober, 1986). Most generally, the turmoil hypothesis purports that conflict and regression are common, if not inescapable, components of adolescence that are a function of the processes initiated during puberty.

Although slow to emerge relative to research on children, results obtained from epidemiological studies have failed to confirm the premises embodied in the turmoil hypothesis. Rather, investigations have found that although there is a moderately high prevalence of symptoms such as dysphoria, irritability, and alienation during adolescence, the frequency of persistently disruptive pathological behaviors is generally low in nonclinical samples and limited to a minority of adolescents who most often evidenced maladjustment prior to the onset of puberty (Strober, 1986).

Findings such as these have helped to shift research on adolescence from a sole focus on risk, to a broader view of adolescence as a time of heightened developmental change. Although adolescence involves many challenges, research has shown that individuals negotiate this period of development with varying degrees of success. For example, Petersen (1988) reports that approximately 11% of young adolescents experience serious chronic difficulties, while 32% manifest only intermittent and situationally based difficulties. In a review of major studies on child and adolescent psychiatric epidemiology, Costello and Angold (1995) conclude that prevalence rates for child and adolescent psychiatric disorder in the range of 10% to 12% have been obtained based on single informant reports. When combining information from different sources, rates of disorder have been found to be in the region of 20%, with adolescents typically reporting more disorders than children (Costello & Angold, 1995).

As the twentieth century draws to a close, professionals as well as the lay public have modified their views of adolescence as a time of inherent stress to one that more accurately reflects the variability in outcome that occurs during this period of development. As such, adolescence can best be conceptualized as a period of heightened risk, as well as a time that embodies opportunities for growth and positive development. In fact, a proliferation over the last twenty years in the number of journals devoted to adolescence, the establishment of specialized research societies and sections, and the number of policy reports addressing the period of adolescence have done much to contribute to a more accurate portrayal of the challenges, as well as of the opportunities that accompany the second decade of life.

On a more sobering note, despite advances in our understanding of adolescence, the 1990s have posed new obstacles that place the adolescent at greater risk than that experienced by any prior generations (Takanishi, 1993). In recent years, more adolescents have been found to experiment with illegal substances prior to the age of 15 (Gans & Blyth, 1990; Jessor, Donovan, & Costa, 1991). Adolescents in the U.S. also are exposed to inordinate amounts of violence (Hammond & Yung, 1993) and are 15 times more likely than English youth to be victims of homicide (Children's Safety Network, 1991). Of note, homicide deaths among African-American males between 15 and 19 years of age increased 111% between 1985 and 1990. Rates of pregnancy among adolescents also have shown marked increases, and adolescent sexual activity now is accompanied by the threat of HIV (Takanishi, 1993). Mental illness also is a significant problem for adolescents. For example, between 7% and

33% of adolescents experience depression (Petersen, Compas, Brooks-Gunn, Stemmler, & Grant, 1993). Moreover, both crime prevalence and incidence peak at approximately age 17 (Moffitt, 1993), placing adolescents in positions to be affected by the consequences of antisocial behavior.

In view of the risks facing adolescents, attributable to intrinsic biological/developmental changes as well as to more external social and situational challenges that they must confront, it is not surprising that the study of adolescent development holds much promise for advancing many of the goals embodied by a developmental psychopathology perspective. A number of issues that have been examined by investigators of adolescent development mirror areas of interest to developmental psychopathologists. These include: 1) the use of interdisciplinary models of development; 2) issues related to the continuity/discontinuity of development across the life course; 3) the boundary and linkages between normal and atypical functioning; 4) the transactions that occur between environmental and more person-specific characteristics; 5) processes associated with risk and resilience; and 6) the integration of basic research into the design and provision of prevention and intervention. In order to place the current volume on adolescence within a developmental psychopathology context, each of these issues and corresponding relevant work on adolescence will be addressed briefly.

Proponents of a developmental psychopathology perspective have consistently advocated the importance of incorporating interdisciplinary perspectives into the study of adaptation and maladaptation across the lifespan (Cicchetti, 1984, 1990, 1993; Cicchetti & Toth, 1991). The utilization of an overarching framework that draws upon the expertise of developmentalists of diverse disciplines holds the greatest potential for elucidating development in multiple domains of functioning and as such is relevant to the study of adolescence (Cicchetti & Cohen, 1995 a, b). Although early conceptualizations of the psychology of adolescence relied on main effects models of development that sought to understand adolescent development as a direct result of either intrapsychic (e.g., Freud, 1958), biological, or environmental factors, recent initiatives increasingly have incorporated more interdisciplinary and transactional models of development (Compas et al., 1995). For example, biopsychosocial models, which recognize the interrelatedness among physical and biological development, cognitive processes, and the social environment (cf. Engel, 1977), have been proffered (e.g., Brooks-Gunn, 1987). Perhaps the most formal call for an interdisciplinary approach to the study of adolescence has been made by Jessor's elucidation of a "developmental behavioral science" (1991, 1992, 1993). The tenets of a developmental behavioral science articulated by Jessor, including the integration of traditionally distinct scientific disciplines, as well as of areas of basic and applied research, are consistent with principles elucidated by proponents of a developmental psychopathology approach to understanding adaptation (Cicchetti, 1984, 1990; Sroufe & Rutter, 1984).

Attention to continuity/discontinuity of development has a long-standing history in investigations conceived within the field of developmental psychopathology (Caspi & Moffitt, 1995; Kohlberg, LaCrosse, & Ricks, 1972; Rutter, 1989, 1995;

Sroufe & Rutter, 1984). Increasingly, investigators of adolescent development have focused their attention on understanding individual differences in the pathways of development from childhood through adolescence and adulthood (Block, 1971, 1993; Cairns & Cairns, 1994; Moffitt, 1993; Rutter, 1989; Sroufe, Carlson & Shulman, 1993). Specifically, research on normal and atypical development during adolescence has identified a number of developmental trajectories. These have been described as including 1) stable adaptive functioning, 2) stable maladaptive functioning, 3) adolescent turnaround or recovery, 4) adolescent decline, and 5) temporary deviation or maladaptation during adolescence (see Compas et al., 1995). Empirical support for these pathways has been derived from a number of investigations, and is perhaps most clearly articulated in Moffitt's (1993) discrimination between adolescence-limited and life-course-persistent antisocial behavior. In general, pathways that involve change during adolescence have been found to be relatively more rare than pathways characterized by stability (Compas et al., 1995).

Because adolescence involves a time of rapid change that requires effective coping if the individual is to function adaptively, research on adolescence has the potential to illuminate theories of development more broadly (Petersen, 1988). With respect to adolescence, investigators have noted the interrelatedness of both healthy and maladaptive behavioral outcomes. For example, connections have been noted in a set of problem behaviors such as substance abuse, delinquency, and drunk driving that reflect a health-compromising lifestyle (Elliott, 1993; Windle, Miller-Tutzauer, & Domenico, 1992). By understanding the effects of the transformations of adolescence on adaptive as well as on maladaptive outcome, theories of development can be examined and evaluated in their full complexity.

In view of the rapid changes in biology that accompany the onset of puberty, as well as the interaction between biology and the adolescent's stimulus value and response to the social environment, adolescence serves as an ideal period in which to examine person/environment transactions. The importance of integrating the study of biology and psychology has been increasingly recognized as a goal of developmental psychopathology (Cicchetti & Cohen, 1995 a,b; Cicchetti & Toth, 1991; Cicchetti & Tucker, 1994). By studying the behavioral correlates of puberty in varied contexts, the investigation of physical characteristic-environment interactions becomes possible (Brooks-Gunn, 1988). For example, Simmons and her colleagues have found that the effects of body image attributable to being an early-maturing sixth grade girl depend on whether the girl attends an elementary or a middle school (Blyth, Simmons, & Zakin, 1985; Simmons, Blythe, & McKinney, 1983). In a similar explication of the interface between physical maturation and the environment, Magnusson, Stattin, and Allen, (1985) found that although early-maturing girls were more likely to drink and to be sexually active than later-maturing girls, the effect was attributable to early-maturing girls who had older friends. Finally, adaptation may be affected by choices of environment made by youngsters prior to the onset of puberty. For example, it has been hypothesized that girls involved in activities such as ballet dancing that devalue pubertal growth may be more adversely affected by early-maturation than girls engaged in activities in which success is not linked to body type (Brooks-Gunn, 1988).

In accord with an overarching interest embodied by developmental psychopathologists in individuals who succeed in the face of adversity, as well as in those who succumb to challenges (Cicchetti & Garmezy, 1993; Rutter, 1990), a recognition of individual differences in pathways of adolescent development is related to an interest in those factors that place adolescents at risk as well as in those factors that may protect them from maladaptive development (Anthony & Cohler, 1987; Garmezy, 1985; Haggerty, Sherrod, Garmezy, & Rutter, 1994). In this regard, the long-term impact of negative life events such as parental divorce, illness, or economic disruption on adolescent development has been found to depend on the effects of the more chronic disruptions associated with these stressors (Compas, 1994).

Finally, the importance accorded to the interface between research and practice by developmental psychopathologists (Cicchetti & Toth, 1992) also is relevant to those invested in promoting positive adolescent development. The transitional nature and disequilibrium of adolescence represents an opportune period for intervention, as times of developmental change may result in a greater receptivity to intervention (Cicchetti & Toth, 1992; Toth & Cicchetti, 1993). Interventions can be divided into prevention and treatment efforts. A number of reviews on prevention attest to the positive outcomes that programs can achieve (Goldston, Yager, Heinicke, & Pynoos, 1990; Institute of Medicine, 1994; Price, Cowen, Lorion, & Ramos-McKay, 1988; Weissberg, Caplan, & Harwood, 1991). In general, there is a consensus that effective programs are likely to stem from an understanding of causes and risk factors that lead to a given problem, to draw on theories and models of human behavior that direct the focus of the intervention, and to monitor intervention implementation carefully (Kazdin, 1993).

A number of issues have been delimited that pose challenges to prevention efforts. For example, although protective factors can be usefully incorporated into the development of preventive interventions, mechanisms that increase resilience are not as well understood as are those that increase risk (Kazdin, 1993; Rutter, 1990). Thus, in order to be optimally effective in informing prevention programs for adolescents, issues such as how focusing on the promotion of competence versus reducing the presence of risks may improve adaptive functioning must be better understood. Most importantly, the development of effective preventive and treatment interventions must rely upon an effective interplay between basic and applied research (Institute of Medicine, 1994; Kazdin, 1993). In order to be optimally effective, interventions for adolescents require a knowledge of developmental trajectories that may lead to maladaptive behavior or psychopathology, as well as to adaptive functioning and to positive mental health.

As this brief review demonstrates, research on adolescent development and psychopathology has burgeoned in recent years. Rather than striving to cover all issues relevant to adolescent risk, psychopathology, and resilience, contributors to this volume were asked to prepare chapters that would be helpful in elucidating what we consider to be a hallmark of adolescence and developmental psychopathology, namely, adolescence as a period encompassed by both risk and opportunity.

The contributors to this volume have examined an array of issues with relevance to adolescent development, including conflict with parents, school functioning, sexuality, suicide, and psychopathology. Cultural considerations also have been addressed, and factors that promote competence as well as hamper positive outcome have been identified. This volume chronicles the many advances that have transpired in our understanding of adolescent development and psychopathology, and provides an agenda for future work in this area. We look forward with great anticipation to the growing contributions that research on adolescence will hold for the discipline of developmental psychopathology.

Dante Cicchetti
Sheree L. Toth

REFERENCES

Anthony, E. J., & Cohler, B. J. (Eds.) (1987). *The invulnerable child.* New York, NY: Guilford Press.

Block, J. (1971). *Lives through time.* Berkeley, CA: Bancroft Books.

Block, J. (1993). Studying personality the long way. In D. Funder, R. Parke, C. Tomlinson-Keasey, and K. Widaman (Eds.), *Studying lives through Time: Personality and development* (pp 9-41). Washington, DC: American Psychological Association.

Blythe, D., Simmons, R., & Zakin, D. (1985). Satisfaction with body image for early adolescent females: The impact of pubertal timing within different school environments. *Journal of Youth and Adolescence, 14,* 207-225.

Brooks-Gunn, J. (1987). Pubertal processes and girls' psychological adaptation. In R. M. Lerner and T. T. Foch (Eds.), *Biological-psychosocial interactions in early adolescence: A lifespan perspective* (pp. 123-153). Hillsdale, NJ: Erlbaum Associates.

Brooks-Gunn, J. (1988). Commentary: Developmental issues in the transition to early adolescence. In M. R. Gunnar and W. A. Collins (Eds.) *Development during the transition to adolescence: The Minnesota Symposia on Child Psychology, Vol. 21* (pp. 189-208). Hillsdale, NJ: Erlbaum Associates.

Brooks-Gunn, J., & Petersen, A. C. (1984). Problems in studying and defining pubertal events. *Journal of Youth and Adolescence, 13,* 181-196.

Brooks-Gunn, J. & Reiter, E. O. (1990). The role of pubertal processes. In S. S. Feldman and G. R. Elliott (Eds.) *At the threshold: The developing adolescent* (pp. 16-53). Cambridge, MA: Harvard University Press.

Cairns, R.B., & Cairns, B.D. (1994) *Lifelines and risks: Pathways of youth in our time.* New York: Cambridge University Press.

Caspi, A., & Moffitt, T. (1995). The continuity of maladaptive behavior: From description to understanding in the study of antisocial behavior. In D. Cicchetti and D. J. Cohen (Eds.) *Developmental Psychopathology, Vol. 1: Theory and Methods* (pp. 472-511). New York: Wiley.

Children's Safety Network (1991). *A data book of child and adolescent injury.* Washington, DC: National Center for Education in Maternal and Child Health

Cicchetti, D. (1984). The emergence of developmental psychopathology. *Child Development*, 55, 1-7.

Cicchetti, D. (1990). An historical perspective on the discipline of developmental psychopathology. In J. Rolf, A. Masten, D. Cicchetti, K. Nuechterlein, and S. Weintraub (Eds.), *Risk and protective factors in the development of psychopathology* (pp. 2-28). New York: Cambridge University Press.

Cicchetti, D. (1993). Developmental psychopathology: Reactions, reflections, projections. *Developmental Review*, 13, 471-502.

Cicchetti, D. & Cohen, D. (Eds.) (1995a). *Developmental Psychopathology, Vol. 1: Theory and Method*. New York: Wiley.

Cicchetti, D. & Cohen, D. (Eds.) (1995b). *Developmental Psychopathology, Vol. 2: Risk, disorder, and adaptation*. New York: Wiley

Cicchetti, D., & Garmezy, N. (1993) (Eds.). Special Issue: Milestones in the development of resilience. *Development and Psychopathology*, 5(4), 497-774.

Cicchetti, D. & Toth, S. L. (1991). The making of a developmental psychopathologist. In D. Cantor, C. Spiker, and L. Lipsitt (Eds.), *Child behavior and development: Training for diversity* (pp. 34-72). Norwood, NJ: Ablex.

Cicchetti, D. & Toth, S. L. (1992). The role of developmental theory in prevention and intervention. *Development and Psychopathology*, 4, 489-493.

Cicchetti, D., & Tucker, D. (1994). Development and self-regulatory structures of the mind. *Development and Psychopathology*, 6, 533-549.

Compas, B. E. (1994). Promoting successful coping during adolescence. In M. Rutter (Ed.) *Psychosocial disturbances in young people: Challenges for prevention*. New York: Cambridge University Press.

Compas, B. E., Hinden, B. R., & Gerhardt, C.A. (1995). Adolescent development: Pathways and processes of risk and resilience. *Annual Review of Psychology*, 46, 265-293.

Costello, E.J., & Angold, A. (1995). Developmental epidemiology. In D. Cicchetti and D. Cohen (Eds.) *Developmental Psychopathology, Vol. 2: Risk, disorder and adaptation* (pp. 23-56). New York: Wiley.

Engel, G. (1977). The need for a new medical model: A challenge for biomedicine. *Science*, 196, 129-135.

Erikson, E. H. (1950). *Childhood and society*. New York: Norton.

Erikson, E. H. (1968). *Identity: Youth and crisis*. New York: Norton.

Freud, A. (1958). Adolescence. In R. Eissler, A. Freud, H. Hartman, and M. Kris (Eds.), *Psychoanalytic study of the child, Vol 13* (pp. 255-278). New York: International Universities Press.

Furstenberg, F. F. (1990). Coming of age in a changing family system. In S. S. Feldman and G. R. Elliott (Eds.) *At the threshold: The developing adolescent* (pp. 147-170). Cambridge, MA: Harvard University Press.

Gans, J. E., & Blyth, D. A. (1990). *America's adolescents: How healthy are they?* (AMA Profiles of Adolescent Health series). Chicago: American Medical Association.

Garmezy, N. (1985). Stress-resistant children: The search for protective factors. In J. Stevenson (Ed.), *Recent research in developmental psychology* (pp. 213-233). Oxford: Pergamon Press (a book supplement to the *Journal of Child Psychology and Psychiatry, Number 4*).

Goldston, S. E., Yager, J., Heinicke, C. M., & Pynoos, R. S. (Eds.). (1990). *Preventing mental health disturbances in childhood*. Washington, DC: American Psychiatric Press.

Haggerty, R. J., Sherrod, L., Garmezy, N., & Rutter, M. (Eds.) (1994). *Risk and resilience in children: Developmental approaches*. New York: Cambridge University Press.

Hall, G. S. (1904). *Adolescence: Its psychology and its relations to physiology, anthropology, sociology, sex, crime, religion, and education.* Englewood Cliffs, NJ: Prentice-Hall.

Hammond, W. R., & Yung, B. (1993). Psychology's role in the public health response to assaultive violence among young African-American men. *American Psychologist, 48,* 142-154.

Harter, S. (1990). Self and identity development. In S. S. Feldman and G. R. Elliott (Eds.) *At the threshold: The developing adolescent* (pp. 352-387). Cambridge, MA: Harvard University Press.

Institute of Medicine (1994). *Reducing risks for mental disorders: Frontiers for preventive intervention research.* Washington, D.C.: National Academy Press.

Jessor, R. (1991). Behavioral science: An emerging paradigm for social inquiry? In R. Jessor (Ed.) *Perspectives on behavioral science: The Colorado lectures* (pp. 309-316). Boulder, CO: Westview.

Jessor, R. (1992). Risk behavior in adolescence: A psychosocial framework for understanding and action. *Developmental Review, 12,* 374-390.

Jessor, R. (1993). Successful adolescent development among youth in high-risk settings. *American Psychologist, 48,* 117-126.

Jessor, R., Donovan, J. E., & Costa, F. M. (1991). *Beyond adolescence: Problem behavior and young adult development.* New York: Cambridge University Press.

Kazdin, A. E. (1993). Adolescent Mental Health: Prevention and treatment Programs. *American Psychologist, 48,* 127-141.

Keating, D. P. (1990). Adolescent thinking. In S. S. Feldman and G. R. Elliott (Eds.) *At the threshold: The developing adolescent* (pp. 54-89). Cambridge, MA: Harvard University Press.

Kohlberg, L., LaCrosse, J. & Ricks, D. (1972). The predictability of adult mental health from child behavior. In B. Wolman (Ed.), *Manual of Child Psychopathology* (pp. 1217-1284) New York: McGraw-Hill.

Magnusson, D., Stattin, H., & Allen, V.L. (1985). A longitudinal study of some adjustment processes from mid-adolescence to adulthood. *Journal of Youth and Adolescence, 14,* 267-283.

Moffitt, T. E. (1993). Adolescence-limited and life-course-persistent antisocial behavior: A developmental taxonomy. *Psychological Review, 100,* 674-701.

Petersen, A. C. (1988). Adolescent development. *Annual Review of Psychology, 38,* 583-607.

Petersen, A. C., Compas, B. E., Brooks-Gunn, J., Stemmler, M., Ey, S., Grant, K. E. (1993). Depression during adolescence. *American Psychologist, 48,* 155-168.

Price, R., Cowen, E., Lorion, R., & Ramos-McKay, J. (Eds.) (1988). *Fourteen ounces of prevention: A casebook for practitioners.* Washington, DC: American Psychological Association.

Rutter, M. (1989). Pathways from childhood to adult life. *Journal of Child Psychology and Psychiatry, 30,* 23-51.

Rutter, M. (1990). Psychosocial resilience and protective mechanisms. In J. Rolf, A.S. Masten, D. Cicchetti, K.H. Nuechterlein, and S. Weintraub (Eds.) *Risk and Protective Factors in the Development of Psychopathology* (pp. 181-214). New York: Cambridge University Press.

Rutter, M. (1995). Relationships between mental disorders in childhood and adulthood. *Acta Psychiatrica Scandinavica, 91,* 73-85.

Simmons, R., Blythe, D., & McKinney, K. (1983). The social and psychological effects of puberty on white females. In J. Brooks-Gunn and A. Petersen (Eds.), *Girls at puberty: Biological and psychosocial perspectives* (pp. 229-272). New York: Plenum.

Sroufe, L. A., Carlson, E., & Shulman, S. (1993). Individuals in relationships: From infancy through adolescence. In D. Funder, R. Parke, C. Tomlinson-Keasey, and K. Widaman (Eds.) *Studying lives through time: Personality and development* (pp. 315-342). Washington, DC: American Pscyhological Association.

Sroufe, L. A. & Rutter, M. (1984). The domain of developmental psychopathology. *Child Development, 55,* 17-29.

Steinberg, L. (1990). Autonomy, conflict, and harmony. In S. S. Feldman and G. R. Elliott (Eds.) *At the threshold: The developing adolescent* (pp. 255-276). Cambridge, MA: Harvard University Press.

Strober, M. (1986). Psychopathology in adolescence revisited. *Clinical Psychology Review, 6,* 199-209.

Takanishi, R. (1993). The opportunities of adolescence - Research, interventions, and policy. *American Psychologist, 48,* 85-87.

Takanishi, R. (1995). Reframing adolescence: From pessimism to possibility. *Contemporary Psychology, 40*(10), 957-958.

Toth, S. L. & Cicchetti, D. (1993). Child Maltreatment: Where do we go from here in our treatment of victims? In D. Cicchetti & S.L. Toth (Eds.) *Child Abuse, Child Development, and Social Policy* (pp. 399-438). Norwood, NJ: Ablex.

Weiner, I. B. (1992). *Psychological disturbance in adolescence* (Second Edition). New York: Wiley.

Weissberg, R. P., Caplan, M., & Harwood, R. L. (1991). Promoting competent young people in competence-enhancing environments: A systems-based perspective on primary prevention. *Journal of Consulting and Clinical Psychology, 59,* 830-841.

Windle, M., Miller-Tutzauer, C., & Domenico, D. (1992). Alcohol use, suicidal behavior, and risky activities among adolescents. *Journal of Research on Adolescence, 2,* 317-330.

1   Adolescent-Parent Conflict: Implications for Adaptive and Maladaptive Development

Judith G. Smetana

Although adolescent-parent conflict has been a topic of consistent interest to theorists and researchers studying the second decade of life, researchers' views of conflict themselves have been conflicting! Whereas some theorists have proposed that conflict between parents and adolescents is normative and the absence of conflict suggestive of disturbance, others have argued that greater conflict is associated with disturbed family relations and problem behaviors in adolescence. Although many researchers now assert that moderate conflict may be adaptive for adolescent development and relationship change (Collins, 1990; Hill, 1987, 1988; Powers, Hauser, & Kilner, 1989; Smetana, 1994), little research has focused on the characteristics of conflict that are functional or dysfunctional for adolescent development.

One strategy for such an endeavor, exemplified by the early work of Freud, would be to construct a theory of normal adolescent-parent relations from studies of adolescent psychopathology. Another strategy, espoused more recently by researchers in the growing field of developmental psychopathology, would be to take a bidirectional approach wherein studies of psychopathology are employed to illuminate normal adolescent development and studies of normal development are used to identify children and adolescents at risk for developmental disturbances and problem behaviors (Cicchetti, 1993).

The strategy adopted in this chapter differs from both of these approaches. In this chapter, variations in adolescent-parent conflict and their correlates in parenting, family interaction, and adolescent development across several samples of normal families are examined. As described in greater detail in subsequent sections, the analyses reveal that these normal families can be characterized by three patterns of adolescent-parent conflict, one of which appears to be most prototypic of adolescent-parent relations, one of which, it will be asserted, appears to be most optimal for adolescent development and relationship change, and the third of which lies at the boundary between normal and pathological family functioning. An examination of parenting practices and family interactions in these families illuminates factors that place adolescents at risk for problem behaviors and psychopathology and also reveals the murky boundaries between normal and atypical development.

I am indebted to Julia Smith and Christyn Dundorf for their assistance with the cluster analyses and to Ayala Gabriel, Barbara Ilardi, Melanie Killen, Larry Nucci, and Maria Santiago Nucci for their comments on an earlier draft.

To provide a context for these arguments, the first section of the chapter presents an overview of different theoretical views of conflict and related research. Then, a social-cognitive perspective that examines different family members' interpretations of conflicts is elaborated. Next, different patterns of conflict in normal families are identified from data drawn from three previously published studies of adolescent-parent conflict, and differences in parenting, parent-adolescent interactions, and adolescent development among these families are discussed, first through quantitative analyses and then through qualitative analyses. In the final section, the implications of these findings for risk and optimal development during adolescence are discussed.

Changing Views of Adolescent-Parent Conflict

Theorists' and researchers' views of normal family relations during adolescence have swung between two extremes. At one extreme is the early view espoused by neopsychoanalytic theorists (e.g., Blos, 1962, 1979; A. Freud, 1937, 1958). According to this view, adolescence is a period of normative developmental disturbance. Healthy adolescence is seen to entail rebellion, storm and stress, turbulent adolescent-parent conflict, and ultimately, detachment from parents. The source of this developmental disturbance, for these theorists, was located in the biological changes of puberty. Pubertal maturation was thought to lead to a resurgence of sexual impulses, accompanied by a reawakening of Oedipal desires that lay dormant during the earlier latency period. The resurgence of these impulses was described as leading to intrapsychic storm and stress, which is expressed as rebellion and conflict with parents. To cope with this stress, these theorists posited that adolescents detach from parents and channel their libidinal impulses in the form of sexual relations with peers. Thus, adolescent-parent conflict was seen as facilitating both adolescent individuation and the development of sexual attachments in late adolescence and young adulthood. Furthermore, conflict and rebellion were seen as normative; those who experience a conflict-free adolescence were described as immature, delaying the developmental tasks of adolescence, and in need of psychological attention. In describing adolescence as a developmental disturbance, this perspective emphasized the fundamental discontinuity between childhood and adolescence. Even-tempered and agreeable children with close relationships to parents were expected to develop into moody, unpredictable, and conflictful adolescents.

This destructive view of conflict is supported in studies comparing normal families with clinically referred families or families of adolescents experiencing behavior problems. For instance, studies indicate higher rates of conflict (as well as greater dominance and lower communicative clarity) in clinically referred than normal families (Jacob, 1975; Krinsley & Bry, 1992; Prinz, Foster, Kent, & O'Leary, 1979; Robin & Foster, 1989). High levels of adolescent-parent conflict also have been found to predict adolescent externalizing behaviors, including marijuana and alcohol use (Baer, Garmezy, McLaughlin, Pokorny, & Wernick, 1987; Kandel, Kessler, & Margulies, 1978), delinquency (Borduin, Pruitt, & Henggeler, 1986; Forehand, Long, & Hendrick, 1987; Patterson, 1982; Patterson & Bank, 1989; Patterson, Capaldi, & Bank, 1991), and premarital sexual relations (Inazu & Fox, 1980), as well as retrospective accounts of running away from home (Adams, Gullotta, & Clancy, 1985). Conflictful or disrupted adolescent-parent relations also have been implicated in such internalizing behaviors as depression (Kashani, Burbach, & Rosenberg, 1988; Puig-Antich, Kaufman, Ryan, & Williamson, 1993; Rutter, Graham, Chadwick, & Yule, 1976) and suicide attempts (Corder, Shorr, & Corder, 1974; Tishler, McKenry, & Morgan, 1981).

Although the view of conflict as normative painted a disturbing portrait of adolescent development in the family (and continues to have an enormous impact on popular views of adolescence), it was based primarily on clinical samples. Empirical research on much more representative samples of adolescents (e.g., Douvan & Adelson, 1966; Kandel & Lesser, 1972; Offer, 1969; Rutter et al., 1976), conducted largely in response to the psychoanalytic view, indicated that for the majority of adolescents, conflict and rebellion were the exception. These studies suggested, in contrast, that close, warm, and supportive family relations during adolescence were the norm. Indeed, these studies were remarkably consistent in demonstrating that among nonclinic families, only 5% to 20% of adolescents experienced emotional turmoil and conflictful relations with parents (Montemayor, 1986; Offer, 1969; Offer, Ostrov, & Howard, 1981; Rutter et al., 1976). Furthermore, many of the adolescents experiencing behavioral problems during adolescence were found to have problems and poor relationships with parents prior to adolescence (Rutter et al., 1976). Thus, the view of parent-adolescent relationships as conflictful and rebellious and development as discontinuous was replaced in the 1960s and 1970s by the view that adolescent-parent relations were normatively close, warm, and supportive, and development in the second decade of life was seen as continuous with childhood.

These classic studies of adolescent-parent conflict in normal families were conducted using questionnaires or interviews and employed global assessment of family closeness, intergenerational tension, or independence. As others (Silverberg, Tennenbaum, & Jacob, 1992; Steinberg, 1987) have noted, they did not assess actual family interactions, nor did they obtain detailed accounts of daily life. These methodological concerns may have lead these researchers to underestimate the prevalence of conflict during adolescence and have lead to further research employing a variety of different and more sophisticated methods, including interviews, questionnaires, standardized scales, and observational techniques (Hill, 1987; Silverberg et al., 1992). The result has been a more moderate perspective on adolescent-parent conflict. Researchers now recognize that emotional bonds between parents and children are maintained during adolescence. However, research also indicates that the quality of adolescents' and parents' relationships and the amount of time they spend together decline in early adolescence (Csikszentmihalyi & Larson, 1984; Montemayor, 1982; Smetana, 1994; Youniss & Smollar, 1985).

Furthermore, research indicates that a significant proportion of American adolescents and their parents experience minor but persistent bickering, squabbling, and conflict over the everyday details of family life, including doing the chores, doing schoolwork, choice of activities and social life, and relationships with friends (for reviews, see Collins, 1990; Hill, 1987; Hill & Holmbeck, 1987; Laursen & Collins, 1994; Montemayor, 1983, 1986; Paikoff & Brooks-Gunn, 1991; Silverberg et al., 1992; Steinberg, 1990). In this more pacific view, conflict is conceptualized as a temporary perturbation leading to transformations or realignments in parent-child relationships (Collins, 1990; Hill, 1987, 1988; Laursen & Collins, 1994; Silverberg et al., 1992; Steinberg, 1990). Furthermore, harkening back to earlier psychoanalytic views, conflict is seen as functional to adolescent development and relationship change (Collins & Laursen, 1992; Grotevant & Cooper, 1985, 1986; Hill, 1988; Smetana, 1988a, 1994). For instance, Hill and his colleagues (Hill, 1988; Hill & Holmbeck, 1987) have asserted that increases in conflict are adaptive because they change the power balance of the family by making disagreements overt, which, in turn, leads to readjustments in family relationships.

In addition, although the interpretation of the research on adolescent problem behaviors has been that conflict leads to deviant behavior, a recent structural model analysis of longitudinal data suggests that relations between adolescent-parent conflict and adolescent behavior problems are bidirectional (Maggs & Galambos, 1993). That is, consistent with other research demonstrating child effects on parent-child relations (Bell, 1968), adolescent-parent conflict was found to predict problem behaviors, but adolescent problem behaviors also predicted subsequent conflict with parents, controlling for prior levels of conflict. Although severe or frequent conflict may adversely affect adolescent development, adolescent behavior problems also affect family functioning.

To summarize, prior research has varied in the extent to which adolescent-parent conflict has been viewed as normative for adolescent development. In the earlier half of the decade, conflict and turbulence in adolescent-parent relations were viewed as normative and necessary for healthy development. This view was replaced in the 1960s and 1970s with the view that conflict was dysfunctional for development. The present view is that moderate conflict, in the context of warm and supportive family relations, is functional for adolescent development and relationship change.

Social-Cognitive Views of Conflict

Ample evidence indicates that although adolescent-parent conflict occurs over mundane, everyday issues, it has much larger significance in the lives of parents and adolescents. As Offer (1969) has stated, "We emphasize that the rebellion has vital and important meaning to the adolescent in this stage of his development. . . . The great majority of parents say that the early adolescent years (12 to 13) are the most difficult time they have in raising their children" (p. 186). The dominant focus of prior research on describing the frequency and intensity of conflict in families with adolescents has led researchers to ignore the meaning of conflict to the participants.

Researchers who have studied close relationships have asserted that to understand individuals' behavior, it is necessary to determine how they represent their circumstances to themselves (Duck & Pond, 1989; Kelley, 1955). Social-cognitive analyses are useful in illuminating the way different family members conceptualize their relationships and interpret conflicts. The focus of my research has been to examine adolescents' and parents' reasoning about conflict and parental authority employing a domain-specific theory of social-cognitive development (Smetana, 1983, in press; Turiel, 1979, 1983; Turiel & Davidson, 1986).

The Domain-Specificity Model

According to this model and as defined in Table 1, social reasoning develops within three conceptually and developmentally distinct domains: the moral, the social-conventional, and the psychological (Turiel, 1978, 1979, 1983). According to this model, morality is defined as prescriptive categorical judgments of right and wrong that pertain to others' welfare, justice, rights, and the fair distribution of resources. Moral concepts are hypothesized to be obligatory, universally applicable (in that they apply to everyone in individual circumstances), impersonal (in that they are not based on personal preferences), and determined by criteria other than agreement, consensus, or institutional convention. Thus, the wrongness of moral transgressions is seen as stemming from their intrinsic features, such as their consequences for others' rights and welfare. Stealing, hitting, hurting, teasing, and lying are all examples of moral issues.

Although morality is viewed as socially constructed, not all social concepts are moral. Accordingly, morality has been distinguished analytically from individuals' understanding of social systems, social organization, and social conventions (Nucci, 1981; Smetana, 1981, 1983; Turiel, 1979, 1983; Turiel & Davidson, 1986). Conventions have been defined as consensually determined, shared behaviors (such as uniformities or rules) that coordinate the interactions of individuals within social systems. They are symbolic elements of social organization and represent shared knowledge among members of society or a social unit. Conventions are arbitrary in that alternative actions could serve the same function; thus they are relative to the social context. Forms of address, dress, manners, greetings, and sex-roles are all examples of social conventions. This distinction between morality and convention is consistent with philosophical analyses of morality (cf. Dworkin, 1978; Gewirth, 1978; J. S. Mill, 1863/1963; Rawls, 1971).

Moral and conventional concepts also have been distinguished conceptually from psychological knowledge. Psychological knowledge pertains to individuals' understanding of self and others as psychological systems and includes an under-

Table 1. Definitions and Examples of Social Knowledge Domains

Domains/Components	Definitions	Examples
Moral	Acts that are prescriptively wrong because they have intrinsic consequences for others' welfare, rights, trust, or the fair distribution of resources	Physical harm (hitting, shoving, pushing, etc.), psychological harm (teasing), stealing, lying, breaking promises
Conventional	Arbitrary and agreed-on behavioral uniformities that structure social interactions in different social systems, including families, cultures	Manners, forms of address, dress, greetings
Psychological	Understanding of self and others as psychological systems	
Personal	Acts that only affect the self and are beyond the bounds of legitimate societal regulation and moral concern; important for the development of self, autonomy, identity	Choice of friends, recreational activities, control over one's body (e.g., choice of clothes, food, appearances)
Prudential	Nonsocial acts that have negative, immediate, and directly perceptible consequences to the self	Health, personal comfort, safety
Psychological	Knowledge of self, identity, personality, and attributions for causes of self's and others' behavior	Concepts of self, self-understanding, causal attributions

standing of self, identity, and personality and attributions regarding the causes of one's own and others' behavior. Individuals' concepts of personal and prudential issue are two aspects of psychological knowledge (see Table 1).

Although concepts of rights and rights claims are an integral aspect of morality, it has been proposed that the notion of rights is grounded in the establishment and maintenance of personal agency (Dworkin, 1978; Gewirth, 1978, 1982; Nucci, 1996; Nucci & Lee, 1993). Individuals may exercise personal agency when asserting control over personal issues. Personal issues have been defined as pertaining only to the actor; therefore, they are considered to be outside the realm of conventional regulation and moral concern (Nucci, 1981, 1996; Nucci & Lee, 1993; Smetana,

1982, 1983; Turiel, 1983; Turiel & Davidson, 1986). Issues of personal choice comprise the private aspects of one's life and entail issues of preference and choice pertaining to friends or activities, the state of one's body, and privacy. The inclusion of actions in the personal domain is hypothesized to represent an important aspect of the individual's autonomy or distinctiveness from others (Nucci, 1981). Thus, in this view, the right to make autonomous decisions is described as an aspect of the self that forms the boundary between the self and the social world. Choice of friends, choice of activities, control over one's body (for instance, in terms of choice of clothes and food), and privacy all are prototypically personal issues.

Moral, conventional, and personal concepts also have been distinguished from prudential issues. Prudential issues have been defined as nonsocial acts pertaining to safety, harm to the self, comfort, and health (Shweder, Turiel, & Much, 1981; Smetana & Asquith, 1994; Tisak, 1993; Tisak & Turiel, 1984). Both moral and prudential rules regulate acts that have physical consequences to persons. However, morality pertains to interactions among people, whereas prudence pertains to acts that have negative consequences to the self which are immediate and directly perceptible to the actor (e.g., harm occurring through carelessness; Shweder et al., 1981). Therefore, prudential issues typically have been judged to be under personal jurisdiction (Nucci, Guerra, & Lee, 1991; Smetana & Asquith, 1994; Tisak, 1993; Tisak & Turiel, 1984). Examples of prudential issues include behaviors regarding one's health, personal comfort, and safety.

These three domains of social knowledge—morality, social convention, and psychological knowledge—are hypothesized to be separate self-regulating systems that are not developmentally ordered. This assumption differs from the models of Piaget (1932) and Kohlberg (Colby & Kohlberg, 1987; Kohlberg, 1971). These theorists have proposed that the process of development involves the gradual differentiation of principles of justice or rights from nonmoral (e.g., conventional, pragmatic, and prudential) concerns. In contrast, in the domain-specificity view, these three domains of social knowledge are hypothesized to coexist from early childhood on.

Social-Cognitive Perspectives on Conflict

Research employing this domain-specificity model has indicated that among American, middle-class families, parents primarily reason about conflicts as issues of social convention (Smetana, 1988b, 1989, 1993; Smetana, Braeges, & Yau, 1991). That is, parents justify their perspective on disputes by appealing to familial or cultural rules and norms; the need for maintaining a system of shared expectations among family members; the negative personal or social consequences of social nonconformity (such as the embarrassment or ridicule acts would cause); the need for politeness, manners, or consideration; the approval of specific authority figures or the existence of rules or laws; or the need to avoid punishment. American middle-class adolescents, in turn, understand but reject their parents' social-conventional interpretations of disputes and appeal instead to exercising or maintaining personal jurisdiction (Smetana, 1988b, 1989, 1993; Smetana, Braeges et al., 1991). These differences in adolescents' and parents' reasoning about conflicts have been found in individual interviews, as well as in structured family interactions (Smetana, Braeges et al., 1991).

Emery (1992) has asserted that family conflicts carry at least two levels of social meaning. The surface meaning refers to its literal content (such as whether or not the adolescent will clean her room, take out the garbage, or be in before curfew). The deep meaning refers to what its process of resolution or outcome conveys about the broader structure of relationships. I have hypothesized that adolescent-parent conflict occurs over the boundaries of adolescents' developing autonomy and that appeals to personal jurisdiction represent a social-cognitive aspect of the individuation process. Consistent with this, research on adolescents' and parents' conceptions of parental authority indicates that adolescents and parents agree that parents have the legitimate authority to regulate moral and conventional issues. However, adolescents and parents disagree as to where the boundary between parents' legitimate authority and adolescents' personal jurisdiction should be drawn (Smetana, 1988a, 1993; Smetana & Asquith, 1994).

Similar findings have been obtained in a recent study of adolescent-parent conflict among lower-class Chinese adolescents in Hong Kong (Yau & Smetana, 1996). Adolescents reported having conflicts with their parents, although the conflicts were less frequent, less severe, and fewer in number than among primarily white, middle-class American adolescents. However, conflicts were over the same types of issues as found in the American sample, and Chinese adolescents, like their American counterparts, reasoned about conflicts primarily as issues of maintaining or exercising personal jurisdiction.

As with other research from the structural-developmental perspective (e.g., Damon, 1977; Kohlberg, 1971; Piaget, 1932; Turiel, 1983), these social-cognitive analyses have focused on identifying normative patterns of adolescent-parent reasoning and adolescent development. Thus, the goals of the research have been consistent with previous research characterizing the incidence and severity of conflict in families with adolescents. Conflict clearly varies across families, however. In some families, relationships are cordial and harmonious, whereas in other families, they are discordant and acrimonious. As Cicchetti (1993) has noted, an understanding of normative developmental processes is essential for understanding developmental psychopathology. Likewise, an interest in developmental psychopathology requires researchers to focus on individual differences and identify factors that are associated with different developmental trajectories. Research on adolescent-parent conflict from social-cognitive or other normative perspectives rarely has examined the characteristics of conflicts held to be functional or dysfunctional for adolescent development and the types of families in which those conflicts occur. As Montemayor (1986) has noted, "It is time to move beyond debate about the 'fundamental' nature of the parent-adolescent relationships . . . . The important question, and one with an empirical answer is, What factors account for variation in parent-adolescent harmony?" (p. 16). Few studies have examined the features that distinguish positive from negative conflicts in normal families. The following sections address this question.

Adolescent-Parent Conflict: An Empirical Investigation

The research reviewed earlier indicated that during different decades over the last century, conflict and rebellion, an absence of conflict, and moderate conflict all have been viewed as normative of adolescent-parent relations. Although variation among these views can be attributed to changes in theoretical perspectives as well as advances in methodology, it is also possible that all these views are, to some degree, correct. That is, there may be different patterns of conflict in normal families that the normative analyses have obscured. The purpose of the analyses to be described here was to characterize conflict in normal families by examining patterns of conflict. Rather than juxtaposing normal with clinically referred families, these analyses examined variation within samples of normal families;[1] the assumption was that examining this variation would help illuminate patterns of optimal functioning, as well as the boundaries between "normality" and "risk."

These analyses were performed on three data sets collected between 1986 and 1993. These data sets were initially collected to address other research questions; therefore, they did not contain all of the measures that would have been optimal to address the questions posed here. Nevertheless, the analyses illustrate one approach to studying conflict in families with adolescents.

The studies used in these analyses focused primarily on social-cognitive aspects of adolescent-parent conflict and authority relations (as described briefly in the previous section). The studies addressed somewhat different questions, but in all three studies, ratings of adolescent-parent conflict were obtained. Table 2 presents an overview of the measures employed in these studies. The aims, samples, methods, and measures for these studies are briefly described below.

Study 1

The data for Study 1, hereafter referred to as the Conflict Study, were collected in Rochester, New York, between 1986 and 1987. Because the study focused on transitions to adolescence, the sample consisted of 102 preadolescents and adolescents (for simplicity, referred to hereafter as adolescents) between the ages of 10 and 18 ($M = 14.10$ years) and their parents (total $N = 306$). All preadolescents and

---

[1] In these studies, normal families were defined as families that were not currently in treatment for mental health or adjustment problems or who had not been clinically referred for treatment. It is possible that the samples included families or individuals who were experiencing problems but were not in treatment.

Table 2. Overview of Measures and Variables in Three Studies

Construct	Conflict Study	Authority Study	Hong Kong Study
Conflict	Intensity (5-point) Frequency (5-point) N of conflicts	Intensity (5-point) Frequency (3-point)	Intensity (5-point) Frequency (5-point) N of conflicts
Parenting		Parenting style	Warmth & control
Conflict resolution	Resolution style	Extent of resolution	Resolution style Solution fairness
Family interactions	Constraining & enabling interactions Summary coding scales	Family decision making style Domain-specific ratings of rules	
Social cognition	Conflict justifications	Rule legitimacy Rule obedience Sorting of items as personal	Justifications
Adolescent outcomes	Social-conventional development School grades	Emotional autonomy	

adolescents came from two-parent families, and both parents participated in the research. The families were primarily white (90%) and middle to upper middle class.

The purpose of this study was to examine adolescents' and parents' reasoning about conflicts from a social-cognitive perspective (see Smetana, 1988b, 1989, for more detail). Accordingly, adolescents and parents were individually interviewed about conflicts. With the interviewer's probing, they generated exhaustive lists of the disagreements and conflicts they had with parents (or with adolescents) and then were interviewed extensively about each disagreement using a semistructured clinical interview. Respondents also rated each conflict for its frequency (on a five-point scale) and intensity (on a five-point scale) and described how each conflict typically was resolved.

Following the individual interviews, adolescents and parents participated in a videotaped 26-minute Family Interaction Task. Families first chose three issues of conflict arising from the individual interviews and then spent 7 minutes discussing each one, stating their positions on the disputes and then working towards resolution (see Smetana, Braeges et al., 1991, or Smetana, Yau, & Hanson, 1991, for more detail). Transcribed videotapes of family interactions were coded microanalytically for enabling interactions, or behaviors that are seen to facilitate or encourage expression of more independent thoughts and perceptions, and constraining interactions, or behaviors that are seen to inhibit or undermine development by resisting adolescents' separation or differentiation from parents (Hauser, Powers, Weiss-Perry,

Follansbee, & Rajanpark, 1985). In addition, the videotapes were rated by trained coders on seven summary scales of family interaction that were composed of 60 five-point Likert Scales assessing affect, communication, conflict, and power in the family (see Smetana, 1994, or Smetana, Yau, Restrepo, & Braeges, 1991, for more detail).

Study 2

The data for Study 2, hereafter referred to as the Authority Study, were collected in Rochester, New York, between 1991 and 1992. Because prior research had indicated that concepts of parental authority shift in the transition to adolescence (Smetana, 1988a), the sample consisted of 110 preadolescents and adolescents (again, referred to subsequently as adolescents) between the ages of 10 and 18 ($M = 13.96$ years) and their parents (total $N = 300$). Most adolescents came from two-parent families (84%), and 108 mothers and 92 fathers participated in the research. The families were primarily white (89%) and lower middle to middle class.

The purpose of this study was to examine adolescents' and parents' conceptions of parental authority in relation to parents' parenting styles (see Smetana, 1995b). Judgments of parental authority in different social-cognitive domains (moral, conventional, prudential, personal, and multifaceted, or acts that overlap the conventional and personal domains) were obtained, and both adolescents and parents rated parents' parenting style using a standardized measure based on Baumrind's (1971) definitions of permissive, authoritarian, and authoritative parental prototypes (Buri, 1989, 1991; Buri, Louiselle, Misukanis, & Mueller, 1988). In addition, adolescents and parents each rated a standard list of 24 issues for frequency of discussion (on three-point scales) and intensity of discussion (on five-point scales). Finally, adolescents reported their family's decision-making style on a checklist based on previous research (Dornbusch, Carlsmith, Bushwall, Ritter, Leiderman, Hastorf, & Gross, 1985) and completed Steinberg and Silverberg's (1986) Emotional Autonomy Scale.

Study 3

The data for Study 3, hereafter referred to as the Hong Kong Study, were collected in Hong Kong between 1992 and 1993. The sample consisted of 121 Chinese adolescents between the ages of 12 and 20 ($M = 15.38$ years). Most adolescents came from two-parent families; unlike the Conflict and Authority Studies, however, parents did not participate in this research. The adolescents were from lower-class families. Fathers had, on average, less than a secondary school education and were employed as skilled manual laborers, whereas mothers had, on average, completed the primary grades and were semiskilled or skilled manual laborers.

The purpose of this study was to examine adolescents' reasoning about conflicts in a different cultural context. Adolescents generated exhaustive lists of the disagreements and conflicts they had with parents and then were individually inter-viewed (in Chinese) about each disagreement. Adolescents also rated each con-flict for its frequency (on a five-point scale) and intensity (on a five-point scale), described how each conflict typically was resolved, and rated the fairness of the solution on a three-point scale (see Yau & Smetana, 1996, for more detail). In addition, adolescents rated their mothers and fathers separately on 11 cultur-ally appropriate items that factored into two scales of warmth and restrictive, dominating control that were developed and normed on populations of Chinese parents in Hong Kong, mainland China, and Taiwan (Berndt, Cheung, Lau, Hau, & Lew, 1993; Lau & Cheung, 1987).

Cluster Analyses

To characterize conflict in these families, cluster analyses (using the squared Euclidean distance) were performed separately on the three data sets. Cluster analysis is a statistical technique for grouping cases (in these analyses, families) based on the relative relations of a set of variables to each other. Cluster analysis also is a holistic and phenomenological approach to grouping families in that the analyses generate different clusters (or profiles) that yield meaningful patterns regarding the combination of variables. This analytic technique was chosen as a way of identifying prototypical patterns of conflict in normal families.

Ratings of conflict frequency and severity were included in the analyses because similar ratings were obtained in all three studies and because these are the compo-nents of conflict that have been examined most often in other research (Monte-mayor, 1986). Furthermore, level of negative affect, in addition to conflict fre-quency, have been found to be good predictors of the effects of conflict (Gottman, 1979; Laursen & Collins, 1994; Laursen & Hartup, 1989).

Because participants in the Conflict and Hong Kong Studies were interviewed about self-generated conflicts, the number of conflicts that families (in the former study) or adolescents (in the latter study) rated for frequency and intensity varied. Participants in the Authority Study all rated the same number of issues, but be-cause the issues were chosen as hypothetical exemplars of different social-cognitive domains, there was great variability in ratings of different issues. Therefore, the variance in families' (or, in the Hong Kong Study, adolescents') ratings of frequency and intensity also were included as separate variables in the cluster analyses to capture information about intrafamily (or intraindividual) variability.

Results of the Cluster Analyses of Families

The results of the three cluster analyses are depicted in Figure 1. As can be seen, three clusters of families emerged in all three analyses.[2] The profiles of families' ratings of conflict frequency and severity were similar across studies, although the

---

[2] To facilitate visual comparisons across studies, the three-point ratings of frequency in the Author-ity study were transformed to a five-point scale in these figures.

Figure 1. Cluster Analyses of Conflict, Authority, and Hong Kong Study Families.

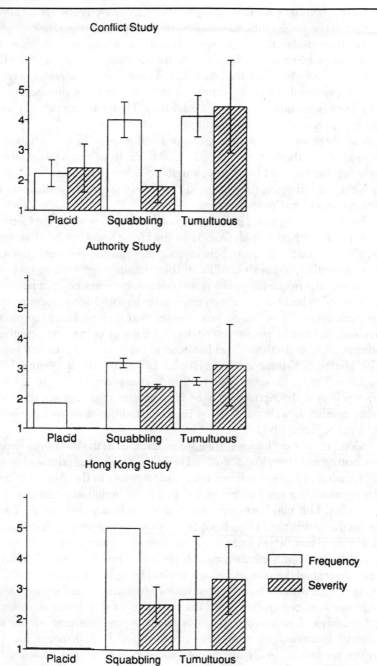

frequencies of families in each cluster varied. (For convenience, participants in the Hong Kong Study will be referred to as families, although in this study, adolescents rated their families.) Moreover, in all three analyses, both the ratings and the variance in the ratings differed significantly across the three groups.

In all three studies, the largest group of families to emerge from the analyses ($n = 48$ in the Conflict Study; $n = 57$ in the Authority Study; $n = 61$ in the Hong Kong Study) were labeled *Frequent Squabblers*. These families reported very frequent conflicts (with only moderate variability in frequency), but the conflicts were low in intensity (with little variability in intensity). Therefore, conflict in these families can best be characterized as bickering.

In all three studies, a second group of families ($n = 17$ in the Conflict Study; $n = 44$ in the Authority Study; $n = 22$ in the Hong Kong Study) were labeled *Placid*. These families reported (with little variability) that they rarely had conflicts, and when they did disagree, they reported (with only moderate variability) that their conflicts were of moderate intensity.

The studies also yielded a final group of families ($n = 36$ in the Conflict Study; $n = 9$ in the Authority Study; $n = 35$ in the Hong Kong Study) that were labeled *Tumultuous*. Like the Frequent Squabblers, these families also were characterized by frequent conflict, although conflict in these families appeared to be less frequent than among the frequent squabblers. In contrast, however, conflict in these families was reported to be, on average, extremely intense, with high variability in intensity. The high frequency of tumultuous families in the Hong Kong sample may seem surprising in light of recent descriptions of Eastern cultures as oriented towards tradition, duty, collectivism, and harmony in interpersonal relations (e.g., Kessen, 1979; Markus & Kitayama, 1991; Shweder, 1986; Shweder & Bourne, 1984; Triandis, 1989, 1990). However, this finding is consistent with a recent analysis of anthropological descriptions of over 160 cultures that suggests that adolescent-parent conflict is widespread, even in more traditional or collectivistic cultures (Schlegel & Barry, 1991).

Indeed, there were far fewer tumultuous families in the Authority Study than in the Conflict and Hong Kong Studies. Furthermore, although the analyses indicated that tumultuous families differed from other families in the Authority Study, conflicts appeared less angry or frequent than among tumultuous families in the other two studies. This may have been due to methodological differences. The conflicts rated in the Conflict and Hong Kong Studies were self-generated; therefore, the assessments tapped conflicts of greatest salience to each family member. In the Authority Study, however, family members rated hypothetical issues of parental authority in different conceptual domains. Although many of these issues are prototypical conflicts among families with adolescents, their salience to different families may have varied.

Groups were constructed based on the results of the cluster analyses, and then further analyses (analyses of variance or, where appropriate, multivariate analyses of variance) examined whether these groups differed on demographic, parenting, conflict resolution, family interaction, social-cognitive, and adolescent outcome variables. The results of these analyses are considered in turn below.

Frequently

☐ Placid
☒ Tumultuous
☐ Squabblers

ry Studies were
e Hollingshead
e latter study)
ttainment was
d families had

nantly intact)
obtained from
rily of intact,
re more likely
lies, whereas
us than were

families did
siblings, or,
wever, Con-
than other

nerated for
long Kong

Married, intact

Divorced / remarried

			%	
Adolescents' age (in years)	Conflict Study	15.45[a]	14.07[b]	10.59
N of conflicts (families)	Conflict Study	8.65[b]	12.03[a]	
N of conflicts (adolescents)	Hong Kong Study	1.23[c]	2.89[a]	1.73[b]

[a, b, c] indicate means that differ significantly.

[d] As scored using Hollingshead's Four Factor Index; scores on the index may range from 10 to 65.

[e] Scored as primary grades or below (1), secondary school (2), postsecondary school (3).

Figure 2. Parenting Styles Among Placid, Tumultuous, and Squabbling Families

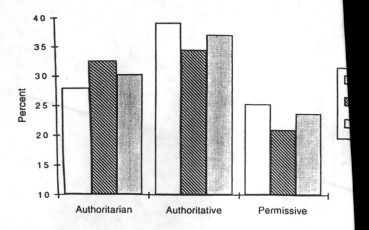

Demographic Characteristics

As shown in Table 3, placid families in the Conflict and Authori significantly higher in socioeconomic status (SES, as measured by the Four Factor Index in the former study and by parental income in the than were tumultuous families. Furthermore, although educational a low among all parents in the Hong Kong Study, mothers from placi more education than other mothers.

The Conflict Study consisted entirely of married (and predomir families, and detailed information on parental marital status was not adolescents in the Hong Kong Study. (The sample consisted prima married families.) However, intact families in the Authority Study we to be placid than were stepparent and divorced, single-parent fam single-parent or stepparent families were more likely to be tumultuo intact, married families (see Table 3).

Across studies, adolescents in tumultuous, squabbling, and placid not differ in gender composition, ethnicity, birth order, number of among adolescents in the Authority and Hong Kong Studies, age. Ho flict Study adolescents from placid families were older, on average, adolescents (see Table 3).

Number of Conflicts

There were significant differences in the number of conflicts ge discussion in the individual interviews in both the Conflict and I

Studies. In both studies, tumultuous families generated more conflicts than did other families. In addition, Hong Kong Study adolescents from frequently squabbling families generated more conflicts than did adolescents from placid families (see Table 3 for means). Although Table 3 suggests that there were many more conflicts among American than Chinese families, this table presents the summed total of conflicts across different family members; across-study differences are smaller when the number of family participants is considered.

Parenting

Parenting Style

Figure 2 depicts family members' (combined) ratings of parents' parenting styles as assessed using Buri's Parental Authority Questionnaire (Buri, 1989, 1991; Buri et al., 1988). Tumultuous families were found to be significantly more authoritarian than placid families, whereas placid and frequently squabbling families were more authoritative than tumultuous families. Placid families also were significantly more permissive than other families.[3]

Parental Warmth and Control

Hong Kong adolescents from placid families rated their parents as higher in parental warmth and lower in restrictive, dominating control (Berndt et al., 1992; Lau & Cheung, 1987) than did adolescents from tumultuous families.

Conflict Resolution

The reports by the Conflict Study families and Hong Kong Study adolescents of how conflicts typically were resolved were coded as parent concedes, adolescent concedes, no resolution, or joint decision making (the "no resolution" category was not employed in the Hong Kong Study). With age controlled,[4] Conflict Study families were found to differ in the proportion of their joint decisionmaking. Although rates were low in all families, there was significantly more joint decision-

---

[3] Previous analyses of these data indicated that adolescents and parents differ considerably in their perceptions of parents' parenting styles (Smetana, 1995b). Both mothers and fathers viewed themselves as more authoritative than did their adolescents, whereas adolescents viewed their parents both as more permissive and more authoritarian than did parents. When the analyses were repeated separately with parents' and adolescents' ratings of parenting style, similar findings were obtained for adolescents' ratings of their parents. When parents rated their own parenting style, however, differences were primarily in their perceptions of permissive parenting; placid parents rated themselves as significantly more permissive than did other families.

[4] Age was treated as a covariate in these analyses because Conflict Study adolescents from placid, tumultuous, and frequently squabbling families differed in age, and mode of conflict resolution was found to be strongly age-related in previous analyses of these data (Smetana et al., 1991).

Figure 3. Conflict Resolution Among Placid, Tumultuous, and Frequently Squabbling Families

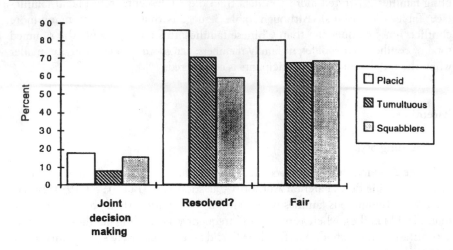

making in squabbling and placid families than in tumultuous families. These differences are depicted graphically in Figure 3. With age controlled, there were no other differences in mode of conflict resolution in this sample, nor were there significant differences in Hong Kong Study adolescents' reports of conflict resolution. However, Hong Kong Study adolescents from placid families viewed conflict solutions as significantly fairer than did other adolescents.

In the Authority Study, adolescents and parents indicated whether or not they resolved actual conflicts over the 24 issues of parental authority. Placid families reported resolving conflicts more than did frequently squabbling families or tumultuous families, and frequently squabbling reported resolving conflicts more than did tumultuous families. These findings also are depicted in Figure 3.

Family Interactions
Conflict Study families did not differ in the frequencies of their cognitive and affective enabling and constraining interactions in the Family Interaction Task, nor did they differ significantly in their family interactions as rated on the summary scales of family interaction.

Family Decision Making
Based on previous research (Dornbusch et al., 1985; Steinberg, Lamborn, Darling, Mounts, & Dornbusch, 1994), Authority Study adolescents' self-reports of family decision making were scored as either youth-alone, joint, or parental decision making. There was a trend towards greater youth-alone decision making among adolescents from tumultuous than placid families.

Table 4. Mean Level of Personal Reasoning and Judgments Among Tumultuous, Placid, and Frequently Squabbling Families

		Family Type		
Variable	Study	Tumultuous	Placid	Frequent Squabblers
Extent of personal rules[c]	Authority Study	2.24[a]	2.00[b]	2.18[a]
Extent of multifaceted (personal / conventional) rules[c]	Authority Study			
Adolescent personal reasoning (%)	Conflict Study	41[b]	56	67[a]
Adolescent personal reasoning (%)	Hong Kong Study	40[b]	46[b]	58[a]
Parental obligation to make rules about personal issues (%)	Authority Study	54[a]	37[b]	36[b]
Fathers' sorting of personal issues as under personal jurisdiction (%)	Authority Study	30[b]	65[a]	59[a]

[a], [b] idicate means that differ significantly.

[c] On a scale ranging from no rules or expectations (1) to firm rules or expectations (3).

Figure 4. Extent of Rules in Placid, Tumultuous, and Frequently Squabbling Families

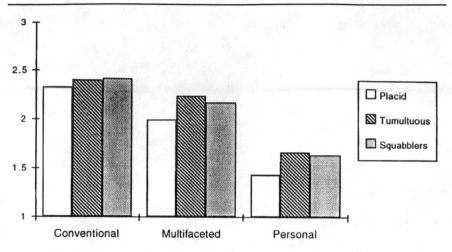

Family Rules

Both tumultuous and frequently squabbling families in the Authority Study reported that they had more rules about multifaceted and personal issues than did placid families (see Table 4 for means). Families did not differ, however, in the extent to which they regulated moral, conventional, or prudential issues.

Social Cognition

Reasoning About Conflict

Adolescents from frequently squabbling families appealed to personal jurisdiction significantly more than did adolescents from tumultuous families (see Table 4 for means). Similar findings were obtained among Hong Kong Study adolescents (see Table 4). In contrast, although moral reasoning about conflicts was infrequent, Chinese adolescents from tumultuous families gave more moral reasons for conflicts (10%) than did either frequently squabbling (3%) or placid (2%) adolescents.

Correspondingly, as depicted in Figure 5, Conflict Study parents in frequently squabbling families invoked social conventions to justify disputes more than did parents in placid families. Thus, the parent-adolescent discrepancies in mode of reasoning about conflicts described in previous research (Smetana, 1988a, 1989) appear to be most characteristic of frequently squabbling families. I have hypothesized elsewhere (Smetana, Braeges et al., 1991) that because they are qualitatively different modes of reasoning, personal and conventional arguments are particularly conflictful perspectives on disputes. The present findings suggest that bickering may persist because these families rely so heavily on these different and difficult to

Figure 5. Placid, Tumultuous, and Squabbling Parents' Reasoning About Conflicts

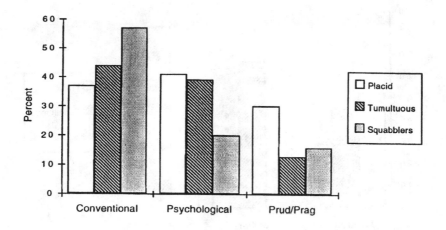

resolve interpretations of disputes. Furthermore, as Figure 5 demonstrates, parents in placid families reasoned about prudential and pragmatic concerns in their individual interviews more than did either tumultuous or frequently squabbling families. As previous research has indicated that these are less conflictful modes of discourse (Smetana & Berent, 1993) than personal, conventional, or moral reasons, these modes of reasoning may be related to the quality of interactions in these families.

Finally, in the Conflict Study, placid and tumultuous parents reasoned about psychological concerns more than did parents in squabbling families.

Reasoning About Parental Authority

Adolescents and parents in the three types of families differed in their judgments of parents' obligation to make rules about different types of issues. As depicted in Figure 6, fathers from tumultuous families viewed parents as having more of an obligation to make rules about all types of issues than did fathers from either squabbling or placid families. Furthermore, parents in squabbling families and fathers in tumultuous families viewed parents as having more of an obligation to make rules about all types of issues than did their adolescents, but parents and adolescents in placid families did not differ.

Judgments of rule obligation and sorting of issues as under personal jurisdiction also differed according to conceptual domain. As shown in Table 4, tumultuous families viewed parents as having more of an obligation to make rules about personal issues than did either squabbling or placid families. Furthermore, fathers from frequently squabbling and placid families more frequently viewed personal issues as under adolescents' personal jurisdiction than did fathers from tumultuous families (see Table 4 for means).

Figure 6. Placid, Tumultuous, and Squabbling Families' Judgments of Parents' Obligation To Make Rules

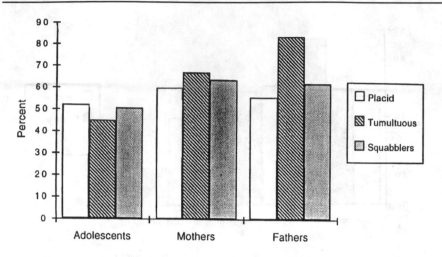

Adolescent Outcomes

Three measures of adolescent development or adolescent outcomes were available for analyses. Authority Study adolescents' emotional autonomy was assessed on Steinberg and Silverberg's (1986) Emotional Autonomy Scale. Additionally, self-reports of Conflict Study adolescents' academic performance (school grades) were obtained from adolescents.[5] Finally, Conflict Study adolescents' developmental level of social-conventional reasoning was assessed using semistructured clinical interviews on two hypothetical dilemmas developed by Turiel (1979, 1983).

Emotional Autonomy

Overall, Authority Study adolescents from tumultuous families scored higher in emotional autonomy than did adolescents from placid families. Recent research has indicated that the Emotional Autonomy Scale assesses detachment from parents rather than healthy individuation (Lamborn, Mounts, Steinberg, & Dornbusch, 1994; Ryan & Lynch, 1989; Steinberg et al., 1994). Therefore, the findings suggest that adolescents from tumultuous families are more detached from parents than are adolescents from placid families.

Academic Performance

Conflict Study adolescents from tumultuous families reported significantly lower academic performance than did adolescents from either placid families or frequently squabbling families.

---

[5] To increase their validity, these reports were obtained in the presence of parents.

Table 5. Levels of Social-Conventional Concepts

Level 1 (ages 6–7). Conventions are viewed as descriptive of uniformities in behavior; the existence of social uniformities is enough to require their maintenance. Children are aware of differences in power status but have no systematic notions of social organization, and uniformities are not understood to serve functions of coordinating interactions within social systems.

Level 2 (ages 8–9). Empirical uniformity is not a sufficient basis for maintaining conventions. Conventional acts are regarded as arbitrary. Conventions are not seen as part of the structure or function of social interaction.

Level 3 (ages 10–11). Conventions are seen as arbitrary and changeable. Adherence to convention is based on concrete rules and authoritative expectations. Conceptions of conventional acts are not coordinated with conceptions of rules.

Level 4 (ages 12–13). Conventions are seen as arbitrary and changeable regardless of rules. Evaluations of rules pertaining to conventional acts are coordinated with evaluations of the act. Conventions are seen as "nothing but" social expectations.

Level 5 (ages 14–16). The emergence of systematic and more integrated conceptions of social systems. Conventions are seen as normative regulation in a system with uniformity, fixed rules, and static hierarchical organization.

Level 6 (ages 17–18). Convention is regarded as codified social standards. Uniformity in convention is not considered to serve the function of maintaining the social system. Conventions are "nothing but" societal standards that exist through habitual use.

Level 7 (ages 18–25). Conventions are uniformities functional in coor-dinating social interactions. Shared knowledge, in the form of conventions, among members of social groups facilitates interaction and operation of the system.

Social-Conventional Development

Turiel (1979, 1983) has proposed that children's understanding of social-conventional concepts develops through a sequence of seven levels of increasingly sophisticated concepts of social convention. In his scheme (characterizing changes in children from approximately 6 to 25 years of age), children's understanding of social conventions is structured by underlying concepts of social organization. Social-conventional knowledge develops through an alternating series of affirmations and then negations of the importance of convention in structuring social life. Each affirmation phase is followed by a negation phase where children reevaluate and reject their understanding at the previous level. These negation phases then lead to a more sophisticated affirmation of the role of conventions. Turiel's (1979, 1983) research demonstrates that development in the conventional domain progresses towards an understanding of social systems as hierarchically organized and a view of social systems as functioning to coordinate social interactions. These levels are more fully described in Table 5.

With age controlled, there were significant differences in Conflict Study adolescents' developmental level of reasoning about social conventions. Adolescents from placid families were significantly more mature in their reasoning (mean developmental level = 5.05) than adolescents from either frequently squabbling families (mean level = 4.04) or tumultuous families (mean level = 3.99). These differences in adolescents' reasoning also may contribute to the varying patterns of conflict found in these families. A salient characteristic of Level 4 thinking (the average developmental level of adolescents from frequently squabbling and tumultuous families) is a rejection of the necessity for many conventions, as well as a criticalness towards the dictates of those in authority. Lacking a more differentiated understanding of social systems, Level 4 adolescents view conventions as "nothing but" social expectations. This description of Level 4 reasoning is consistent with Offer's (1969) description of adolescent rebellion as negation: "He defines what he does in term of what his parents do not want him to do. In periods like this, it becomes obvious that the adolescents' decisions are in reality based on the negative of the parents' wishes . . . the adolescent becomes a general irritant to his parents" (p. 187).

Level 5 represents a significant advance in thinking, as well as a change from a negation phase to an affirmation phase in the development of social conventional concepts. Level 5, which was most characteristic of adolecents from placid families, represents the emergence of systematic concepts of social organization. At this level, adolescents develop an understanding of the role of individuals within social units and a view of social conventions as normative regulation in a system with uniformity, fixed roles, and hierarchical organization. Thus, adolescents at this level are likely to affirm the validity of parents' expectations, although they may still have conflicts with parents because the expectations are poorly timed or may conflict with adolescents' desires.

Qualitative Analyses

The videotaped family interactions in the Conflict Study further illuminate differences among the three types of families as well as underline the limitations of this categorization. In many cases, the family interactions were virtually indistinguishable in tone and content. This is not surprising, as the quantitative analyses had revealed significant differences on some variables but not others; in particular, the quantitative analyses had failed to reveal statistically significant differences in the family interactions. Although these findings could indicate method variance, as most of the significant differences were obtained on self-report measures, it is also possible that the coding of the family interactions did not tap the salient dimensions on which these families differed. Placid, tumultuous, and frequently squabbling families all found conflicts to discuss, as they were instructed to do, and the types of issues raised in each group varied from trivial to serious. Regardless of family categorization, discussions were sometimes contentious, but laughing, humor, and signs of affection also were evident across family classifications. At the same time, some qualitative differences in the ways families handled disputes emerged.

Placid Families

Although placid families clearly found issues to discuss, and some disagreements were over very serious issues, the interactions were characterized by respect for and willingness to listen to others' points of view. The following examples, excerpted from much longer discussions, illustrate this.

TEEN: OK, I see it like this—that I feel that I'm responsible enough, you know, my friends. You know basically who they are, you've met some of them. And . . . I'm talking about a party without supervision, or having people after school. To me, I don't see a difference between having Tim over after school, or having Tim and Vivian over after school. Because it's not like I'm having a major emotional relationship. I've known Tim forever.

MOM: Mm-hmm (*agreeing*).

TEEN: It's just, I mean, he's my friend, and I'm friends with him the same way I'm friends with Vivian. It's not like it's anything physical or it's this love affair or anything like that I don't—

FATHER: I think that you have a good point in that Mom and I have made a distinction between having girls over and having boys over and maybe you're right and maybe that shouldn't be a rule and that's probably our fear of what can develop . . . when you have friends over we're not in the room with you so it, it's something that could happen in one part of the house and not in another is what I'm saying. But the thing is that we feel that there should be parental supervision nearby.

MOTHER: And that's something that's important, okay, that's something that's an important issue for us, liability, care, and well being of whoever's in our home. That's one issue for us. Second issue also is, you know, when you talk about the idea of Tim, of a boy being your friend as a girl and I know that's true for Tim— I don't doubt that . . . . How will I know when those relationships change and they become emotional relationships, which are of concern to me because there is a right and wrong moral behavior that will take place in my house? . . .

TEEN: But don't you think I'd tell you? I always—I've told you in the past when things have changed, I mean—. . . . I think, I think you should be able to trust me enough to know I'm gonna make the decision, I'm gonna—

MOTHER: But I also know from experiences that I've had growing up, that it's very difficult to say no to a boy, that if a boy knows that no one is in the house, that he's got the physical strength to empower you, whether or not you want to be empowered. Okay? That's the difference, that men have a strength . . . .

FATHER: Do you know where Mommy and I are coming from?

TEEN: Yeah, but I don't agree with it.

FATHER: But do you understand?

Although the discussion concerned a serious issue and the adolescent and her parent took opposing views, the interactions occurred in a climate that allowed differences of opinion to be expressed. This was also evident in the following discussion.

TEEN: . . . And calling her [the sister] names, names don't mean much to me.

FATHER: Yeah, but she gets riled up by that—you call her names; [it's] just not very nice.

TEEN: Why?

FATHER: Because that's not what she is.

TEEN: So?

FATHER: So why should you do it? It's unfair.

TEEN: (sarcastically): I feel terrible!

MOTHER: She's a sensitive kid, too.

FATHER: Well, I don't know if she's that sensitive, but she's a tough kid, anyway. Do you call anyone else "bitch" in this world?

TEEN: Oh, definitely.

MOTHER: Who?

TEEN: Girls at school call me asshole, and I call them bitches. But it's a friendly—

MOTHER: Yes, but with your sister, it also has the intonation of unfriendliness.

TEEN: Oh, should I sit her down and explain?

MOTHER: Well, anyway, you have a good vocabulary, right? So you could call her other words—

TEEN: I forgot all my vocabulary words. You saw that on my SAT score!

MOTHER: You could try other words that are more colorful and not so horrible.

TEEN: Truculent young rebel—

MOTHER: That's better, because then she will not think you are calling her a bitch anymore, and she will cool down, right?

TEEN: . . . that doesn't sound as good as "little bitch"!

FATHER: No, the idea of calling her names is that she responds by yelling, and then that encourages him.

TEEN: [I like] the antagonism . . . . Well, I don't do it that often [hurt her feelings] do I? . . . I'm a good attention getter, see?

In the final example, this family had just finished discussing several conflicts and then turned to the larger issue of the adolescent's view of her parents.

FATHER: That's an interesting question which is a little off the topic, but we have time. What issue, or to what degree do you listen to us or try to live up to our expectations?

TEEN: Oh no, God, I try to live up to your expectations!

FATHER: What?

TEEN: Are you kidding?

FATHER: What? Maybe I didn't hear what you said.

TEEN: I listen to you very much.

FATHER: Oh, okay, I understand. That's right, that's why we don't have any arguments.

TEEN: I know, I shouldn't [argue].

FATHER: Alright, let me ask you this then: Do we impose too much on you? I mean, we don't have any rules, really.

TEEN: I know. No, I mean . . . I mean, I certainly don't drink and all that.

FATHER: That's what I said [to the interviewer], too. I said, you know, we don't have any problems.

TEEN: I know.

FATHER: We got this on tape.

TEEN: No, but listen to me. (*Everyone is laughing.*) Now you made me turn bright red . . . . No, I mean, I listen to you, but I don't have a need to . . . I don't have a need to disagree with you. Except for clothes . . .

Frequently Squabbling Families

In many of the frequently squabbling families, there was a similar tone of warmth, good humor, and respect for differences. At the same time, interactions differed in several ways. In some families, the conflicts were left unresolved, discussions got derailed, or the families shifted focus from one issue to another. In the following example, the family had just completed a lengthy discussion of whether the adolescent would begin to clean up after the pet ferret, who was allowed to run around free in the house.

FATHER: Yes, and I will assist you in assisting him to assist you [in cleaning up]. It's resolved.

TEEN: The next subject is my room. Now I don't believe it's messy; I just believe it's not entirely neat. I mean, I put my clothes away and I vacuum it every week.

FATHER: I agree with you.

TEEN: I don't know how you can say I'm messy.

MOTHER: Because, why do you drop clothes on the floor?

TEEN: Mom, I go to bed and I put clothes on the floor and then I wake up in the morning and I throw them down. I mean, is it such a big deal?

MOTHER: But you don't.

TEEN: I do.

MOTHER: Sometimes I see piles of clothes.

TEEN: Those are my clean clothes that I haven't gotten around to putting away yet.

FATHER: How are we going to resolve it?

TEEN: This one will take us the full 7 minutes . . . . We've already resolved the ferret poops. It's just—I don't know.

FATHER: I know how we can resolve the ferret poops. I get to wear shoes and not go barefooted because it upsets me when my foot steps in it bare. So I will promise not to walk around the house barefooted and I will get Dan [the brother] to assist you in cleaning them up . . . .

TEEN: Now, what do you think about my room, Mom?

MOM: I just want you to clean it . . . .

FATHER: I think we ought to just say we don't have any conflicts and leave. We don't. We don't have conflicts. These aren't conflicting issues. This is ridiculous. It's trivial.

Thus, although the issue of the ferret was resolved, apparently to everyone's satisfaction, it was revisited again. Similar examples were evident in many of the frequently squabbling families' interactions; issues that appeared to be resolved (or that were left unresolved) resurfaced throughout their discussions. In addition, although the issues may have been mundane, this father's claim that his family had no disagreements was belied by the number of issues this family discussed, their evident disagreement with each other, and their many statements indicating that these issues were ongoing sources of family discord.

Another aspect of frequently squabbling families' interactions was that the issues were not clearly demarcated. Conflicts over one issue shaded into other disputes, so that the content appeared to be continually shifting, as the following example illustrates.

TEEN: There's nothing wrong with what I wear, just because everyone else wears these clothes.

MOTHER: Well, I think you're doing a better job than you have been of putting them together.

TEEN: You know mom, it's not my fault that I can't match clothes, I mean, what am I supposed to do?

MOTHER: I know. Well, I think that what you said to me a long time ago is true . . . that what you wear is not a reflection on me.

TEEN: That's right. I keep telling you that. Mom, every one of my friends say, "Your mom is such a prep." I'm like, yeah. (Father laughs.)

MOTHER: I guess that's something you'll have to live with.

TEEN: What's so bad. Mom, this is what people wear all the time. You just don't understand . . . .

MOTHER: You know, I expected this feminine little girl who was going to run around in all coordinated clothes and rouge on her cheeks . . . .

TEEN: W-R-O-N-G!

FATHER: Always be dressed in Polly Flanders dresses. (laughs).

TEEN: Mom, do you want me to dress like Lana _____ and Jane _____?

FATHER: How do they dress?

TEEN: They wear the tightest things, they look like hookers.

MOTHER: No, I want you to wear the long skirts with the pretty knit sweaters over the blouses.

TEEN: Good, buy me some $60 vests and I'll wear them . . . .

FATHER: I'm amazed when you say you're not dressing in a certain style 'cause you don't figure you want to spend the money to get those clothes . . . .

Although this family's discussions were good-humored, the issues were constantly were transformed into new issues. In the above example, the focus shifted from the adolescents' choice of clothes, to money spent on clothes, and then to more general money issues, without resolving any of them, so that the issues became more complex and often difficult to resolve. The following family appeared to value "understanding" more than resolution.

TEEN: Yeah, messy room. This shouldn't take as long.

MOTHER: Messy room.

FATHER: We ought to ask ourselves, before we move to messy room, we should ask ourselves if we, if this thing is in the category of resolved or is an open issue or more insight, but we have to keep on working on it.

TEEN: I think it's still just as open because you know next week it'll be just the same as it always was.

FATHER: Okay, fine. That's the important thing. We need to understand where we are.

Tumultuous Families

Tumultuous families, like frequently squabbling families, often left conflicts unresolved, and like frequent squabblers, the grounds of the disputes appeared to shift constantly. But these families differed from the other families in that their disputes appeared to be more contentious. Despite the fact that tumultuous parents were well educated and articulate (as were their adolescents) and despite their attempts to justify their perspectives on disputes, these families appeared to be less tolerant of differences and made less of an effort to take the other's perspective, as the following example illustrates.

MOTHER: Why don't we start with academic achievement. We have a little less direction there.

FATHER: Why don't you go ahead and elaborate a bit more.

MOTHER: Okay, in the course of the conversation that I was having with the interviewer we were talking about academic achievement as a source of conflict, and I was trying to figure out why academic achievement was important to me. It wasn't the grades per se, it was what the grades were going to buy Diana later on that bothered me . . . . That was my concern.

FATHER: Is that a problem to you?

TEEN: What?

FATHER: Long-term consequences of—do you feel that you're involved fully in your school work?

TEEN (*sarcastically*): No, Dad, I don't feel I'm involved fully in my school work.

FATHER: Okay, is it a problem that you're not?

TEEN: Should it be?

FATHER: No, Mommy has said that to her it's a problem.

MOTHER: It is to me. It's one of my conflicts.

FATHER: Now I'm asking if it's a problem to you? . . . I mean, do you think very much of what is going to happen several years down the road? Do you think at all about that?

TEEN: No, dad.

MOTHER: And? What's going to happen?

TEEN: I don't know, Mom. I guess you know how to sum it up according to your point of view.

Disputes in these families often appeared to escalate in intensity, perhaps because of this lack of understanding, as the following example (excerpted heavily because of it's length) illustrates.

TEEN: Alright, what about after school and stuff.

MOTHER: What do you mean?

TEEN: Like I have to come home all the time. I can never stay after school and stay for a meeting.

MOTHER: You always stay after school, Tara, you go to basketball games.

TEEN: Once in a while.

MOTHER: You go whenever you want to go.

TEEN: No!

MOTHER: When do you other than on Tuesdays?

TEEN: A couple of times they've been on Fridays . . .

MOTHER: I don't know how many times you've stayed after school.

FATHER: We've never stopped you from staying after school.

TEEN: But then I can't do anything after school because I have to come home and watch Jeff.

MOTHER: But what is it you want to do?

TEEN: I couldn't go to a friend's house after school if I wanted to.

MOTHER: Tara, in all honesty—

TEEN: I can't.

MOTHER: Have you been asked to a friend's house? You have not said a word to me that you want to go to a friend's house.

TEEN: I couldn't if I wanted to.

MOTHER: You have not said—why do you assume?

FATHER: Tara, we will make arrangements with Jeffrey, and I will tell the other two girls how to act to make sure there is no fighting.

MOTHER: Or if they came to the—Tara, you have never said one word.

TEEN: Yes I have . . . . I can't invite a friend over to the house—

MOTHER: —I'm telling you that is not true.

TEEN: They can only stay until 4:30.

MOTHER: They can stay later . . . .

In this family, as with other tumultuous families, the interactions became increasingly heated as parents and adolescents sought but did not find common grounds for agreement. The quantitative analyses also indicated that tumultuous families were more authoritarian than other families, and this was evident in many of the interactions.

FATHER: I will tell the kids, "Let me use the phone for a minute or two and when I get off you can recall your friends back." For me to sit there in a chair and wait for Lisa to get off the phone for 15 minutes, 20 minutes, or a half an hour . . . I think that . . . as a parent, I have a right to do that.

MOTHER: Because it's your house.

FATHER: Exactly. Even though the children live there, they have to abide by our rules. And there has to be some order, okay? Standards of order set up, and then I think they have to be abided by, lived to, and democratically—

TEEN: Yeah, but we never sit down and discuss anything, so come on!

FATHER: —reason we really [don't] sit down and discuss anything is because we don't get along too well.

TEEN: Right.

FATHER: When it comes to communicating or sharing our ideas, beliefs—

MOTHER: Well, children are not going to agree totally with parents.

TEEN: Just stay on the issue! The phone thing!

The lack of agreement, as well as parents' more authoritarian view of parenting, is also evident in the following family.

MOTHER: There has to be some leadership, and we're providing the leadership as parents.

TEEN: Well, I think it can be shared and helped.

MOTHER: In many ways it can be.

TEEN (*slamming her hand down*): Well, this is one of the ways that it can be! It's such a stupid thing to fight over.

MOTHER: This is one thing that we've decided is a "NO" in our home.

TEEN (*puts her hand to her eyes*): This is—Oh God, I can't wait until I move.

Summary

To summarize, three clusters of families emerged from these analyses of normal, primarily middle-class white American families and normal, primarily lower-class Chinese adolescents from Hong Kong: frequently squabbling, tumultuous, and placid families. The same patterns of family conflict were replicated in three separate analyses utilizing data from two different cultures and different socioeconomic statuses. These analyses included over 700 participants in over 300 families and employed different assessments of conflict frequency and severity; this provides strong support for the validity of the clusters described here. Furthermore, these three clusters of families differed in patterns of family interaction, parenting, and adolescent development.

Tumultuous families experienced frequent and variable but intense conflict and, in both American and Chinese samples, generated more conflicts for discussion than did other families. Parents from tumultuous families were more likely to be divorced or remarried and were lower in income and SES than other parents; adolescents from tumultuous families also had poorer academic performance than other adolescents. Furthermore, tumultuous parents in the Hong Kong Study were rated higher in restrictive, dominating parental control and lower in warmth than were parents from placid families.

Although there was strong endorsement of authoritative parenting among tumultuous American families, they were more authoritarian, had more rules, and were more restrictive in their judgments of adolescents' personal jurisdiction than other families. Emery (1992) has noted that more frequent and more intense conflicts are expected when boundaries depart significantly from developmental norms (for instance, when restrictiveness is great) or when boundaries are deeply penetrated (for instance, when parents attempt to regulate behavior that has long been under the child's control). This appears to be the case for tumultuous families, who appear to intrude more deeply than is developmentally appropriate in their adolescents' personal domains. Tumultuous families viewed parents as having more of an obligation to regulate personal issues and viewed personal issues as under adolescents' personal jurisdiction less than did other families. Furthermore, tumultuous families reported having more rules about multifaceted and personal issues than did placid families. Therefore, it appears that tumultuous parents' greater authoritarianism and intrusion into adolescents' personal domains was met by greater adolescent detachment, as assessed by scores on Steinberg and Silverberg's (1986) Emotional Autonomy Scale and adolescents' reports of family decision making.

In contrast, conflict in placid families was low in frequency and intensity. Placid families in the Conflict Study and adolescents from placid families in the Hong Kong Study generated fewer conflicts for discussion than did other families or adolescents, respectively, but with an average number of nearly 9 conflicts generated per American family and at least one conflict generated by Chinese adolescents, these families clearly were not free of conflict. This was also evident in the qualitative analyses; placid families did not lack for issues to discuss in the Family Interaction Task. Placid parents were higher in SES, more likely to be professionally employed, and more likely to be in intact marriages than were parents from tumultuous families. Placid families were more authoritative and engaged in more joint decision making than tumultuous families, but they were also more permissive and reported resolving conflicts more than did all other families. Furthermore, Chinese adolescents from placid families rated their parents as higher in warmth and lower in restrictive, dominating control than did tumultuous adolescents. Although Chinese adolescents from placid families did not differ from other adolescents in their modes of conflict resolution, they viewed conflict solutions as fairer than did other adolescents. Parents from placid families used less conflictful (e.g., prudential or pragmatic) reasons to justify their perspectives on conflicts than did other parents, which may have contributed to the placid nature of relations in these families. Finally, adolescents from placid families were significantly older and more developmentally mature in their understanding of social-conventional concepts than other ado- lescents and significantly lower in emotional autonomy (and thus, less detached, in an unhealthy way, from parents) than adolescents from tumultuous families.

Squabbling families, like tumultuous families, also had frequent conflicts, but their conflicts were of low intensity. Squabbling families were characterized by the type of bickering that other researchers (e.g., Hill, 1987, 1988) have viewed as typical of the adolescent years. Consistent with this observation, frequent squabbling occurred with the greatest frequency in all three of the data sets examined here. Although frequent squabblers were less clearly differentiated than either placid or tumultuous families, in many respects they resembled placid families. They did not differ from placid families in the extent to which they endorsed authoritative parenting or joint decision making, and frequently squabbling adolescents did not differ from their placid counterparts in their academic performance. They differed from adolescents from placid families (and resembled adolescents from tumultuous families), however, in the developmental maturity of their social-conventional concepts.

The most distinctive pattern observed in frequently squabbling families was in their reasoning about conflict. Frequently squabbling parents offered more social-conventional justifications for disputes than did other parents. In both American and Chinese families, adolescents from frequently squabbling families offered more personal justifications than did other adolescents. Thus, the normative pattern of reasoning about adolescent-parent conflict that has been described in previous research (Smetana, 1988a, 1989, 1993) seems to be most characteristic of these families.

There are interesting parallels between these three clusters of families described here and the patterns of psychological growth identified in Offer and Offer (1975) in their landmark longitudinal study of adolescent males. Offer and Offer's continuous growth group, which comprised 23% of their sample, resemble the adolescents from placid families identified here (although the personality and interpersonal assessments obtained by Offer and Offer were much more extensive than the assessments available here). Offer and Offer described these adolescents as extremely well-adjusted and as displaying many of the qualities that define positive mental health. Parents of these adolescents encouraged their sons' independence, and parent-adolescent relations were characterized by mutual trust, respect, and affection. Offer and Offer (1975) also identified a surgent growth group, comprising 35% of their sample, who resemble the frequent squabblers discussed here. Their relations with parents were characterized by some conflict and disagreement. Although these adolescents were often as well-adjusted and successful as the continuous growth group, they were slightly more prone to depression. Finally, Offer and Offer also identified a tumultuous group, comprising 21% of their sample. These adolescents' demographic backgrounds and patterns of family interaction were very similar to the tumultuous families described in this chapter. As described by Offer and Offer (1975): "These are the students who go through adolescence with much internal turmoil that manifests itself in overt behavioral problems in school and at home. These adolescents have been observed to have . . . escalating conflicts with parents" (p. 45).

That similar patterns of conflict, as well as reasoning about conflicts, were observed among lower-class Chinese adolescents from Hong Kong as compared to middle-class American adolescents also deserves comment. First, although these analyses highlight similarities across cultures, this should not be taken as an indication that there are no cultural differences in adolescent-parent conflict or adolescent development. The cultural issues are complex and beyond the scope of this chapter; however, other analyses reported elsewhere (Yau & Smetana, 1996) indicated both similarities and differences between Chinese and American adolescents' reasoning about conflict and in the expression of those conflicts. Furthermore, although there are variations in how personhood is defined across cultures (Geertz, 1975; Markus & Kitayama, 1991; Shweder & Bourne, 1984), it has been asserted, and surveys of the anthropological literature indicate (LeVine & White, 1986), that notions of self and personhood are basic human concepts (Damon & Hart, 1988; Geertz, 1975). This suggests that although there may be cultural variations in the boundaries that constrain or define the personal domain (Nucci, 1996; Nucci & Lee, 1993), and thus, what is considered to be under the individual's personal jurisdiction, all cultures must treat some issues as under personal control and necessary for the establishment of the self. In turn, this suggests that variations in the conceptual boundaries between the personal and conventional domains may be directly related to the occurrence of adolescent-parent conflict. That is, conflict may be greater in cultures (such as ours) that view the individual as exercising personal jurisdiction over a broad range of issues—as having what Goffman (1971) terms numerous "territories of the self." Conflict also may be greater in cultures (such as ours) where adolescents clearly have transitional status in society (Benedict, 1938) and where the boundaries between the personal and the conventional are sometimes ambiguously defined. Although adolescent-parent conflict may be more muted in cultures where the boundaries of the personal domain are drawn both more clearly and more restrictively and where issues that in this culture are considered to be under personal jurisdiction are seen as regulated by society and social convention, conflict may serve the same function in liberating the adolescent from parental constraint (Smetana, 1995a) . Thus, the content of the personal domain may be canalized by culture (Nucci, 1996; Nucci & Lee, 1993), but how cultures draw the boundaries between individuals' personal jurisdiction and societal convention and social order may be continually renegotiated during adolescence within the context of the family.

Implications for Developmental Risk and Resilience

Adolescence has been described as a developmental period that entails special risks; there are a variety of behaviors that adolescents engage in (such as anti-social acts, substance use, and school dysfunction) that impede the quality of functioning (Kazdin, 1993). Thus, psychiatric diagnostic categories do not adequately describe the range of conditions that are of concern during adolescence or that impair adolescent functioning. In this final section, the implications of the analyses presented here for the development of problem behaviors and psychopathology as well as optimal development are considered.

Risk

The findings suggest that adolescents from tumultuous families may be most at risk for developmental disturbance and problem behaviors. This hypothesis is clearly speculative, as these analyses were performed on samples of normal families, and little information on the incidence of problem behaviors was obtained. Furthermore, previous research has indicated that well-functioning families are overrepresented in family research, especially when active informed consent procedures are employed (Weinberger, Tublin, Ford, & Feldman, 1990), as was done in the studies described here.

However, several findings within these data, as well as the congruence between these findings and a variety of other research, lend some credence to this speculation. First, there was a trend towards more frequent conflicts over prudential issues of drug, alcohol, and cigarette use among tumultuous families in the Authority Study, as compared to placid families, and conflicts over these issues were significantly more intense than in other families.[6] Second, in another section of the Authority Interview, Authority Study adolescents and parents were asked to endorse the legitimacy of different authority sources, such as adolescents, friends, parents, teachers, religious institutions, and the law, to govern different types of acts. Adolescents from tumultuous families were significantly more oriented towards friends as legitimate authority sources for multifaceted issues than were other adolescents. Although this measure of peer influence is indirect and significant findings were obtained in only one conceptual domain, they are notable in that others have found that susceptibility to peer influence is an important factor in problem behavior (e.g., McCord, 1990). Multifaceted issues are, by definition, ambiguous with respect to whose authority is legitimate, and it is interesting that adolescents from tumultuous families were more oriented towards their peers regarding these issues than were other adolescents.

---

[6] This is a bit circular, of course, as the clusters were initially defined on the basis of ratings of conflict frequency and intensity, but these findings are domain-specific, whereas the classifications were based on global ratings.

Furthermore, the high levels of conflict and low levels of conflict resolution observed in tumultuous families are consistent with the descriptions of factors predicting a range of problem behaviors, including drug and alcohol use (Baer et al., 1987; Kandel et al., 1978), delinquency (Borduin et al., 1986; Forehand et al., 1987; Patterson, 1982; Patterson et al., 1991), and depression (Kashani et al., 1988; Puio-Antich et al., 1993; Rutter et al., 1976). In addition, others have found that marital disruption and divorce, combined with other family stressors, lead to deterioration in adolescent functioning (Forehand, Wierson, Thomas, Armistead, Kempton, & Neighbors, 1991; Simmons, Burgeson, Carlton-Ford, & Blyth, 1987). Although other stressors were not assessed, tumultuous families were more likely than other families to have experienced at least one type of disruption (marital dissolution) that has been found to predict adolescent dysfunction. Finally, there are interesting parallels between Offer and Offer's (1975) description of tumultuous adolescents and the tumultuous families described here. In Offer and Offer's (1975) study, psychopathology was more prevalent in this group than in their other two groups.

Although the analyses described here suggest that adolescents from tumultuous families are at greater risk for maladaptation than other adolescents, the present analyses do not illuminate the processes through which tumultuous families are created. It is unclear whether this pattern of family conflict has its origins in parenting style, adolescents' personality characteristics or developmental needs for autonomy, or patterns of family interaction. Although longitudinal research would be necessary to address this question, it is likely that tumultuous family relations are created out of a complex interaction of all these factors. For instance, Patterson (1992) has described a coercive cycle in the development of aggressive behavior. Likewise, authoritarian parenting may clash with adolescents' developmental needs for autonomy to create a pattern of unhealthy family interaction, which may, in turn, lead to adolescent problem behavior.

Experiencing conflict, or exposure to others' conflicts, may not necessarily be harmful for development; indeed, it has been proposed that conflict may be adaptive for development (Collins, 1990; Hill, 1987, 1988; Powers et al., 1989; Smetana, 1994) and may lead to the development of conflict resolution skills (Grych & Fincham, 1993). Therefore, it is important to understand the conditions under which conflict has negative effects on adolescents. As both tumultuous and frequently squabbling families had very frequent conflicts, tumultuous families were distinguished primarily by the intensity of their conflicts. The findings from several lines of recent research suggest that it is this conflict intensity that may place adolescents at greater risk of dysfunction over time. A recent study indicates that negative affect is one of the most salient characteristics in adolescents' reports of "important" conflicts from the previous day and is associated with disengagement and lack of conflict resolution (Laursen & Koplas, 1995). Likewise, Patterson (1982) has implicated negative affect as an important aspect of a coercive cycle that provokes aggressive behavior. Finally, previous research has demonstrated the negative effects of interparental conflict for children's adjustment (see Grych & Fincham, 1990 for a review) and has suggested that conflict intensity, in particular, may lead to adjustment problems. For instance, more aggressive, intense conflicts among parents is related to greater anger, fear, and sadness in children (Cummings, Vogel, Cummings, & El-Sheikh, 1989; Cummings, Zahn-Waxler, & Radke-Yarrow, 1981; Grych & Fincham, 1993;). Therefore, the role of negative affect in the development of adolescent maladaptation needs systematic examination in future research.

More speculatively still, the findings suggest intriguing connections between tumultuous families' social judgments and potential risks for maladjustment. The personal domain has been defined as the set of actions that are beyond the bounds of justifiable social regulation and moral concern. Thus, control over personal issues serves to establish the boundary between the self and others. This definition of the personal domain is consistent with what William James (1899) described as an awareness of the "I," that aspect of the self that organizes and interprets experiences as reflective awareness of one's agency, continuity, and distinctness. Theorists from a variety of theoretical perspectives (Baldwin, 1906; Damon & Hart, 1988; Erikson, 1968; Kohut, 1978; Mahler, 1979; Selman, 1980) as well as researchers examining the development of self and personhood across cultural contexts (Geertz, 1975; Markus & Kitayama, 1991; Shweder & Bourne, 1984) have argued for a close connection between personal autonomy and the formation of the individual. Consistent with these theorists, Nucci (1996) has asserted that the establishment of an arena of personal freedom is necessary for the formation of the social individual.

The child establishes personal concepts through negotiation with parents and other agents of society (Nucci, 1996). For instance, Nucci and Weber (1995) have reported observations of 3- and 5-year old children's interactions with mothers that suggest that the personal domain is socially constructed through reciprocal parent-child interactions. Likewise, the research on adolescent-parent conflict discussed here suggests that conflict forces parents to reevaluate the limits of their authority, as adolescents actively attempt to negotiate the boundaries of their personal jurisdiction. Adolescents and parents negotiate from very different positions in the social structure. Thus, parents' appeals to social convention, adolescents' rejection of their parents' perspective, and their reinterpretation of con- ventions as legitimately under their personal jurisdiction form a continual dialectic in which the boundaries of parental authority are subtly transformed. Parents shift from viewing a variety of conflicts as conventional and legitimately subject to their authority to granting the adolescent increasing personal jurisdiction over these issues. Adolescents' appeals to personal jurisdiction serve to increase their agency, or enlarge their sphere of personal action, and thus serve to individuate adolescents from their parents.

As Nucci (1996) has pointed out, clinical data demonstrate that disruption in the formation of these personal boundaries can damage individual psychological health (Kernberg, 1975; Kohut, 1978; Mahler, 1979). There appears to be basic psychological limits to the extent to which others can impinge on others' private lives. In this light, it is especially interesting to note that one characteristic of tumultuous families was their restrictive judgments regarding adolescents' personal domain. Tumultuous families viewed rules about personal issues as more obligatory than did other families, and fathers from tumultuous families granted adolescents less personal jurisdiction over personal issues than did other fathers. Although these findings were obtained in normal families, the theory and research just discussed suggest that there may be connections among tumultuous families' treatment of personal issues, adolescents' formation of their personal domains, and risks for psychological maladjustment that deserve further investigation. For instance, Nucci (1996) presents an illuminating clinical case study of a young woman with a borderline personality disorder. Although psychological disorders cannot be reduced simply to an issue of personal autonomy, this case suggests intriguing links between psychopathology and extreme parental control over and intrusion into the personal spheres of the young woman's life, including issues of eating, personal friendships, and recreational activities (all prototypical aspects of the personal domain). This speculation is compatible with Noam (1993), who has elaborated a view of how delays in self and ego development may lead to vulnerabilities in adolescent development.

Optimal Development

It has been noted that adolescent mental health includes not only the absence of dysfunction, but also *optimal* social and psychological functioning (Kazdin, 1993). Thus, in addition to identifying families or adolescents at risk, it is important to identify the patterns of interaction that lead to optimal functioning. Several findings from the analyses reported here suggest that placid families promote adolescent development. Placid adolescents were lower in emotional autonomy (detachment from parents), had higher academic achievement, and were more mature in their developmental level of reasoning about social conventions than were other adolescents. Furthermore, the description of placid families and adolescents is consistent with Offer and Offer's (1975) description of adolescent males in their continuous growth group; these adolescents were seen as displaying many of the qualities identified with positive mental health. However, the age differences observed between adolescents from placid and frequently squabbling families in the analyses presented here and the lack of differentiation between these families on many measures suggests another possibility: that frequent squabbling is a developmental phase on the route to more placid family relations. This is an intriguing possibility that must be addressed through longitudinal research, but it is consistent with the view that conflict is a temporary perturbation, characteristic primarily of early adolescence, that leads to transformations or realignments in parent-child relationships (Collins, 1990; Hill, 1987, 1988; Silverberg et al., 1992; Smetana, 1994; Steinberg, 1990).

The hypothesis that placid family relations promote optimal adolescent development appears, at first glance, to be at odds with the current consensus that moderate amounts of conflict, in the context of familial warmth and support, are adaptive for adolescent development and relationship change (Collins & Laursen, 1992; Grotevant & Cooper, 1985, 1986; Hill, 1988; Smetana, 1988b, 1994). For instance, I (Smetana, 1994, 1995a), as well as others (Collins & Laursen, 1992; Hill, 1987, 1988), have hypothesized that conflict provides a context for transformations in parental authority to occur. However, the qualitative analyses indicated clearly that placid families did have conflicts (although fewer in number than tumultuous families); rather, they appeared to negotiate and resolve them more effectively. It is also possible that researchers' focus on conflict has obscured our understanding of the types of interactions that lead to relationship change.

Several programs of research have provided information about the types of constructive family interactions that facilitate adolescent development. Hauser and his colleagues (Hauser, Powers, Noam, Jacobson, Weiss, & Follansbee, 1984; Leaper, Hauser, Kremen, Powers, Jacobson, Noam, Weiss-Perry, & Follansbee, 1989; Powers, 1988) have found that enabling interactions, entailing acceptance and empathy, as well as focusing, problem solving, and curiosity in both adolescents and parents facilitate adolescents' ego and moral development. Cluster analyses of these same data revealed greater adolescent ego development in families where differences of opinion in a revealed differences task were discussed in a noncompetitive atmosphere (Powers, Hauser, Schwartz, Noam, & Jacobson, 1983). Competitive, challenging behaviors also led to high levels of ego development when there was little avoidance, rejection, or distortion in communication. These results are consistent with Grotevant and Cooper (Cooper, Grotevant, & Condon, 1983; Grotevant & Cooper, 1985, 1986), who report that both separation (indexed partially by both direct and indirect disagreements and warmth) and connectedness with parents, assessed in the context of a family interaction task, facilitate adolescents' identity development and role-taking skills. Furthermore, greater compromise during conflicts with peers was associated with higher levels of self-esteem (Cooper & Cooper, 1992). Thus, the findings from these research programs suggest that optimal development in the emotional realm entails an interplay between individuation and connectedness and that adolescents (at least white, middle-class, American adolescents) fare best in an environment that facilitates an open exchange of views in the context of warmth, supportiveness, and trust. Furthermore, optimal development in the social-cognitive realm appears to entail a willingness to negotiate the boundaries between adolescents' personal jurisdiction and parental and societal conventions. This is consistent with the picture of placid families that emerged from the analyses described in this chapter.

If the analyses and speculations presented here are correct, bickering may be the most characteristic pattern of normative family relations in Western, middle-class, white families with early and mid adolescents. Adolescents in tumultuous families may be most at risk for developing problem behaviors and in more extreme forms, disruptions in the development of self that may lead to psychopathology. Finally, placidity in family relations, possibly following a developmental period of frequent squabbling, may facilate adolescent development and protect against maladaptation. It may be time for researchers studying normal adolescent development to shift from studying adolescent-parent conflict to determining how to promote the types of healthy interactions, including clear communication, developmentally appropriate boundaries between parental authority and adolescent jurisdiction, and negotiation and compromise, that appear to promote optimal adolescent development in the context of the family.

REFERENCES

Adams, G. R., Gulotta, T., & Clancy, M. N. (1985). Homeless adolescents: A descriptive study of similarities and differences between runaways and throwaways. *Adolescence, 20*, 715–724.

Baer, P. E., Garmezy, L. B., McLaughlin, R. J., Pokorny, A. D., & Wernick, M. J. (1987). Stress, coping, family conflict, and adolescent alcohol use. *Journal of Behavioral Medicine, 10*, 449–466.

Baldwin, J. M. (1906). *Thought and things* (Vol. I). London: Swan Sonnenschen.

Baumrind, D. (1971). Current patterns of parental authority. *Developmental Psychology Monographs, 4* (1, Pt. 2).

Bell, R. Q. (1968). A reinterpretation of the direction of effects in studies of socialization. *Psychological Review, 75*, 81–95.

Benedict, R. (1938). Continuities and discontinuities in cultural conditioning. *Psychiatry, 1*, 161–167.

Berndt, T. J., Cheung, P. C., Lau, S., Hau, K.-T., & Lew, W. J. F. (1993). Perceptions of parenting in mainland China, Taiwan, and Hong Kong: Sex differences and societal differences. *Developmental Psychology, 29*, 156–164.

Blos, P. (1962). *On adolescence.* New York: Free Press.

Blos, P. (1979). *The adolescent passage.* Madison, CT: International Universities Press.

Borduin, C. M., Pruitt, J. A., & Henggeler, S. W. (1986). Family interactions in black, lower-class families with delinquent and nondelinquent boys. *Journal of Genetic Psychology, 147*, 333–342.

Buri, J. R. (1989). Self-esteem and appraisals of parental behavior. *Journal of Adolescent Research, 4*, 33–49.

Buri, J. R. (1991). Parental authority questionnaire. *Journal of Personality Assessment, 57*, 110–119.

Buri, J. R., Louiselle, P. A., Misukanis, T. M., & Mueller, R. A. (1988). Effects of parental authoritarianism and authoritativeness on self-esteem. *Personality and Social Psychology Bulletin, 14*, 271–282.

Cicchetti, D. (1993). Developmental psychopathology: Reactions, reflections, and projections. *Developmental Review, 13*, 471-502.

Colby, A., & Kohlberg, K. (Eds.). (1987). *The measurement of moral judgment* (Vols. 1–2). New York: Cambridge University Press.

Collins, W. A. (1990). Parent-child relationships in the transition to adolescence: Continuity and change in interaction, affect, and cognition. In R. Montemayor, G. R. Adams, & T. P. Gulotta (Eds.), *From childhood to adolescence: A transitional period?* (pp. 85–106). Newbury Park, CA: Sage.

Collins, W. A., & Laursen, B. (1992). Conflict and relationships during adolescence. In C. U. Shantz & W. W. Hartup (Eds.), *Conflict in child and adolescent development* (pp. 216–241). Cambridge, England: Cambridge University Press.

Cooper, C. R., & Cooper, R. G., Jr. (1992). Links between adolescents' relationships with their parents and peers: Models, evidence, and mechanisms. In R. D. Parke & G. W. Ladd (Eds.), *Family-peer relationships: Modes of linkage* (pp. 135–158). Hillsdale, NJ: Erlbaum.

Cooper, C. R., Grotevant, H. D., & Condon, S. M. (1983). Individuality and connectedness as a context for adolescent identity formation and role-taking skill. In H. D. Grotevant & C. R. Cooper (Eds.), *New directions for child development: Adolescent development in the family* (pp. 43–59). San Francisco: Jossey-Bass.

Corder, B. F., Shorr, W., & Corder, R. F. (1974). A study of social and psychological characteristics of adolescent suicide attempters in an urban, disadvantaged area. *Adolescence, 9,* 1–6.

Csikszentmihalyi, M., & Larson, R. (1984). *Being adolescent.* New York: Basic Books.

Cummings, E. M., Vogel, D., Cummings, J. S., & El-Sheikh, M. (1989). Children's responses to different forms of expression of anger between adults. *Child Development, 60,* 1392–1404.

Cummings, E. M., Zahn-Waxler, C., & Radke-Yarrow, M. (1981). Young children's responses to expression of anger and affection by others in the family. *Child Development, 52,* 1274–1281.

Damon, W. (1977). *The social world of the child.* San Francisco: Jossey-Bass.

Damon, W., & Hart, D. (1988). *Self-understanding in childhood and adolescence.* Cambridge, England: Cambridge University Press.

Dornbusch, S. M., Carlsmith, J. M., Bushwall, S. J., Ritter, P. L., Leiderman, H., Hastorf, A. H., & Gross, R. T. (1985). Single-parents, extended households, and control of adolescents. *Child Development, 56,* 326–341.

Douvan, E., & Adelson, J. (1966). *The adolescent experience.* New York: Wiley.

Duck, S., & Pond, K. (1989). Friends, Romans, and countrymen, lend me your retrospections: Rhetoric and reality in personal relationships. In C. Hendrick (Ed.), *Close relationships* (pp. 17–38). Newbury Park, CA: Sage.

Dworkin, R. (1978). *Taking rights seriously.* Cambridge, England: Harvard University Press.

Emery, R. E. (1992). Family conflicts and their developmental implications: A conceptual analysis of meanings for the structure of relationships. In C. U. Shantz & W. W. Hartup (Eds.), *Conflict in child and adolescent development* (pp. 270–298). Cambridge, England: Cambridge University Press.

Erikson, E. H. (1968). *Identity, youth, and crisis.* New York: Norton.

Forehand, R., Long, N., & Hendrick, M. (1987). Family characteristics of adolescents who display overt and covert behavior problems. *Journal of Behavior Therapy and Experimental Psychiatry, 18,* 325–328.

Forehand, R., Wierson, M., Thomas, A. M., Armistead, L., Kempton, K., & Neighbors, B. (1991). The role of family stressors and parent relationships on adolescent functioning. *Journal of the American Academy of Child and Adolescent Psychiatry, 30,* 316–322.

Freud, A. (1937). *The ego and the mechanisms of defense.* London: Hogarth Press.

Freud, A. (1958). Adolescence. *Psychoanalytic Study of the Child, 13,* 255–278.

Geertz, C. (1975). On the nature of anthropological understanding. *American Scientist, 63,* 47–53.

Gewirth, A. (1978). *Reason and morality.* Chicago: University of Chicago.

Gewirth, A. (1982). *Human rights: Essays on justification and applications.* Chicago: University of Chicago Press.

Goffman, E. (1971). *Relations in public.* New York: Harper & Row.

Gottman, J. M. (1979). *Marital interaction: Experimental investigations.* San Diego, CA: Academic Press.

Grotevant, H. D., & Cooper, C. R. (1985). Patterns of interaction in family relationships and the development of identity exploration in adolescence. *Child Development, 56,* 415–428.

Grotevant, H. D., & Cooper, C. R. (1986). Individuation in family relationships. *Human Development, 29*, 82–100.

Grych, J. H., & Fincham, F. D. (1990). Marital conflict and children's adjustment: A cognitive-contextual framework. *Psychological Bulletin, 108*, 267–290.

Grych, J. H., & Fincham, F. D. (1993). Children's appraisals of marital conflict: Initial investigations of the cognitive-contextual framework. *Child Development, 64*, 215–230.

Hauser, S. T., Powers, S. I., Noam, G. G., Jacobson, A. M., Weiss, B., & Follansbee, D. (1984). Familial contexts of adolescent ego development. *Child Development, 55*, 195–213.

Hauser, S. T., Powers, S. I., Weiss-Perry, B., Follansbee, D. , & Rajanpark, D. C. (1985). *Constraining and Enabling Family Coding System.* Unpublished manuscript, Harvard University.

Hill, J. (1987). Research on adolescents and their families: Past and prospect. In C. E. Irwin, Jr. (Ed.), *New directions for child development: Vol. 27, Adolescent social behavior and health* (pp. 13–31). San Francisco: Jossey-Bass.

Hill, J. (1988). Adapting to menarche: Familial control and conflict. In M. R. Gunnar & W. A. Collins (Eds.), *Minnesota Symposia on Child Psychology: Development during the transition to adolescence,* vol. 21 (pp. 43–78). Hillsdale, NJ: Erlbaum.

Hill, J., & Holmbeck, G. (1987). Familial adaptation to pubertal change during adolescence. In R. M. Lerner & T. Foch (Eds.), *Biological-psychosocial interactions in early adolescence: A life-span perspective* (pp. 207–223). Hillsdale, NJ: Erlbaum.

Inazu, J. K., & Fox, G. L. (1980). Maternal influence on the sexual behavior of teenage daughters. *Journal of Family Issues, 1*, 81–102.

Jacob, T. (1975). Family interaction in disturbed and normal families: A methodological and substantive review. *Psychological Bulletin, 82*, 33–65.

James, W. (1899). *The principles of psychology.* London: Macmillan.

Kandel, D. B., Kessler, R. C., & Marguilies, R. Z. (1978). Antecedents of adolescent initiation into stages of drug use: A developmental analysis. In D. B. Kandel (Ed.), *Longitudinal research on drug use: Empirical findings and methodological issues* (pp. 73–99). New York: Wiley.

Kandel, D. B., & Lesser, G. S. (1972). *Youth in two worlds.* San Francisco: Jossey-Bass.

Kashani, J. H., Burbach, D. J., & Rosenberg, T. K. (1988). Perception of family conflict resolution and depressive symptomatology in adolescents. *Journal of the American Academy of Child and Adolescent Psychiatry, 27*, 42–48.

Kazdin, A. E. (1993). Adolescent mental health: Prevention and treatment programs. *American Psychologist, 48*, 127–141.

Kelley, G. A. (1955). *A psychology of personal constructs.* New York: Norton.

Kernberg, O. F. (1975). *Borderline conditions and pathological narcissism.* New York: J. Aronson.

Kessen, W. (1979). The American child and other cultural inventions. *American Psychologist, 34*, 815–820.

Kohlberg, L. (1971). From is to ought: How to commit the naturalistic fallacy and get away with it in the study of moral development. In T. Mischel (Ed.), *Cognitive development and epistemology* (pp. 151–235). New York: Academic Press.

Kohut, H. (1970). *The search for the self: Selected writings (1950–1978)* New York: International University Press.

Krinsley, K. D., & Bry, B. J. (1992). Sequential analyses of adolescent, mother, and father behaviors in distressed and nondistressed families. *Child and Family Behavior Therapy, 13*, 45–62.

Lamborn, S. D., Mounts, N. S., Steinberg, L., & Dornbusch, S. M. (1994). Patterns of competence and adjustment among adolescents from authoritative, authoritarian, indulgent, and neglectful homes. *Child Development, 62*, 1049–1065.

Lau, S., & Cheung, P. C. (1987). Relations between Chinese adolescents' perception of parental control and organization and their perception of parental warmth. *Developmental Psychology, 23*, 726–729.

Laursen, B., & Collins, W. A. (1994). Interpersonal conflict during adolescence. *Psychological Bulletin, 115*, 197–209.

Laursen, B., & Hartup, W. W. (1989). The dynamics of preschool children's conflicts. *Merrill-Palmer Quarterly, 35*, 281–297.

Laursen, B., & Koplas, A. L. (1995). What's important about important conflicts: Adolescents' perceptions of daily disagreements. *Merrill-Palmer Quarterly, 41*, 536-553.

Leaper, C., Hauser, S. T., Kremen, A., Powers, S. I., Jacobson, A. M., Noam, G. G., Weiss-Perry, B., & Follansbee, D. J. (1989). Adolescent-parent interaction in relation to adolescents' gender and ego development pathway: A longitudinal study. *Journal of Early Adolescence, 9*, 335–361.

LeVine, R., & White, M. (1986). *Human conditions: The cultural basis for educational development.* New York: Routledge & Kegan Paul.

Maggs, J. L., & Galambos, N. L. (1993). Alternative structural models for understanding adolescent problem behavior in two-earner families. *Journal of Early Adolescence, 13*, 79–101.

Mahler, M. S. (1979). *The selected papers of Margaret S. Mahler: Vols. I & II.* New York: J. Aronson.

Markus, H. R., & Kitayama, S. (1991). Culture and the self: Implications for cognition, emotion, and motivation. *Psychological Bulletin, 98*, 224–253.

McCord, J. (1990). Problem behaviors. In S. S. Feldman & G. R. Elliot (Eds.), *At the threshhold: The developing adolescent* (pp. 414–430). Cambridge, MA: Harvard University Press.

Mill, J. S. (1968). *Utilitarianism.* New York: Washington Square Press. (Original work published 1863).

Montemayor, R. (1982). The relationship between parent-adolescent conflict and the amount of time adolescents spend alone and with parents and peers. *Child Development, 53*, 1512–1519.

Montemayor, R. (1983). Parents and adolescents in conflict: All families some of the time and some families most of the time. *Journal of Early Adolescence, 3*, 83–103.

Montemayor, R. (1986). Family variation in storm and stress. *Journal of Adolescent Research, 1*, 15–31.

Noam, G. G. (1993). "Normative vulnerabilities" of the self and their transformations in moral action. In G. G. Noam and T. E. Wren (Eds.), *The moral self: Building a better paradigm* (pp. 209–238). Cambridge, MA: MIT Press.

Nucci, L. (1981). The development of personal concepts: A domain distinct from moral or societal concepts. *Child Development, 52*, 114–121.

Nucci, L. P. (1996). Morality and the personal sphere of actions. In E. S. Reed, E. Turiel, & T. Brown (Eds.), *Values and Knowledge* (pp. 41-60) Mahwah, NJ: Erlbaum.

Nucci, L. P., Guerra, N., & Lee, J. (1991). Adolescent judgments of the personal, prudential, and normative aspects of drug usage. *Developmental Psychology, 27*, 841–848.

Nucci, L. P., & Lee, J. (1993). Morality and personal autonomy. In G. G. Noam & T. Wren (Eds.), *The moral self: Building a better paradigm* (pp. 123–148). Cambridge, MA: MIT Press.

Nucci, L. P., & Weber, E. (1995). Social interactions in the home and the development of young children's conceptions of the personal. *Child Development, 66*, 1438-1452.

Offer, D. (1969). *The psychological world of the teenager.* New York: Basic Books.

Offer, D., & Offer, J. B. (1975). *From teenage to young manhood: A psychological study.* New York: Basic Books.

Offer, D., Ostrov, E., & Howard, K. (1981). *The adolescent: A psychological self-portrait*. New York: Basic Books.

Paikoff, R. L., & Brooks-Gunn, J. (1991). Do parent-child relationships change during puberty? *Psychological Bulletin, 110*, 47–66.

Patterson, G. R. (1982). *Coercive family process*. Eugene, OR: Castalia Publishing.

Patterson, G. R., & Bank, L. (1989). Some amplifying mechanisms for pathologic process in families. In M. R. Gunnar & E. Thelen (Eds.), *Minnesota Symposia on Child Psychology*: Vol. 22. (pp. 167–209). Hillsdale, NJ: Erlbaum.

Patterson, G. R., Capabaldi, D., & Bank, L. (1991). An early starter model for predicting delinquency. In D. J. Peplier & K. Rubin (Eds.), *The development and treatment of childhood aggression* (pp. 139–168). Hillsdale, NJ: Erlbaum.

Piaget, J. (1965). *The moral judgment of the child*. New York: Free Press. (Original work published 1932).

Powers, S. I. (1988). Moral judgment development in the family. *Journal of Moral Education, 17*, 209–219.

Powers, S. I., Hauser, S. T., & Kilner, L. (1989). Adolescent mental health. *American Psychologist, 44*, 200—208.

Powers, S., Hauser, S. T., Schwartz, J. M., Noam, G. G., & Jacobson, A. M. (1983). Adolescent ego development and family interaction: A structural-developmental perspective. In H. D. Grotevant & C. R. Cooper (Eds.), *New directions for child development: Adolescent development in the family* (pp. 5–25). San Francisco: Jossey-Bass.

Prinz, R. J., Foster, S. L., Kent, R. N., & O'Leary, K. D. (1979). Multivariate assessment of conflict in distressed and non-distressed mother-adolescent dyads. *Journal of Applied Behavior Analysis, 12*, 691–700.

Puig-Antich, J., Kaufman, J., Ryan, N. D., & Williamson, E. E. (1993). The psychosocial functioning and family environment of depressed adolescents. *Journal of the American Academy of Child and Adolescent Psychiatry, 32*, 244–253.

Rawls, J. (1971). *A theory of justice*. Cambridge, MA: Harvard University Press.

Robin, A. L., & Foster, S. L. (1989). *Negotiating parent-adolescent conflict: A behavioral-family systems approach*. New York: Guilford.

Rutter, M., Graham, P., Chadwick, O. F. D., & Yule, W. (1976). Adolescent turmoil: fact or fiction? *Journal of Child Psychology and Psychiatry, 17*, 35–56.

Ryan, R. M., & Lynch, J. H. (1989). Emotional autonomy versus detachment: Revisiting the vicissitudes of adolescence and young adulthood. *Child Development, 60*, 340–346.

Schlegel, A., & Barry, H., 3rd. (1991). *Adolescence: An anthropological inquiry*. New York: Free Press.

Selman, R. (1980). *The growth of interpersonal understanding: Developmental and clinical analyses*. New York: Academic Press.

Shweder, R. A. (1986). Divergent rationalities: In D. W. Fiske & R. A. Shweder (Eds.), *Metatheory in social science: Pluralisms and subjectivities* (pp. 163–196). Chicago: University of Chicago Press.

Shweder, R. A., & Bourne, E. J. (1984). Does the concept of the person vary cross-culturally? In R. A. Shweder & R. A. Levine (Eds.), *Culture theory: Essays on mind, self, and emotion* (pp. 158–199). Cambridge, England: Cambridge University Press.

Shweder, R. A., Turiel, E., & Much, N. (1981). The moral intuitions of the child. In J. H. Flavell & L. Ross (Eds.), *Social-cognitive development: Frontiers and possible futures* (pp. 288–305). Cambridge: Cambridge University Press.

Silverberg, S. B., Tennenbaum, D. L. , & Jacob, T. (1992). Adolescence and family interaction. In V. B. Van Hasselt & M. Hersen (Eds.), *Handbook of social development: A lifespan perspective* (pp. 347–370). New York: Plenum Press.

Simmons, R. G., Burgeson, R., Carlton-Ford, S., & Blyth, D. A. (1987). The impact of cumulative change in early adolescence. *Child Development, 58,* 1220–1234.

Smetana, J. G. (1981). Preschool children's conceptions of moral and social rules. *Child Development, 52,* 1333–1336.

Smetana, J. G. (1982). *Concepts of self and morality: Women's reasoning about abortion.* New York: Praeger.

Smetana, J. G. (1983). Social-cognitive development: Domain distinctions and coordinations. *Developmental Review, 3,* 131–147.

Smetana, J. G. (1988a). Adolescents' and parents' conceptions of parental authority. *Child Development, 59,* 321–335.

Smetana, J. G. (1988b). Concepts of self and social convention: Adolescents' and parents' reasoning about hypothetical and actual family conflicts. In M. R. Gunnar & W. A. Collins (Eds.), vol. 21 *Minnesota Symposia on Child Psychology: Development during the transition to adolescence* (pp. 79–122). Hillsdale, NJ: Erlbaum.

Smetana, J. G. (1989). Adolescents' and parents' reasoning about actual family conflicts. *Child Development, 60,* 1052–1067.

Smetana, J. G. (1993). Conceptions of parental authority in divorced and married mothers and their adolescents. *Journal of Research in Adolescence, 3,* 19–40.

Smetana, J. G. (1994). Conflict and coordination in adolescent-parent relationships. In S. Shulman (Ed.), *Close relationships and socioemotional development* (pp. 155–184). Norwood, NJ: Ablex.

Smetana, J. G. (1995a). Context, conflict, and constraint in parent-adolescent authority relationships. In M. Killen & D. Hart (Eds.), *Morality in everyday life: Developmental perspectives.* (pp. 225-255). Cambridge, MA: Cambridge University Press.

Smetana, J. G. (1995b). Parenting styles and conceptions of parental authority during adolescence. *Child Development, 66,* 299–316.

Smetana, J. G., & Asquith, P. (1994). Adolescents' and parents' conceptions of parental authority and adolescent autonomy. *Child Development, 65,* 1143–1158.

Smetana, J. G., & Berent, R. (1993). Adolescents' and mothers' evaluations of justifications for disputes. *Journal of Adolescent Research, 8,* 252–273.

Smetana, J. G., Braeges, J. L., & Yau, J. (1991). Doing what you say and saying what you do: Reasoning about adolescent-parent conflict in interviews and interactions. *Journal of Adolescent Research, 6,* 276–295.

Smetana, J. G., Yau, J., & Hanson, S. (1991). Conflict resolution in families with adolescents. *Journal of Research on Adolescence, 1,* 189–206.

Smetana, J. G., Yau, J., Restrepo, A., & Braeges, J. (1991). Adolescent-parent conflict in married and divorced families. *Developmental Psychology, 27,* 1000–1010.

Steinberg, L. (1987). Recent research on the family at adolescence: The extent and nature of sex differences. *Journal of Youth and Adolescence, 3,* 191–197.

Steinberg, L. (1990). Interdependence in the family: Autonomy, conflict, and harmony in the parent-adolescent relationship. In S. S. Feldman & G. R. Elliot (Eds.), *At the threshhold: The developing adolescent* (pp. 255–276). Cambridge, MA.: Harvard University Press.

Steinberg, L., Lamborn, S. D., Darling, N. S., Mounts, N. S., & Dornbusch, S. M. (1994). Over-time changes in adjustment and competence among adolescents from authoritative, authoritarian, indulgent, and neglectful families. *Child Development, 65,* 754–770.

Steinberg, L., & Silverberg, S. B. (1986). The vicissitudes of autonomy in early adolescence. *Child Development, 57*, 841–851.

Tisak, M. (1993). Preschool children's judgments of moral and personal events involving physical harm and property damage. *Merrill-Palmer Quarterly, 39*, 375–390.

Tisak, M. S., & Turiel, E. (1984). Children's conceptions of moral and prudential rules. *Child Development, 55*, 1030–1039.

Tishler, C. L., McKenry, P. C., & Morgan, K. C. (1981). Adolescent suicide attempts: Some significant factors. *Suicide and Life-Threatening Behavior, 11*, 86–92.

Triandis, H. C. (1989). The self and social behavior in differing cultural contexts. *Psychological Review, 96*, 508–520.

Triandis, H. C. (1990). Cross-cultural studies of individualism and collectivism. In J. J. Berman (Eds.), *1989 Nebraska Symposium on Motivation: Vol 37. Cross cultural perspectives* (pp. 41–133). Lincoln: University of Nebraska Press.

Turiel, E. (1978). Social regulation and domains of social concepts. In W. Damon (Ed.), *New Directions for Child Development Vol. I: Social cognition* (pp. 45–74). San Francisco: Jossey-Bass.

Turiel, E. (1979). Distinct conceptual and developmental domains: Social convention and morality. In C. B. Keasey (Ed.), *Nebraska Symposium on Motivation* (pp. 77–116). Lincoln: University of Nebraska.

Turiel, E. (1983). *The development of social knowledge: Morality and convention.* Cambridge: Cambridge University Press.

Turiel, E., & Davidson, P. (1986). Heterogeneity, inconsistency, and asynchrony in the development of cognitive structures. In I. Levin (Ed.), *Stage and structure: Reopening the debate* (pp. 106–143). Norwood, NJ: Ablex.

Weinberger, D., Tublin, S., Ford, M., & Feldman, S. (1990). Preadolescents' social-emotional adjustment and selective attrition in family research. *Child Development, 61*, 1374–1386.

Yau, J., & Smetana, J. G. (1996). *Adolescent-parent conflict among Chinese adolescents in Hong Kong.* Child Development, 67, 1262-1275.

Youniss, J., & Smollar, J. (1985). *Adolescents' relations with mothers, fathers, and friends.* Chicago: University of Chicago Press.

II   Round Holes, Square Pegs, Rocky Roads, and Sore Feet: The Impact of Stage-Environment Fit on Young Adolescents' Experiences in Schools and Families

Jacquelynne S. Eccles, Sarah E. Lord, & Robert W. Roeser

There has been a dramatic increase over the last 10 to 15 years in research attention to adolescence. Adolescence is a particularly good developmental period to study the ontogeny of individual differences in adjustment given the multitude of changes that can co-occur during this period. Very few developmental periods are characterized by so many changes on so many different levels—changes due to pubertal development, social role redefinitions, cognitive development, school transitions, and the emergence of sexuality. The core developmental tasks that accompany these changes include achieving a greater sense of autonomy from parents, confronting issues related to sexuality and expanded peer relationships, and establishing a coherent sense of personal competence and identity. In addition, the early adolescent is often faced with adjusting to the demands of a new school environment that places more emphasis on extrinsic motivation and relative ability, teacher control, and involvement in a variety of extracurricular activities. All of these changes can challenge adolescents' evolving sense of self and can offer newfound opportunities for identity exploration and social role definition. These concurrent changes in both the individual and in his or her surrounding social contexts during the period of adolescence make it an ideal focus for the study of human development.

With the nature and pace of the challenges associated with the developmental changes of adolescence also come opportunities for adolescents to embark on different developmental trajectories—trajectories that can result in either positive or negative developmental outcomes. While for many early adolescents the challenges

The research discussed in this chapter was made possible by grants from the National Institute of Child Health and Human Development, the National Science Foundation, the Spencer Foundation, and the MacArthur Foundation Research Network on Successful Adolescent Development. We wish to thank all of our colleagues for assistance in designing, running, and analyzing the data from the studies reported herein. Special thanks go to Christy Miller Buchanan, Harriet Feldlaufer, Constance Flanagan, Jan Jacobs, Douglas Mac Iver, Carol Midgley, Allan Wigfield, David Reuman for their work on MSALT and Elaine Belansky, Diane Early, Kari Fraser, Ariel Kalil, Linda Kuhn, Karen McCarthy, and Sherri Steele for their work on MAGICS. We would also like to thank all of the teachers, school personnel, families, and students who agreed to participate in these studies. Copies can be obtained from the first author at the Institute for Social Research, University of Michigan, Ann Arbor, MI 48106-1248.

of this period can promote positive growth and adjustment, for others the multiple changes can render them vulnerable to self-esteem and mental health difficulties, as well as to peer pressure for involvement in a variety of risky and problem behaviors. As a result, a substantial portion of America's adolescents are not doing very well: between 15 and 30% (depending on ethnic group) drop out of school before completing high school; adolescents have the highest arrest rate of any age group; and increasing numbers of adolescents consume alcohol and other drugs on a regular basis (Office of Educational Research and Improvement, 1988). In addition, studies have shown academic failure to co-occur with many of the other problem behaviors that are increasingly manifest during adolescence, such as disruptive behavior, truancy, delinquency, substance abuse, and teenage pregnancy (Donovan & Jessor, 1985; Dryfoos, 1990; Rosenbaum, 1976).

In addition to an increased incidence in behavioral problems, the prevalence of several types of clinical dysfunctions also increase during this developmental period (Carnegie Council on Adolescent Development, 1989; Kazdin, 1993). For example, there is an increase in the prevalence of depression and eating disorders, particularly among females (e.g., Kazdin, 1993; Lewinsohn, Hops, Roberts, & Seeley, 1993; Petersen, Compas, Brooks-Gunn, Stemmler, Ey, & Grant, 1993). Perhaps most serious, the incidence of attempted and completed suicide increases dramatically with the onset of adolescence (e.g., Hawton, 1982).

Given the fact that many of these problems begin during the early adolescent years (Carnegie Council on Adolescent Development, 1989), the question of Why? arises. Is there something unique about this developmental period that puts individuals at greater risk for difficulty as they pass through it? This paper focuses on these questions and advances the hypothesis that some of the "negative" psychological and behavioral changes associated with adolescent development result from a mismatch between the needs of the developing adolescent and the opportunities afforded them by their social environments. Consistent with the view elaborated by Higgins and Parsons (1983), we suggest that the unique transitional nature of early adolescence results, at least in part, from an interaction between the developmental characteristics of early adolescents and the social contexts in which they live. We assert that the manner in which these changes are negotiated by the adolescent and the other members of their social worlds can either facilitate or undermine positive adjustment. In particular, we advance the hypothesis that some of the negative psychological and behavioral changes associated with adolescent development result from a mismatch between the needs of developing adolescents and the opportunities afforded them by their social environments. We provide examples of how this mismatch develops and operates in two particularly relevant social environments: the school and the home.

Before discussing how changes in these social contexts influence changes in adolescent functioning and behavior, we first discuss how our studies of "normative" development during this period can be framed within a developmental psychopathology framework. We then review the evidence on the "problematic" psychological and behavioral changes associated with the period of early adolescence.

Normative Adolescent Development as a Domain of Developmental
Psychopathology

In this chapter, we have adopted a developmental psychopathology organizational framework to address the question of why early adolescence seems to be a time of heightened risk of maladjustment for some young people. From this perspective, it is assumed that information about individual differences in patterns of normative adolescent development can provide useful information for understanding the processes associated with the development of more serious psychopathology. Patterns of positive and negative adjustment within a normative sample can be used to identify potential risk and protective factors for the development of more serious problems, and may provide clues to the etiology and processes associated with both normative and pathological development. Such information can be used not only to inform theory but, more practically, to guide prevention and intervention strategies. While the individual differences discussed in this chapter may not necessarily reflect functional impairments at a clinical level, some of the negative changes noted in early adolescence are quite debilitating (e.g., school dropout, drug/alcohol addiction, teen pregnancy, AIDS), and potentially have long-term detrimental consequences for an adolescent's future.

Problematic Changes Associated with
Early Adolescent Development

Research on academic motivation, achievement, and mental health issues during adolescence has expanded considerably over the last two decades (Cicchetti & Toth, 1993; Eccles & Midgley, 1989; Simmons & Blyth, 1987; Zaslow & Takanishi, 1993). While many negotiate this period with few difficulties, evidence does suggest that adolescents today face greater risks to their current and future well-being than at any time previously documented (Barber & Crockett, 1993; Takanishi, 1993). Early adolescence marks the beginning of a downward spiral for some that eventuates in academic failure, school dropout, delinquency, and substance abuse (see Dryfoos, 1990; Finn, 1989; Simmons & Blyth, 1987). The research documenting these negative changes is reviewed in the next section.

Changes in Academic Motivation and Achievement

Evidence from several sources suggests that early adolescence is marked by important changes in individuals' psychological functioning, particularly with regard to school adjustment. For example, developmental declines have been documented for such motivational constructs as interest and feelings of belonging in school (Epstein & McPartland, 1976; Roeser, Midgley, & Maehr, 1994); intrinsic motivation (Harter, 1981); self-concepts / self-perceptions (Eccles, Midgley, &

Adler, 1984; Harter, 1982; Simmons, Blyth, Van Cleave, & Bush, 1979), and confidence in one's intellectual abilities, especially following failure (Parsons & Ruble, 1977). There are also reports of age-related increases during early adolescence in such negative motivational and behavioral characteristics as test anxiety (Hill, 1980) and general academic worries (Schulenberg, Asp, & Petersen, 1984; McGuire, Mitic, & Neumann, 1987), learned helpless responses to failure (Rholes, Blackwell, Jordan, & Walters, 1980), and a focus on self-evaluation rather than task mastery (Nicholls, 1980). Finally, Simmons and Blyth (1987) found a marked decline in some early adolescents' school grades as they moved into junior high school, with the magnitude of this decline being predictive of subsequent school failure and dropout.

In sum, these studies, using data from normative adolescent samples, suggest a negative pattern of changes for a large number of young adolescents. These changes are problematic because scholastic anxiety and stress, loneliness and isolation, and feelings of low academic competence are all known to be risk factors for the psychological well-being of early adolescents as they make the transition to a new school environment and need to contend with new social and academic demands (e.g., Harter, Whitesell, & Kowalski, 1992; Kennedy, 1993; Lord, Eccles, & McCarthy, 1994). While these changes do not necessarily indicate a negative developmental trajectory characterizing all adolescents, such changes do suggest that substantial numbers of young people are at risk for poor adjustment and, potentially, for more serious negative outcomes in terms of mental health. These changes may be especially problematic among those who were most at risk before they entered the adolescent years.

Changes in Prevalence of Mental Health and Problem and Risky Behaviors

While the evidence on the prevalence of clinical dysfunctions during adolescence is not extensive, some studies do report negative changes in the psychological adjustment and well-being of some adolescents during this period. For instance, there is an increase in depressive symptomatology and disorders, especially among young women (Cohen, Cohen, Kasen, Velez, Hartmark, Johnson, Rojas, Brook, & Streuning, 1993; Elliott, Huizinga, & Menard, 1989; Kazdin, 1993; Petersen et al., 1993). While Elliott et al. (1989) found no increase in emotional problems other than depression across adolescence, other studies document an increase in eating disorders, particularly in females (e.g. Casper, Eckert, Halmi, Goldberg, & Davis, 1980; Garfinkel & Garner, 1982; Lewinsohn et al., 1993; Petersen et al., 1993).

In terms of behavioral problems, there is evidence of increases in the prevalence rates of both conduct disorders for boys and oppositional defiant disorders for both boys and girls that meet the DSM-3R criteria during middle adolescence (Cohen et al., 1993). Other evidence suggests that the rates of specific risky and problem behaviors also rise (Masten, 1988), with the prevalence of behaviors such as lying and cheating becoming especially high (Offord & Boyle, 1988). In addition, the number of youth who report current use of substances such as cigarettes, alcohol, marijuana, and cocaine also show significant increases across adolescence (e.g.,

Achenbach, Howell, Quay, & Conners, 1991; Newcomb & Bentler, 1988). Moreover, for a small percentage of adolescents, heavy use of substances such as marijuana, alcohol, and cocaine also increase.

Finally, studies show that school-related problem behaviors become more prevalent in adolescence (e.g., Achenbach et al., 1991; Dryfoos, 1990). According to Dryfoos (1990), rates of school truancy and suspension increase considerably from early to late adolescence. Among 10-14 year olds, 7% report being suspended and 18% report truanting school in a one year span. Among 15-17 year olds, these figures increase to 15% and 53% respectively. Boys have slightly higher rates of these behaviors than girls, and whites are more likely to be truant than are blacks (Dryfoos, 1990). These findings parallel other studies that have shown increases in school-related misconduct during adolescence, with boys reporting more antisocial behavior (Buchanan, Eccles, & Becker, 1992; Offord & Boyle, 1988; Simmons & Blyth, 1987).

Although these types of changes are not extreme for most adolescents, there is sufficient evidence of a gradual decline in various indicators of mental health, self-perceptions, academic motivation, performance, and positive school behaviors and a gradual increase in school problems and antisocial behaviors over the early adolescent years to make one wonder what is happening (see Eccles & Midgley, 1989, for review). It seems plausible that the problems of academic alienation, poor school performance, and minor delinquency evident in early adolescence are linked to more negative outcomes such as depression and other emotional difficulties, substance abuse, and school dropout that are manifest later in adolescence (e.g., Dryfoos, 1990; Eggert, Seyl, & Nicholas, 1990; Finn, 1989; Hawkins, Catalano, & Miller, 1992; Newcomb & Bentler, 1988). But, in fact, few studies have addressed these associations. As Finn (1989; p. 118) notes, "the educational processes linking school failure to behavior problems have received very little attention . . . leaving unanswered the question: 'What mechanisms cause these two sets of outcomes to co-occur?' " Because we know that achievement-related affects, cognitions, and behaviors (e.g., Ames, 1992a, 1992b; Nicholls, 1984), and problem behaviors all cluster together (e.g., Donovan & Jessor, 1985; Jessor & Jessor, 1977), we can extend Finn's question to inquire into the processes underlying the co-occurrence of maladaptive patterns of school motivation, mental health, and behavior during early adolescence.

While the relation of academic motivation, achievement, and mental health during adolescence is not highly developed in the literature, it seems clear that academic engagement and achievement is likely to be critical to continued patterns of personal adjustment during this period of development and may serve as a protective factor against many of the negative outcomes that are manifest later in adolescence, including problem behaviors and mental health problems (Bloom, 1976; Cowen, 1991; Rae-Grant, Thomas, Offord, & Boyle, 1989). Conversely, academic underachievement and alienation in early adolescence are likely to be risk factors for later adjustment. Promising lines of inquiry linking adolescents' school experiences with their general mental health are being undertaken. For instance, several authors have discussed the relation between negative self-perceptions of academic and social competence and internalized and externalized distress (e.g., Achenbach et al., 1991; Cole, 1991; Gold & Mann, 1985; Kennedy, 1993; Lord et al., 1994).

Changes in Family Interactions

In considering studies documenting changes in family interactions during adolescence, similar types of declines have been noted. Again, although the findings are neither universal nor indicative of major disruptions for most adolescents and their families, evidence suggests that there is a temporary increase in family conflict, particularly over issues related to autonomy and control, during the early adolescent years (see Buchanan et al., 1992; Collins, 1990; Hauser, Powers, & Noam, 1991; Hill, 1988; Montemayor, 1986; Paikoff & Brooks-Gunn, 1991; Smetana, 1988a, 1988b, 1989; and Steinberg, 1990, for recent reviews). For example, Hill and Steinberg, in both their observational and self-report studies, have found increased conflict between mothers and their sons and daughters during the early and middle adolescent years, particularly for early maturing adolescents (e.g., Hill, 1988; Steinberg, 1981, 1987, 1988).

Unraveling the Processes Related to
Negative Changes During Adolescence

A variety of explanations have been offered to explain the negative changes often associated with adolescence. For example, Simmons and her colleagues have suggested that the concurrent timing of the junior high school transition and pubertal development accounts for the declines in the school-related measures and self-esteem (e.g., Blyth, Simmons & Carlton-Ford, 1983; Simmons & Blyth, 1987). Drawing upon cumulative stress theory, they suggest that declines in motivation and mental health indicators occur because so many young adolescents must cope with at least two major transitions: pubertal change and the move to middle or junior high school. To test this hypothesis, Simmons and her colleagues compared the pattern of change on early school-related outcomes for adolescents who moved from sixth to seventh grade in a K–8, 9–12 system with the pattern of change for adolescents who made the same grade transition in a K–6, 7–9, 10–12 school system. This work unconfounds the conjoint effects of age and school transition operating in most developmental studies of this age period. These researchers find clear evidence, especially among girls, of greater negative change among adolescents making the junior high school transition between grades 6-7 than among adolescents remaining in the same school setting across these grades. But are these differences due to the cumulative impact of school transition and pubertal change for girls who moved to a junior high school at grade seven or are they due to differences in the nature of the school environments in these two educational structures? Or are the differences due to both of these sets of experiences? Simmons and her colleagues (see Simmons and Blyth, 1987) now argue for the latter.

Similarly, Eccles and her colleagues have suggested that the change in the nature of the learning environment associated with the junior high school transition is a plausible explanation for the declines in the school-related measures associated with the junior high school transition (Eccles et al., 1984; Eccles & Midgley, 1989). Drawing upon person-environment fit theory (see Hunt, 1975), Eccles and Midgley (1989) proposed that the motivational and behavioral declines evident during early adolescence could result from the fact that junior high schools are not providing appropriate educational and social environments for early adolescents. According to person-environment fit theory, behavior, motivation, and mental health are influenced by the fit between the characteristics individuals bring to their social environments and the characteristics of these social environments. Individuals are not likely to do very well, or be very motivated, if they are in social environments that do not meet their psychological needs. If the academic and social environments in the typical junior high school do not fit with the psychological needs of adolescents, then person-environment fit theory predicts a decline in motivation, interest, performance, and behavior as adolescents move into and through this environment. This is the perspective elaborated next in this paper. This perspective is then extended to the family context, focusing on the possible mismatch between adolescents' need for greater autonomy from parental control and the opportunities for such autonomy provided by adolescents' parents.

Stage-Environment Fit and School-Related Changes

Various explanations have been offered for the declines in academic motivation and general mental health associated with early adolescence. In this section, the possible role that the school context may play in precipitating these declines is discussed. To understand this role, two types of evidence regarding school effects are presented: evidence drawn from studies that follow the standard environmental influences approach and evidence from studies that adopt a developmental variant on the person-environment fit paradigm, or as Eccles and Midgley have termed it, the "stage / environment fit" approach (see Eccles & Midgley, 1989).

General Environmental Influences

Work in a variety of areas has documented the impact of various classroom and school environmental characteristics on motivation. For example, the big school/ small schools literature has demonstrated the motivational advantages of small schools, especially for marginal students (Barker & Gump, 1964). Similarly, the teacher efficacy and teacher-student relationship literatures document the importance of high teacher efficacy and positive teacher-student relations for positive teacher and student motivation (Brookover, Beady, Flood, Schweitzer, & Wisenbaker,

1979; Fraser & Fisher, 1982; Moos, 1979, 1991). Finally, work in motivational psychology has demonstrated the important influence of academic learning environments that stress task mastery and self-improvement (Ames, 1992b; Maehr & Midgley, 1991; Nicholls, 1984) and student participation and self-control (de-Charms, 1980; Deci & Ryan, 1985, 1987) on student's motivational beliefs, affect, learning strategies, and achievement.

While this list is by no means exhaustive, the main point is that there may be systematic differences in some of these characteristics of educational environments between typical elementary classrooms and schools, and typical junior high classrooms and schools, and that these differences may account for some of the motivational and behavioral changes seen among early adolescents as they make the transition into middle or junior high school. Indeed, evidence is emerging that suggests that with increasing grade level, especially around the transition to middle-level schools, the characteristics of school environments become less facilitative of continuing achievement and positive personal development. For example, there is diminished support for the development of competence, for feelings of belongingness, and for opportunities for autonomy in these environments (Higgins & Parsons, 1983). If the school environment does become less facilitative of development as the adolescent progresses into middle-level schools, then some of the motivational and mental health problems seen at early adolescence may be a consequence of these negative changes in the school environment, rather than characteristics of the developmental period per se (see Higgins & Parsons, 1983, for a full elaboration of this argument).

A Stage-Environment Fit Approach

A slightly different analysis of the possible environmental causes of the motivational and mental health changes associated with the junior high school transition draws on the idea of person-environment fit. Such a perspective leads one to expect negative consequences for individuals when they are in environments that do not fit well with their needs (Hunt, 1975; Lewin, 1935). At the most basic level, this perspective suggests the importance of looking at the fit between the needs of early adolescents and the opportunities afforded them in the traditional junior high school environment. A poor fit would help explain the declines in motivation and mental health associated with the transition to junior high school.

An even more interesting way to use the person-environment fit perspective is to put it into a developmental framework. Hunt (1975) argued for the importance of adopting a developmental perspective on person-environment fit in the classroom:

Maintaining a developmental perspective becomes very important in implementing person-environment matching because a teacher should not only take account of a student's contemporaneous needs by providing whatever structure he presently requires, but also view his present need for structure on a developmental continuum along which growth toward independence and less need for structure is the long-term objective. (Hunt, 1975, p. 221)

Here, Hunt was suggesting that teachers should provide the optimal level of structure for children's current levels of maturity, while at the same time providing a sufficiently challenging environment to pull the children along a developmental path towards higher levels of cognitive maturity.

While Hunt's consideration of person-environment fit was geared towards facilitation of optimal learning style and cognitive adaptation, there is good reason to believe that a developmentally appropriate environment will promote positive emotional development as well. Indeed, as at any developmental period, developmental and social psychologists have long argued for the importance of integration of the cognitive and emotional aspects of individuals' psychologies for optimal self-development and adjustment. It could be argued that emotional development is of equal importance to the overall well-being of early adolescents and yet support for such emotional development is largely ignored in the designs of classrooms in which early adolescents spend a substantial amount of time. While the importance of the integration of cognition and affect has been addressed in work pertaining to the mental health and psychological well-being of infants and young children (e.g., Beeghly & Cicchetti, 1987; Hesse & Cicchetti, 1982), little work has focused on the link between these two dimensions in adolescence. Much as Vygotsky employed the notion of a "zone of proximal development" for describing the role of adult scaffolding for facilitating the cognitive and affective development of the toddler, adults maintain a similarly important role in continuing to scaffold both cognitive and emotional development during the adolescent years.

What we find especially intriguing about this suggestion is its application to an analysis of the motivational and mental health declines associated with the junior high school transition. If it is true that different types of educational environments may be needed for different age groups to meet developmental needs and to foster continued positive growth, then it is also possible that some types of changes in educational environments may be especially inappropriate at certain stages of development, such as the early adolescent period. In fact, some types of changes in the educational environment may be developmentally regressive. Exposure to such changes is likely to lead to a particularly poor person-environment fit, and this lack of fit could account for some of the declines in motivation seen at this developmental period.

In essence, then, we are suggesting that it is the fit between the developmental needs of the adolescent and the educational environment that is important. Imagine two trajectories: one a developmental trajectory of early adolescent growth, the other a trajectory of environmental change across the school years. We believe there will be positive motivational and mental health consequences when these two trajectories are in synchrony with one another; that is, when the environment is both responsive to the changing needs of the individual and offers the kinds of stimulation that will propel continued positive growth. In other words, transition to a facilitative and developmentally appropriate environment, even at this vulnerable age, should have a positive impact on children's perceptions of themselves and their educational environment. In contrast, negative motivational and mental health consequences will result if the two trajectories are out of synchrony. In this case, transition into a developmentally inappropriate educational environment should result in the types of motivational and psychological declines that have been identified as occurring with the transition into junior high school. This should be particularly true if the environment is developmentally regressive; that is, if it doesn't provide the adolescent with appropriate opportunities for continued positive growth.

This analysis suggests a set of researchable theoretical and descriptive questions. First, what are the developmental needs of the early adolescent? Second, what kind of educational environment would be developmentally appropriate in terms of both meeting these needs and stimulating further development? Third, what are the most common changes experienced by young adolescents as they move into middle school or junior high school? Finally, and most important, are these changes compatible with the physiological, cognitive, psychological, and social changes early adolescents are experiencing? Or is there a developmental mismatch between maturing early adolescents and the classroom environments they experience before and after the transition to the junior high school—a mismatch that results in a deterioration in academic motivation, performance, and general psychological adjustment for some adolescents?

Systematic Changes in School Environments With the Transition into Junior High School

We believe that there are developmentally inappropriate changes in a cluster of classroom organizational, instructional, and climate variables, including task structure, task complexity, grouping practices, evaluation techniques, motivational strategies, locus of responsibility for learning, and quality of teacher-student and student-student relationships. Additionally, we believe that these changes contribute to the negative change in students' motivation and achievement-related beliefs assumed to coincide with the transition into junior high school. Although relatively little research has been done, the little that exists supports these suggestions.

Remarkably few empirical studies have focused on differences in the classroom or school environment across grades or school levels. Nonetheless, the emerging evidence points to a developmentally regressive change in several aspects of the junior high school environment, including less perceived social support and more of an emphasis on grades and competition (Harter et al., 1992; Midgley, Anderman, & Hicks, 1995; Roeser, Midgley & Maehr, 1994; Roeser, Urdan, & Midgley, 1993). In addition, many of the comparative studies have focused on school-level characteristics, such as school size, degree of departmentalization, and extent of bureaucratization, and not on variables more proximally related to students' beliefs, attitudes, and behaviors. Although differences on these macro-characteristics can have important effects on teacher beliefs and practices and, consequently, on adolescent motivation, well-being and achievement, until quite recently these links have rarely been studied explicitly. Most attempts to assess the classroom environment have included only one grade level and have related between classroom differences in the environment to student outcomes, particularly scores on achievement tests. Little research has focused on systematic differences in the classroom environments across grade level or between elementary and junior high schools. But looking across the various relevant studies, six patterns emerge with a fair degree of consistency.

First, junior high school classrooms, as compared to elementary school classrooms, are characterized by a greater emphasis on teacher control and discipline and fewer opportunities for student decision making, choice, and self-management (e.g., Brophy & Evertson, 1978; Deci & Ryan, 1987; Midgley & Feldlaufer, 1987; Midgley, Feldlaufer, & Eccles, 1988b; Moos, 1979). For example, Brophy, Evertson, and their colleagues have found consistent evidence that junior high school teachers spend more time maintaining order and less time actually teaching than elementary school teachers (Brophy & Evertson, 1978). In our own work, sixth grade elementary school math teachers reported less concern with controlling and disciplining their students than these same students' seventh grade junior high school math teachers reported one year later (Midgley et al., 1988b).

Similar differences emerge on indicators of student opportunity to participate in decision making regarding their own learning. For example, Ward and his colleagues found that upper elementary school students are given more opportunities to take responsibility for various aspects of their schoolwork than seventh grader students in a traditional junior high school (Ward, Mergendoller, Tikunoff, Rounds, Dadey, & Mitman, 1982). In our work, both seventh graders and their teachers in the first year of junior high school reported less opportunity for students to participate in classroom decision making than did these same students and their sixth grade elementary school teachers one year earlier. In addition, using a measure developed by Lee, Statuto, and Kedar-Voivodes (1983) to assess the congruence between the adolescents' desire for participation in decision making and their perception of the opportunities for such participation, Midgley and Feldlaufer (1987) found a greater discrepancy when the adolescents were in their first year in junior high school than when these same adolescents were in their last year in elementary school. The fit between the adolescents' desire for autonomy and their perception of the extent to which their classroom afforded them opportunities to engage in autonomous behavior had decreased over the junior high school transition. Such discrepancies are likely to be particularly problematic for the early adolescent, given that a key developmental task of this period is to establish a sense of autonomy and personal efficacy, yet these young people are not in environments that afford such opportunities. One can imagine the adolescents' frustration with such an environment, frustration that can adversely impact on both their motivation for academic engagement and achievement and their overall psychological well-being.

Second, junior high school classrooms, as compared to elementary school classrooms, are characterized by less personal and positive teacher-student relationships (see Eccles & Midgley, 1989). For example, in Trebilco, Atkinson, and Atkinson

(1977), students reported less favorable interpersonal relations with their teachers after the transition to secondary school than before. Similarly, in our work, both students and observers rated junior high school math teachers as less friendly, less supportive, and less caring than the teachers these students had one year earlier in the last year of elementary school (Feldlaufer, Midgley, & Eccles, 1988). In addition, the seventh grade teachers in this study also reported that they trusted the students less than did these students' sixth grade teachers (Midgley et al., 1988b). So, at a time period when most early adolescents are confronted with an uncertainty about themselves that derives from the often daunting tasks of establishing a sense of coherent personal identity and negotiating newfound social roles in the face of a myriad of changes, they are met with distrust by the very people who could provide support for them during the negotiation of these tasks. It is no wonder that a significant proportion of early adolescents come to feel alienated from adults and institutions at this time period.

Third, the shift to junior high school is associated with an increase in practices such as whole-class task organization, between classroom ability grouping, and public evaluation of the correctness of work (see Eccles & Midgley, 1989). For example, in the study by Ward and his colleagues, whole-group instruction was the norm in the seventh grade, small-group instruction was rare, and individualized instruction was not observed at all. In contrast, the sixth grade teachers mixed whole- and small-group instruction within and across subjects areas (Rounds & Osaki, 1982). Similar shifts towards increased use of whole-class instruction with most students working on the same assignments at the same time, using the same textbooks, and completing the same homework assignments were evident in our study of the junior high school transition (Feldlaufer et al. 1988). In addition, several reports have documented the increased use of between class ability grouping beginning at junior high school (e.g., Oakes, 1981).

Changes such as these are likely to increase social comparison, concerns about evaluation, and competitiveness (e.g., Eccles et al., 1984; Marshall & Weinstein, 1984; Rosenholtz & Simpson, 1984). They may also increase the likelihood that teachers will use normative grading criteria and more public forms of evaluation, both of which may have a negative impact on many early adolescents' self-perceptions and motivation. These changes may also make aptitude differences more salient to both teachers and students, leading to increased teacher expectancy effects and decreased feelings of efficacy among teachers. In fact, recent work by Midgley, Maehr, and their colleagues does suggest that both teachers and students in middle schools report more emphasis placed on relative ability, competition, and extrinsic rewards in their schools than teachers and students in elementary school settings (Midgley et al., 1995; Roeser, Midgley et al., 1994). Thus, at a time when young people are by nature of the developmental period acutely sensitive to their competencies relative to others, their social environments highlight these comparisons. For many, such competitive environments can lead to grade declines and increasing anxiety about one's competence and, perhaps for some, may help to crystallize a downward spiral toward eventual school dropout.

Fourth, junior high school teachers appear to use a higher standard in judging students' competence and in grading their performance than do elementary school teachers (see Eccles & Midgley, 1989). There is no stronger predictor of students' self-confidence and sense of efficacy than the grades they receive. If grades change, then we would expect to see a concomitant shift in adolescents' self-perceptions and academic motivation. There is evidence that junior high school teachers use stricter and more social comparison-based standards than elementary school teachers to assess student competency and to evaluate student performance, leading to a drop in grades for many early adolescents as they make the junior high school transition. For example, Finger and Silverman (1966) found that 54% of the students in New York State schools experienced a decline in their grades when they moved into junior high school. Similarly, Simmons and Blyth (1987) found a greater drop in grades between sixth and seventh grade for adolescents making the junior high school transition than for adolescents who remained in K–8 schools. Interestingly, the decline in grades is not accompanied by a similar decline in the adolescents' scores on standardized achievement tests, which suggests that the decline reflects a change in grading practices rather than a change in the rate of the students' learning (Kavrell & Petersen, 1984). Imagine what this decline in grades might do to young adolescents' self-confidence, especially in light of the evidence suggesting that the material may be less intellectually challenging than the work in the elementary school grades.

Finally, junior high school teachers feel less effective as teachers, especially for low-ability students. This was one of largest differences we found between sixth and seventh grade teachers. In mathematics, seventh grade teachers in traditional junior high schools report much less confidence in their teaching efficacy than sixth grade elementary school teachers in the same school districts (Midgley et al., 1988b). This is true in spite of the fact that the seventh grade math teachers were more likely to be math specialists than the sixth grade math teachers. Here again, at a time when early adolescents are working to establish a sense of personal competence and efficacy and are in need of adult role models for this task, significant adults in their lives do not themselves feel efficacious in their roles. It is likely that teachers' feelings of low efficacy can adversely affect both the academic motivation and mental health of adolescents not only by means of negative social role modeling, but also in terms of the environments that these teachers then provide for their students. The environments created by teachers who do not feel efficacious may likely reflect an accommodation to the teachers' own insecurities and consequent needs for excessive structure and control—needs that are likely to be in direct contrast to the developmental needs of the adolescents themselves.

The Adolescent in Context: A Developmental Mismatch

Changes such as those noted above are likely to have a negative effect on many adolescents' motivational orientation towards school and mental health at any grade level. But we believe these types of school environment changes are particularly harmful at early adolescence given what is known about psychological development during this stage of life. As noted earlier, evidence from a variety of sources indicates that early adolescent development is characterized by increases in desire for autonomy and self-determination, peer orientation, self-focus and self-consciousness, salience of identity issues, concern over sexuality, and capacity for abstract cognitive activity (see Simmons & Blyth, 1987).

Simmons & Blyth (1987) have argued that adolescents need a reasonably safe, as well as an intellectually challenging, environment to adapt to these shifts—an environment that provides a "zone of comfort," as well as challenging new opportunities for growth. In light of these needs, the environmental changes often associated with transition to junior high school seem especially harmful in that they emphasize competition, social comparison, and ability self-assessment at a time of heightened self-focus; they decrease decision making and choice at a time when the desire for control is growing; and they disrupt both social networks at a time when adolescents are especially concerned with peer relationships and may be in special need of close adult relationships outside of the home. We believe the nature of these environmental changes, coupled with the normal course of individual development, results in a developmental mismatch so that the fit between the early adolescent and the classroom environment is particularly poor, increasing the risk of negative motivational and psychological outcomes, especially for adolescents who are having difficulty succeeding in school academically. In the next section, we review research findings relevant to these predictions.

Before turning to our studies, however, it is important to step back and consider briefly why junior high school classrooms might have these characteristics. Several sources have suggested that these characteristics result, in part, from the size and bureaucratic nature of the junior high school as an institution (e.g., Barker & Gump, 1964; Bryk, Lee, & Smith, 1989; Carnegie Council on Adolescent Development, 1989; Simmons & Blyth, 1987). For example, it is likely that such school characteristics as large size, minimal connections to the community, and a rigidly hierarchical system of governance, as well as organizational characteristics such as departmentalized teaching, ability grouping, normative grading, and large student load, undermine the motivation of both teachers and students. It is difficult for teachers to maintain warm, positive relationships with students if they have to teach 25 to 30 different students each hour of the day. Similarly, it is hard for teachers to feel efficacious about their ability to monitor and help all of these students. Finally, it seems likely that teachers will resort to more controlling strategies when they have to supervise such a large number of students.

The consequences of the size and organization of traditional junior high schools on teachers' motivation are likely to be exacerbated by the negative stereotypes about adolescents propagated in this culture by both presumed experts and the mass media (see Miller, Eccles, Flanagan, Midgley, Feldlaufer, & Harold, 1990; Offer, Ostrov, & Howard, 1981). Such stereotypes characterize adolescence as a period of "storm and stress." The adolescent is often portrayed as impulsive, out of control, confused, self-absorbed, angst-ridden, rebellious, violent, alienated, and involved in drugs, alcohol, and unremitting sex. Due to these characterizations, the popular view that adolescents need to be tightly controlled until they ultimately "snap out of it" and settle into their adult roles is widespread in this culture. Interestingly, rarely does the media portray the potential role of the social environment in creating a world of alienation for some adolescents. There is also limited discussion in the media of the ways in which the adults in young adolescents' lives could help and support them through this life passage, or ways in which adults could give adolescents socially responsible tasks to engage in so as to exercise their need for establishing adult roles. Finally, there is relatively little attention in the media, or elsewhere, to the many positive things done by early adolescents in their communities and schools. Such coverage is desperately needed to counterbalance the prevalence of sensational reports of adolescents' problem behaviors.

In sum, the characteristics of the school environments in which early adolescents spend a substantial portion of time are not likely to be conducive to meeting the developmental needs of early adolescents (e.g., the need for increasing autonomy, for a sense of competence, and for continued relatedness). Indeed, such environments could be considered to be developmentally regressive. As a consequence, we predict that these environments are likely to undermine the cognitive and emotional development of young adolescents and have negative implications for both their academic motivation and overall psychological well-being. More specifically, we predict that, as adolescents move into classrooms characterized by more teacher control and discipline and less student participation in decision making, their values and attitudes towards academics will suffer. Similarly, when adolescents encounter junior high classrooms that are perceived as less supportive and warm compared to their elementary school environments, their attitudes and values should become more negative. Finally, we hypothesize that the increases in instructional practices that stress social comparison and tougher grading standards and the decreases in teachers' feelings of efficacy between elementary and junior high environments will have a detrimental impact on adolescents' ability self- concepts.

The Impact of Classroom Environmental Changes on Early Adolescents' Motivation: The Michigan Study of Adolescent Life Transitions (MSALT)

To test the predictions outlined above, we conducted a large-scale two-year, four wave longitudinal study of the impact of changes in the school and classroom environments on early adolescents' achievement-related beliefs, motives, values, and behaviors (Michigan Study of Adolescent Life Transitions, MSALT). The sample was drawn from 12 school districts located in middle-income communities in southeastern Michigan. Approximately 1500 early adolescents participated at all four waves of the study. These adolescents moved from the sixth grade in an elementary school into the seventh grade in a junior high school during the course of the study. As is typically the case, the students did not move as a group into the junior high school—they were assigned to various different classes when they arrived at the junior high school. Questionnaires were administered in the students' math classes at school during the fall and spring terms of the two consecutive school years. A few of our key findings are reviewed below.

Teacher Efficacy

One of the largest differences we found between the sixth and seventh grade teachers was in their confidence in their teaching efficacy: The seventh grade teachers reported less confidence than the sixth grade teachers. Although the relation between teacher efficacy and student beliefs and attitudes is yet to be firmly established, Brookover et al. (1979), using schools as the unit of analysis, found positive correlations between teachers' sense of efficacy and students' self-concept of ability and self-reliance. Given these associations, differences in teachers' sense of efficacy before and after the transition to junior high school could contribute to the decline in early adolescents' beliefs about their academic competency and potential.

To test this hypothesis, we divided our adolescent sample into four groups based on median splits of their math teachers' ratings of their personal teaching efficacy (see Midgley, Feldlaufer, & Eccles, 1989, for a full description of this study). The largest group (559 of the 1329 included in these analyses) moved from a high efficacy sixth grade math teacher to a low efficacy seventh grade math teacher. Another 474 adolescents had low efficacy teachers both years, 117 moved from low to high efficacy teachers, and 179 had high efficacy teachers both years. Thus, fully 78% of the children in our sample moved to a low teacher efficacy math classroom in the seventh grade. The potential impact of such a shift on the motivation and self-perceptions of early adolescents, especially those having difficulty mastering the academic material is frightening. We know, in particular, that low teacher expectations for students undermine the motivation and performance of low achieving students (Eccles & Wigfield, 1985). Moving from a high to a low efficacious teacher may produce a similar effect.

As predicted, the adolescents who moved from high efficacy to low efficacy math teachers during the transition (the most common pattern) ended their first year in junior high school with lower expectancies for themselves in math, lower perceptions of their performance in math, and higher perceptions of the difficulty of math than the adolescents who experienced no change in teacher efficacy or who moved from low to high efficacy teachers. Also as predicted, teacher efficacy beliefs had a stronger impact on the low achieving adolescents' beliefs than on the high achieving adolescents' beliefs. By the end of the junior high school year, low achieving adolescents who had moved from high to low efficacy math teachers suffered a dramatic decline in their confidence in their ability to master mathematics. This drop may signal the beginning of the downward spiral in school motivation that eventually leads to school dropout among so many low achieving adolescents. It is important to note, however, that this same decline was not characteristic of the low achieving adolescents who moved to high efficacy seventh grade math teachers, suggesting that the decline is not a general feature of early adolescent development but rather a consequence of the learning environment experienced by so many early adolescents as they make the junior high school transition. Whether a similar pattern characterizes other subject areas remains to be demonstrated.

Teacher-Student Relationships

As reported earlier, we also found that student-teacher relationships deteriorate after the transition to junior high school. Research on the effects of classroom climate indicates that the quality of student-teacher relationships is associated with students' academic motivation and attitudes toward school (e.g., Fraser & Fisher, 1982; Moos, 1979; Trickett & Moos, 1974). Consequently, there is reason to believe that transition into a less supportive classroom will have a negative impact on early adolescents' interest in the subject matter being taught in that classroom. In a sample of 1300 students, we looked at the effect of differences in perceived teacher support before and after the transition to junior high school on the value early adolescents attach to mathematics (see Midgley, Feldlaufer, & Eccles, 1988a, for a full description of this study). As predicted, the early adolescents who moved from elementary teachers they perceived to be low in support to junior high school math teachers they perceived to be high in support showed an increase in the value they attached to math. In contrast, the early adolescents who moved from teachers they perceived to be high in support to teachers they perceived to be low in support showed a decline in the value they attached to mathematics. Again we found evidence that low achieving students are particularly at risk when they move to less facilitative classroom environments after the transition.

Both of these studies show that the declines often reported in studies of early adolescents' motivational orientation are not inevitable. Instead, these declines are associated with specific types of changes in the nature of the classroom environment experienced by many early adolescents as they make the junior high school transition. The studies also show that a transition into more facilitative classrooms can induce positive changes in early adolescents' motivation and self-perceptions. Unfortunately for all adolescents, but especially for low achieving adolescents, our findings also indicate that most adolescents experience a negative change in their classroom experiences as they make the junior high school transition.

Person-Environment Fit in Classroom Decision Making

Neither of these studies, however, directly tested our stage-environment fit hypothesis. To do this, one must directly assess person-environment fit and relate this fit to changes in adolescents' self-perceptions and motivation. Data from MSALT provide an opportunity to do this analysis. Both the adolescents and the teachers in this study were asked to rate whether students were allowed to have input into classroom decisions regarding seating arrangements, classwork, homework, class rules, and what to do next and whether students *ought* to have input into each of these decisions. (These items were developed by Lee et al., 1983). These questions can be used in the following ways: (1) to plot the developmental changes in adolescents' preferences for decision making opportunities in the classroom, (2) to determine changes in the opportunity for them to participate in decision making, and (3) to determine the extent of match or mismatch between their preferences and the opportunities actually afforded them in the school environment. Grade-related changes in this match can then be related to developmental changes in the adolescents' self-perceptions and school-related motivation.

Developmental and Grade-Related Changes in Fit

As noted earlier, both early adolescents and their teachers reported that there was less opportunity for participation in classroom decision making at the seventh grade than at the sixth grade. In contrast, there was an increase both across the school transition and over time during the seventh grade year in the early adolescents' desires for more participation in classroom decision making. As a consequence of these two divergent patterns, the congruence between early adolescents' desires for participation in classroom decision making and their perceptions of the opportunities available to them was lower in the seventh grade than in the sixth grade (Midgley & Feldlaufer, 1987).

The Motivational Consequences of a Poor Developmental Stage-Environment Fit

As outlined earlier, person-environment fit theory suggests that the mismatch between young adolescents' desires for autonomy and control and their perceptions of the opportunities in their learning environments should result in a decline in the adolescents' intrinsic motivation and interest in school. But more importantly from a developmental perspective, the exact nature of the mismatch should also be important. As noted earlier, given the appropriate developmental progression towards increased desire for independence and autonomy during the early adolescent period, adolescents who experience decreased opportunities for participation in classroom decision making along with an increased desire for greater participation in such decisions (i.e., a "can't but should be able to" mismatch) should be more at risk for negative motivational outcomes than adolescents experiencing other forms of mismatch (such as the "can but shouldn't be able to" mismatch).

In a longitudinal analysis of the Lee et al. (1983) items, Mac Iver and Reuman (1988) provided some support for this prediction. They compared the changes in intrinsic interest in mathematics for adolescents reporting different longitudinal patterns in their responses to the actual and preferred decision-making items across the four waves of data. Consistent with the prediction, it was the adolescents who perceived their seventh grade math classrooms as putting greater constraints on their preferred level of participation in classroom decision making than their sixth grade math classrooms who evidenced the largest and most consistent declines in their intrinsic interest in math as they moved from the sixth grade into the seventh grade. These are the students who are experiencing the type of developmental mismatch we outlined in our discussion of stage-environment fit.

Person-Environment Mismatch as Lack of Attunement: Implications for Psychological Well-Being

Another way to think about the issue of person-environment fit is in terms of adult attunement, or sensitivity, to the autonomy, competence, and relatedness needs of the developing adolescent. Attunement to these needs can be conceptualized in terms of providing appropriate environments for supporting the developmental needs of young adolescents. Such attunement can also be conceptualized in terms of sensitivity to the psychological experiences and emotional well-being of the adolescents in these contexts. While a central task of early adolescent development is to establish a greater sense of autonomy, there is also a continued need for support and relatedness from significant adults. Indeed, adult support and responsiveness may be particularly important at this period of development for two reasons: (1) It is a time of multiples changes, and exposure to multiple changes increases vulnerability self-esteem and mental health difficulties; and (2) it is a time of increasing salience of the peer group and increasing exposure to opportunities for involvement in risky behaviors. However, although the peer culture becomes increasingly important at adolescence relative to the elementary years, adult figures can retain their significance as bases and scaffolds from which the adolescent can explore new identities and negotiate newfound expectations for more adult roles and responsibilities if they are available and responsive to the needs of the adolescents (e.g., Eccles, Midgley, Wigfield, Buchanan, Reuman, & Mac Iver, 1993; Hartup, 1989; Ryan & Lynch, 1989). The literature in stress and coping has pointed to a positive relationship with an adult as a key protective factor for children at risk for psychosocial problems (see Garmezy, 1983; Rutter, 1981).

The discussion to this point has highlighted the available evidence for the negative impact of a poor developmental stage-environment fit on constructs relevant to students' academic competence and values. There is also evidence to suggest that the characteristics of these school environments can undermine teachers' ability to be attuned to the emotional needs of their adolescent students. This could decrease the probability of teachers being a protective influence during this difficult transitional period. For example, in our study of MSALT adolescents who made the transition to a seventh grade junior high school, we found that over and above prior achievement level and self-esteem in the sixth grade, adolescents' perceptions of competence in academic, social, physical appearance, and athletic domains were facilitative of gains in self-esteem across the transition to junior high school, while worries and self-consciousness about performance in these domains were predictive of declines in self-esteem across this transition (Lord et al., 1994). Are the teachers aware of the importance of these psychological predictors of adjustment to junior high school? Do they know which adolescents are and which are not adapting well to the transition? Our findings suggest not. Unlike the results for the adolescents' report of self-esteem, students' worries and self-consciousness about academics were positively rather than negatively related to teachers' ratings of adolescents' adjustment to junior high. Apparently, indicators of anxiety such as academic nervousness and self-consciousness that are related to declines in adolescents' self-esteem are seen as indicators of good adjustment by the teachers. Perhaps the teachers interpret students' nervousness and self-consciousness about their performance as indicators of positive motivation and concern on the part of the student. These findings suggest that early adolescents may be receiving positive feedback from their teachers for the very characteristics that are linked to declines in self-esteem. It is likely that receiving positive feedback for characteristics that are internally distressful could increase adolescents' psychological distress.

Another interesting discrepancy emerged in the results for teachers and adolescents in the MSALT data: In contrast to the pattern of predictors of change in the adolescent reports of their self-esteem, sixth grade academic performance was a much stronger predictor of teachers' ratings of the adolescents' adjustment than was sixth grade self-esteem. Apparently, teachers' view of early adolescents' adjustment to junior high school is primarily related to the adolescents' academic performance. In contrast, the strongest predictors of increases in adolescents' reports of their self-esteem following the junior high school transition were self-concepts, lack of self-consciousness, and developmentally appropriate authority relations at home. These findings are of concern since it appears that teachers think that a student is adjusting well to the junior high school transition as long she or he is doing well academically. But the student data indicate that achievement level is not a good predictor of adjustment to this school transition. Since teachers are in a very good position to identify adolescents at psychological risk, it is unfortunate that they appear so unaware of their students' psychological state until the problem begins to affect the students' academic achievement.

Although these results are disappointing, they are not surprising. The finding that fewer of the students' psychological characteristics predicted to the teachers' rating of the students' adjustment to junior high probably reflects the increased student anonymity and decreased student-teacher connections implicit in the junior high setting. Because of the increased number of students and the departmentalized classrooms in these settings, teachers are not likely to get to know each of the students on a personal basis. It follows that teachers' criteria for assessing students' adjustment would be limited and would be based on their knowledge of the students' performance level. Again, while adolescents' need for empathic and developmentally responsive support from teachers increases, opportunities for such supportive contact decline.

While this diminished sensitivity is probably in large part due to the structure of large, departmentalized junior high environments, the low attunement of teachers to their students may also reflect the teachers' lack of understanding about what early adolescents are up against at this period and about what adolescents need from the adults in their lives. Currently, teachers rarely receive any special certification or education about adolescence as a developmental period (Braddock & McPartland, 1992). Teacher education reform initiatives have indeed recommended such education and training in order to better train teachers for working with this age group, yet such reform has been slow in the making.

In sum, these results are cause for concern since they indicate that many teachers of early adolescents are not very well attuned to the experiences and needs of adolescents making the transition to junior high school. In addition to having negative psychological implications for adolescents, this lack of attunement also reduces the likelihood that teachers can identify adolescents at risk in time to help them get the additional services and support they may need. Thus, teachers aren't as able to provide the safety net that early adolescents need as they move through this challenging developmental period and as they adapt to new educational settings. Consequently, adolescents may not be receiving what they need emotionally from the adults in their school environments during the transition to junior high school.

The Consequences of Lack of Fit / Nonattunement in School for Early Adolescents' Competence, Motivation, and Mental Health

The research discussed thus far highlights the importance of understanding early adolescents' development in terms of the fit between their psychological needs and the school context in which they spend a good portion of their waking hours. A key challenge for adolescents is to develop a coherent personal identity that integrates personal competencies with the expanding social roles and experiences that accompany this developmental period (Erikson, 1959). The contexts in which adolescents develop can either facilitate, or undermine, this process of identity development. We have presented evidence that school environ- ments that are not attuned to the developmental needs of the adolescent have implications not only for adolescents' motivation for school achievement, but also for their psychological well-being. Indeed, middle and junior high school environments can be developmentally regressive and can thus inhibit the growth of personal competencies that are intimately tied with an individual's affective self-appraisals and experiences. As we noted earlier, academic achievement is thought to be critical to continued patterns of personal adjustment during this time and may indeed help to serve as a protective factor against negative outcomes later in adolescence, including problem behaviors and mental health problems (Bloom, 1976; Cowen, 1991; Rae-Grant et al., 1989). Conversely, academic underachievement and alienation in early adolescence are risk factors for later adjustment. While few studies have directly looked at the relations among academic learning environments, academic engagement, and mental health measures (e.g., Maehr, Midgley, Pintrich, & De Groot, 1992), the results we have presented thus far suggest that school environments that do not provide for the needs of adolescents may be related to patterns of alienation and school disengagement that eventuate in more serious negative life trajectories.

We have recently begun to examine more closely how academic engagement, motivation, and psychological well-being are interrelated during the period of early adolescence in order to gain a better understanding of the patterns of factors that could put early adolescents at risk for later problems (Roeser, Lord, & Eccles, 1994). For these analyses, we used data from the first wave of a longitudinal study of approximately 1400 early adolescents and their families living in a large mid-Atlantic county. This study (The Maryland Adolescent Growth in Context Study, MAGICS; Principal Investigator: Jacquelynne Eccles) is a collaborative effort of the MacArthur Foundation Research Network on Successful Adolescent Development (Chair: Richard Jessor). It was designed to assess how parents, schools, neighborhoods, and peers influence the academic, psychological, and social development of adolescents. The target youths, primary caregivers (approximately 92% mothers), and secondary caregivers (approximately 75% fathers) were interviewed in their homes. Of the target youth, 50.9% were male and 62% were African American. The sample includes a wide socioeconomic range of families with mean income of $45,000–$49,999. Both the interview and the questionnaire contained items assessing parent and adolescent perceptions of parent involvement in school and at home; parenting style; family environment and relationships; self-concepts in school, social, and extracurricular domains; and parent and adolescent adjustment.

Using the MAGICS data, we examined the profiles of academic competence, motivation, and psychological well-being for groups of adolescents who displayed varying levels of academic alienation (Roeser et al., 1994). School disengagement, or alienation, was operationalized using both psychological and behavioral measures. Psychological alienation was assessed by negative attitude towards school (not liking school), school disengagement (e.g., classes are boring, school is a waste of time) and devaluing the importance of a good education. Behavioral alienation was assessed by student and parent reports of the adolescents' school problem behaviors, including skipping classes and being suspended or expelled from school. Students in the upper quartile of the psychological alienation from school scale who also evidenced high rates of school problems were designated as the most alienated. Another group of alienated adolescents was created from those in the upper quartile of the psychological alienation scale who did not evidence as many school problems. Finally, a low alienation group was comprised of those adolescents in the lower quartile of the psychological alienation from school scale. We then tested the association between the level of school alienation and an array of indicators of self-perceptions of com-

Table 1. Group Comparisons[1] by Level of School Alienation for Motivation and Achievement, School Behaviors, and Mental Health Measures

Descriptive Variables	Low Psychological Alienation From School	High Psychological Alienation From School		Significance
	Few School Problems Group 1 (N = 342)	Some School Problems Group 2 (N = 306)	School Problems Group 3 (N = 61)	
Motivation and achievement				
Academic self-concept	5.97a	4.74b	4.42c	F(2,696) = 153.59**
Academic importance[2]	6.55a	4.26b	4.15b	F(2,700) = 485.55**
Academic liking [2]	4.28a	2.63b	2.66b	F(2,701) = 700.91**
Feelings of belonging in school	4.37a	3.68b	3.54b	F(2,699) = 67.85**
School is important now / future	4.41a	2.63b	2.66b	F(2,701) = 137.37**
Grades	4.33a	3.73b	3.60c	F(2,696) = 61.32**
Parent's concern about child's education / performance[3]	−0.22a	0.06b	0.87c	F(2,694) = 84.92**
School behaviors				
Percent skipping school	4a	14b	33c	F(2,691) = 28.36**
Percent suspended from school[4]	11a	12a	100b	F(2,704) = 228.65**
Percent failing a class[4]	14a	27b	100c	F(2,704) = 131.44**
Mental Health				
Self-worth	3.98a	3.36b	3.35b	F(2,704) = 55.42**
Resourcefulness	3.90a	3.26b	3.34b	F(2,701) = 60.35**
Depression	1.71a	2.03b	2.22b	F(2,701) = 23.36**
Anger	2.04a	2.65b	3.09c	F(2,704) = 62.77*

[1] Tukey HSD mean comparisons were used to test all possible pairs. Different superscripts for a particular variable across groups indicates a significant mean difference at the $p \leq .05$ level for all possible pairs. / [2] These variables were part of the "Psychological Alienation Scale" used to designate groups, and are presented for descriptive purposes. / [3] This parent report measure was used to validate the levels of academic alienation groupings. It represents a measure of parent concern about their child's academic life. / [4] These variables were part of the "School Problems Scale" used to designate groups and are presented for descriptive purposes. / * $p \leq .05$ / * $p \leq .01$.

petence, academic values and feelings, academic grades, and mental health indicators. Means and significance levels for the three groups on both student and parent reports of the adolescent's motivation and mental health are presented in Table 1.

Overall, the results from both the adolescent and parent reports suggest a strong association between school alienation and competence, motivation and mental health during early adolescence. Relative to those seventh graders who were not alienated, the profile of measures that differentiated adolescents' who were highly alienated from school included lower academic values and ability self-concepts, lower self-esteem and less personal resourcefulness, and higher reported anger and depressive symptomatology than less alienated adolescents. Highly alienated students also reported less teacher social support, poor overall evaluations of the quality of their schools, and more negative teacher expectancies for them in school. While our data do not allow us to draw causal conclusions, these results are consistent with findings from other work indicating that at-risk students increasingly encounter alienating environments as they progress through school (e.g., Eccles et al., 1993; Finn, 1989; Kagan, 1990; Strahan, 1988).

Another worrisome set of differences also distinguished the three groups: Relative to low alienated students, the high alienated adolescents reported being involved in peer groups characterized by more antisocial characteristics (e.g., involvement in drugs, vandalism, gangs, unprotected sex) and fewer prosocial characteristics (e.g., value school, good education). These findings are particularly troubling given that there is evidence that adolescents who are alienated from conventional groups (e.g., school and family) often establish strong social bonds with antisocial peer groups in order to obtain a sense of belonging (see Elliott et al., 1989; Fuligni & Eccles, 1992). Elliott and his colleagues (1989) showed that the social rewards (e.g., sense of belonging) gained from bonding to antisocial persons or groups can serve as positive feedback for continued involvement with the group, and they concluded that those adolescents who were alienated from conventional bonds and bonded to antisocial peer groups were at most risk for involvement in serious delinquency and substance abuse. Given the increased peer pressure during early adolescence, and the higher vulnerability of the developing adolescents to such pressure, it is likely that involvement in these peer groups may put these young adolescents on a developmental trajectory towards further negative outcomes, such as school dropout, gang and drug involvement, teen pregnancy, or death.

Overall, the work reviewed above and our own studies begin to address the issue of the intertwined nature of academic motivation, achievement, and personal adjustment during adolescence and highlight the importance of the school context for influencing adolescents' motivational, psychological, social, and behavioral well-being. The results suggest the developmentally inappropriate school environments during early adolescence can contribute to putting substantial numbers of young people at risk for the negative trajectories that some lives take during this critical period. We turn now to a discussion of the role of another powerful social context for the adolescent: the family.

Stage-Environment Fit in Perceived Control in the Family

Evidence from several investigators suggests that adolescents' relationships with their parents also undergo a stressful period during early and middle adolescence (e.g., Eccles et al., 1993; Hill, 1988; Montemayor, 1983; Steinberg, 1990). Because adolescence is a time for renegotiating the power and authority relationships within the family, this stress is often focused on issues of control and autonomy within the family. When children are young, their relationships with their parents are asymmetrical in terms of power and authority. But as children in this culture mature, they need to take more and more responsibility for themselves until, ultimately, they leave their natal home and take primary responsibility for their own lives. In the optimal situation in this culture, parents will reinforce and stimulate this process of developing autonomy, while at the same time providing sufficient structure and control to protect their adolescent from the dangers and risks in their social worlds. But it is very likely that the renegotiation processes associated with these developmental trajectories will not be smooth. It is not easy for parents to determine the optimal balance of autonomy versus control for their children at all ages. Accordingly, from a stage-environment fit perspective, one would predict strained relationships whenever there is a poor fit between the child's desire for increasing autonomy and the opportunities for independence and autonomy provided by the child's parents.

Early adolescence seems a likely developmental period for asynchrony in the desire for and opportunities for autonomy to emerge within the family context. Many of the developmental changes during early adolescence can precipitate increased family conflict around issues of control and autonomy within the family. These changes include the early adolescents' expanding social contacts with same-age peers and the families of their new friends, their cognitive maturation and ability to understand social roles in a more complex way, and their emerging sexuality. Evidence suggests that increased parent-child stress may be related to each of these developmental changes in adolescence.

Social changes in the world of adolescents substantially increase the opportunity for them to experience independence outside the home. The transition to junior high school, and cultural beliefs regarding "appropriate" amounts of adult supervision for children of different ages lead to a dramatic increase in the amount of unsupervised age-mate contact during this developmental period (Higgins & Parsons, 1983). This increase creates the opportunity for adolescents to spend a lot of time in relationships that are likely to be more symmetrical in terms of interpersonal power and authority. The opportunity to be exposed to a broader range of families is also likely to increase with the junior high school transition because these schools are typically larger and draw their attendance from a more diverse range of neighborhoods and communities. This broadened exposure, in turn, may lead early adolescents to question the legitimacy of their parents' rules (Higgins & Parsons, 1983; Laupa & Turiel, 1986; Smetana, 1988a & b, 1989; Tisak, 1986).

These new social experiences, together with the increasing cognitive capacities to understand, integrate, and coordinate diverse social perspectives, may lead early adolescents to question their parents' authority and to push for a more symmetrical relationship with their parents. Finally, parents, in response to their children's emerging sexuality and increased involvement with opposite sex peers, may become more concerned about their children's safety and may actually become more restrictive than they were during the period of middle childhood. This enhanced restrictiveness, in turn, might further exacerbate the perceived asynchrony in the child's mind between desired autonomy and actual provisions within the family. However, as the family adjusts to these changes, one would expect new authority relationships to emerge and the strain to decrease over the adolescent years (see Montemayor, 1983).

Perhaps the best support for this analysis comes from the work of Smetana (1988a, 1988b, 1989). Drawing on evidence regarding age changes in children's understanding of moral versus social conventional reasoning, and their understanding of the legitimacy of adult authority, Smetana has conducted in-depth interviews with adolescents and their parents about authority relationships within the family and about the nature and origin of conflicts in the family. Like others, she finds that most parent-adolescent conflicts focus on day-to-day mundane issues, like cleaning one's room, curfew, and so forth. The conflicts often result because adolescents now define these issues as personal issues (e.g., issues that the individual should decide), while the parents still define these issues as conventional issues (e.g., issues for which parents have some right to establish "rules"). In cross-sectional comparisons, Smetana found a linear age-related increase in the adolescents' view that most such issues are personal rather than conventional. Shifts in the parents' views were less systematic. But most importantly for the stage-environment fit hypothesis, the greatest increase in mismatch between the adolescents' and their parents' views occurred during the early adolescent period (grades 5–8) and mirrored increases in reported conflict (Smetana, 1989).

We are in the process of examining similar issues in our study of adolescent development (the MSALT study described earlier). We assessed family decision making in two ways: Both the adolescents and their parents responded to two items derived from the Epstein and McPartland (1977) scale of family decision making (e.g., "In general, how do you and your child arrive at decisions?" [1 = I tell my child just what to do; 3 = We discuss it and then we decide; 5 = I usually let my child decide]; and "How often does your child take part in family decisions that concern herself or himself?" [1 = never; 4 = always]). The adolescents were also asked to rate how they thought decisions ought to be made in their family, and the extent to which they think "their parents treated them more like a kid than like an adult."

Consistent with the analyses reported earlier for schools, we found both an increase over time in adolescents' desire for greater participation in family decision making and positive associations between the extent of the adolescents' participation in family decision making and indicators of both intrinsic school motivation and positive self-esteem (Flanagan, 1986, 1989; Lord et al., 1994; Miller & Taylor, 1986; Yee, 1986, 1987; Yee & Flanagan, 1985). Even more interesting from the stage-environment fit perspective, the parents reported that they included their children more in family decision making than the children perceived to be true (Flanagan, 1986; Yee, 1987). Furthermore, for girls in particular, the discrepancy between the adolescents' and the parents' perception of the opportunities for the adolescents to participate in family decision making increased over the four waves of our study (Yee, 1987). Finally, and most important, the pattern of changes in early adolescents' self-esteem and intrinsic versus extrinsic motivation for school work were systematically, and predictably, related to changes in their perceptions of the opportunity to participate in family decision making at home. As our developmental stage-environment fit perspective on adult control implies, the adolescents who reported decreasing opportunities to participate in family decision making showed decreases in their self-esteem and intrinsic motivation over the period of this study; the opposite pattern of change occurred for the adolescents who reported increasing opportunities to participate (Flanagan, 1989; Yee, 1987).

The opportunity to participate in family decision making also predicted positive adjustment to the junior high school transition (Eccles, McCarthy, Lord, Harold, Wigfield, & Aberbach, 1990; Lord et al., 1994). Sixth graders who perceived their families as supportive of their involvement in decision making and as nonintrusive showed increases in their self-esteem across the transition to junior high, while the self-esteem of those adolescents who felt their parent(s) did not provide them with opportunities to be involved in decision making and thwarted their efforts to be autonomous declined across this transition. These effects were over and above the influence of sixth grade self-esteem, actual academic ability, ability self-concepts, and worries about adjustment to the transition (Lord et al., 1994). Thus, not only may a mismatch between authority relationships in the home precipitate increased conflict, it may also be detrimental to the adolescents' school-related motivation, self-esteem, and overall psychological well-being.

We have demonstrated the salience of support for autonomy in decision making for the psychological well-being of adolescents in two other studies that represent more ethnically diverse populations. For example, in a cross-sectional study of 11–15-year-old adolescents living with their family in low- to low-middle socioeconomic neighborhoods in inner-city Philadelphia, relative to other dimensions of parenting, parents' support of decision-making opportunities and parent-adolescent mutuality in problem-solving situations were the strongest predictors of adolescent psychological adjustment (Lord, 1994). In addition, in the MAGICS study described earlier, we have demonstrated patterns of negative associations between early adolescents' perceptions of lack of involvement in family decision making and a broad range of indicators of adolescents' psychological well-being (Eccles & Lord, 1993). In this study, seventh grade adolescents were grouped according to the degree of congruence between their perceived involvement in family decision making and the degree to which they thought they should be involved in such decision making.

Figure 1 illustrates these results: Relative to those seventh graders who were satisfied with their involvement in decision making, adolescents who reported that

Figure 1. Main Effect of Decision-Making Synchrony Versus Mismatch for Adolescent Adjustment Measures: Maryland Adolescent Growth in Context Study, Grade 7

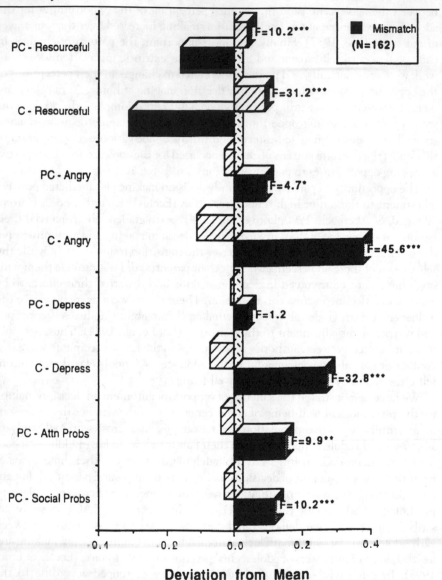

Deviation from Mean

Note: C = Child Report
       PC = Parent Report

they were not as involved in family decision making as they thought they should be reported lower self-esteem, more depressive symptomatology (particularly girls), more anger, and less personal resourcefulness. These differences were found for both adolescents' self-report and parents' report of their adolescent's adjustment. Furthermore, those adolescents who reported a poor fit between their desire for opportunities to participate in decision making and the opportunities provided also reported that their parents were more intrusive and overprotective (e.g., "treat me more like a kid than an adult," "tries to protect me too much"). Finally, as can be seen in Figure 2, multivariate analyses revealed that these two constructs, decision-making match and perceived parental intrusiveness, independently contributed to adolescents' self-reported adjustment indicators when both were entered into a multiple regression equation.

These results suggest that decision-making fit and perceptions of parental intrusiveness tap unique, though related, aspects of adolescents' developmental autonomy needs. While the desire to be involved in decision making reflects adolescents' growing need for establishing a sense of personal efficacy within their environment, adolescents' perceptions of parent intrusiveness represents their need for personal space and clear relational boundaries. It is interesting to note that in our analysis of the academic alienation among these same early adolescents described earlier (Roeser et al., 1994), we found that the academically alienated adolescents were also more likely to report poor decision-making fit in their families than their peers.

Several qualifications to the results presented above are warranted. First, the one-point-in-time, correlational nature of data from MAGICS restricts our ability to model the causal direction of the relation between stage-environment mismatch and adolescent mental health. For instance, while the results are consistent with our hypothesis that a stage-environment mismatch results in a negative change in self-esteem over time, they are also consistent with the hypothesis that characteristics of the child create the mismatch. It may be, for example, that parents are indeed more protective and controlling of their children (and thereby are perceived as more intrusive) if they are worried about their child's development (e.g., if the child is already becoming involved in risk taking or problem behavior, or if the child is depressed). Since we are collecting longitudinal data on this sample, we will be able to model these two directional hypotheses in the near future.

Overall, these results on different samples provide good preliminary support for the hypothesis that a misfit, or lack of attunement, between parents and their adolescents is negatively related to both the adolescents' academic motivation and their psychological well-being. A developmentally responsive environment can help adolescents develop certain competencies (such as autonomy, psychosocial maturity and competence, and high self-esteem) that can, in turn, serve as protective factors for successfully coping with change and adversity. Parents who are able to support adolescent needs for developmentally and contextually appropriate levels of autonomy likely exert a facilitative effect on their adolescents' psychological functioning and academic motivation. By providing such opportunities for participation in decision making, parents help their adolescents develop a sense of

Figure 2. Percent of Variance Accounted for in Adolescent Mental Health Indicators by Parent Intrusion and Decision Making Fit: Maryland Adolescent Growth in Context Study, Grade 7.

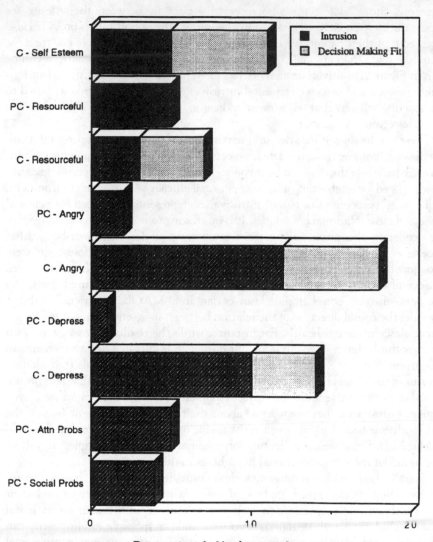

Percent of Variance Accounted For

Note: C = Child Report
       PC = Parent Report

personal efficacy in part because adolescents can see that their parents respect their opinions and trust them to be involved in important decisions. Parents who use more controlling strategies when interacting with their children may convey to them that they are not to be trusted to undertake activities independently and that they are not considered competent (Maccoby and Martin, 1983).

The Interplay Among Biological, Psychological, and Social Context Factors: Biology and the Notion of Fit

Another way to look at developmental change is to look for interindividual differences at the same time point between same-aged children of different maturational levels. During early adolescence, the extent of pubertal development provides a good indicator of individual differences in physical maturation, particularly for females. There is some evidence to suggest that pubertal maturation is associated with stage-environment mismatch effects within both schools and families.

Maturational Differences in the Desire for Autonomy in School

Using data from our MSALT study, we related an indicator of physical maturation to female adolescents' desire for input into classroom decisions using the Lee et al. (1983) items. Consistent with the intraindividual longitudinal pattern of age-related change reported above, the more physically mature female adolescents expressed a greater desire for input into classroom decision making than did their less physically mature female peers (Miller, 1986). Unfortunately, as was true for the longitudinal results, the more physically mature females did not perceive greater opportunities for participation in classroom decision making. Although the females with varying degrees of pubertal development were in the same classrooms, the more physically mature females (i.e., the early developers) reported *fewer* opportunities for participation in classroom decision making than did their less mature peers (i.e., the on-time and late developers).

These maturational differences were even more striking when we looked at the within-year changes in these female adolescents' perceptions of the opportunities they had to participate in classroom decision making. We calculated the mean change in their perceptions of opportunities from the fall to the spring testing waves. We then looked at this change as a function of their pubertal status. The early-maturing females reported less opportunity to participate in classroom decision making in the spring term than they had reported in the *previous* fall term. In contrast, the late-maturing females in these same classrooms showed an increase over the course of the school year in these opportunities (Miller, 1986). How can this be, given that these adolescents were in the same classrooms? Did the teachers actually treat these adolescent females differently (i.e., did the teachers respond to earlier physical maturity with more controlling behavior)? Or did the adolescents perceive a similar environment differently (i.e., did the early maturing adolescents perceive the same level of adult control as providing less opportunity for self-determination than did the later maturing adolescents)?

Evidence from educational psychology, developmental psychology, and general psychology suggests that either or both of these explanations could be accurate: Teachers do respond differently to various children in the same classroom depending on a variety of characteristics (Brophy & Evertson, 1976), and people do perceive similar environments differently depending either on their cognitive or motivational orientation or both (see Baron & Graziano, 1991; Ryan & Grolnick, 1986). More detailed classroom observations are needed to determine the exact nature of the relation between teachers' behavior and adolescents' perceptions.

But more important for the issues central to this chapter, the degree of mismatch between these female adolescents' desires for input and their perceptions of these opportunities in their classroom environment was related to their pubertal status: There was a greater degree of mismatch among the more physically mature female adolescents than among the less mature. In fact, by the end of the school year, almost twice as many early-maturing females reported experiencing the "can't but should" type of mismatch (e.g., answering "no" to the question "Do you get to help decide what math you work on during math class?" but "yes" to the question "Should you have a say about this?") as did their less physically mature classmates.

We find this last set of results especially interesting in light of the findings of Simmons and her colleagues (e.g., Simmons & Blyth, 1987; Simmons et al., 1979). They found that the pubertal status of female adolescents at the time of the junior high school transition is related to changes in their self-esteem and their self-reports of truancy and school misconduct: The more physically mature females reported the highest amount of truancy and school misconduct after they made the junior high school transition. Simmons and her colleagues suggest that experiencing both school and pubertal transitions simultaneously puts these girls at particular risk for negative outcomes. Additionally, it is possible that the size of the mismatch between their desire for a less controlling adult environment and their perceptions of the actual opportunities for participation puts these females at risk for the most negative motivational outcomes.

Similar results characterize our data on interindividual pubertal status effects. Miller and Taylor (1986) tested the relation between female pubertal status and self-esteem. Consistent with other studies (e.g., Simmons & Blyth, 1987), the early-maturing sixth grade girls reported lower self-esteem than their less physically mature classmates. But, consistent with the person-environment fit perspective outlined in this paper, it was only the early maturing girls who also felt they had relatively little opportunity to participate in family decision making who reported lower self-esteem. There was no effect of pubertal status on self-esteem among those sixth grade girls who reported relatively high opportunity to participate in their family's decision making.

Maturational Differences in the Desire for Autonomy in Families

We have just begun to examine the influence of pubertal development on person-environment fit and psychological well-being in the MAGICS study (Lord & Eccles, 1994). Relative to their on-time and late-maturing female peers, early-maturing seventh grade girls felt that their parents were more intrusive. Whether or not this perception represented the adolescents' dissatisfaction with their parent, or a real characteristic of the situation, cannot be ascertained from the cross-sectional data. Indeed, the parents of more mature adolescent females may be more protective of and intrusive with their daughters because of the increased risks that can accompany such maturity (dating, unprotected sex, etc.).

There is also some evidence in MAGICS that the influence of decision making fit and pubertal status may interact to influence indicators of adolescents' psychological well-being. For example, the negative impact of decision making misfit on these adolescents' reports of depressive symptomatology was strongest for the early-maturing females. That is, relative to on-time and late-maturing females, the early-maturing females who also perceived a decision-making misfit in their families reported the highest levels of depressive symptomatology (Lord & Eccles, 1994). While others have documented the important relationships between changing self-images and perceptions (e.g., Petersen et al., 1993) and multiple concurrent life changes (Simmons & Blyth, 1987) for depression in young females, our results implicate stage-environment mismatch as another relevant focus of study for understanding depression in adolescence.

Conclusion

In this chapter, we adopted a developmental psychopathology organizational framework to address the question of why adolescence is a time of heightened risk for maladjustment among some adolescents. We discussed patterns of positive and negative adjustment within normative samples of adolescents, and identified potential risk and protective factors for both academic motivation and general mental health. We highlighted the importance of understanding these patterns of development within the context of home and school environments. In doing so, we argued that optimal development takes place when there is good stage-environment fit, attunement, and synchrony between the needs of developing adolescents and the opportunities afforded them by their social spheres of experience.

In particular, we have provided evidence of the negative effects of the diminished quality of personal relationships with teachers after the transition to junior high school, and have argued that this decline in the quality of relationships is especially problematic during early adolescence when young people are in special need of close relationships with adults outside of their homes. We have also noted the increase in ability grouping and comparative and public evaluation at a time when young adolescents have a heightened concern about their status in relation to their peers. Finally, we discussed, and provided evidence where available, the negative consequences of these kinds of developmentally inappropriate environmental changes on early adolescents' school motivation, academic self-concepts, and mental health.

We also discussed the role of opportunities for self-determination and participation in rule making, pointing out the particularly important need for a match between the individual's increasing desires for autonomy and self-determination and the opportunities for such autonomy provided both in school and in the home. Although adolescents desire more freedom from adult control than children, they do not want total freedom and they do not want to be emotionally detached from their parents. Instead they desire a gradual increase in the opportunity for self-determination and participation in decision and rule making. Furthermore, evidence suggests that adolescents develop best when these increasing opportunities occur in environments that are emotionally supportive (Baumrind, 1971; Lord, 1994; Ryan & Lynch, 1989).

Unfortunately, our research suggests that many early adolescents do not have these experiences in either the school or the home. After the transition to junior high school in particular, early adolescents are often confronted with a regressive environmental change. Many early adolescents experience a decrease in the opportunity to participate in classroom decision making when they move into junior high school. Not surprisingly, there is also a decrease in intrinsic motivation and an increase in school misbehavior associated with this transition, as well as a decline in indicators of mental health. These changes are most apparent among the adolescents who report experiencing the greatest mismatch between their needs and the opportunities afforded them to participate in classroom decision making. Such motivational and mental health changes are not apparent in adolescents who report the more developmentally appropriate increase in opportunities for participation in classroom decision making.

We have also found evidence suggesting that a similar process is occurring in the family. Excessive parental control is linked to lower intrinsic school motivation, to more negative changes in self-esteem and mental health following the junior high school transition, to more school misbehavior, and to relatively greater investment in peer social attachments. However, since some of the evidence we have presented is correlational, it is possible that excessive parental control is the consequence rather than the cause of these negative adolescent outcomes. Our preliminary longitudinal analyses do suggest, however, that the causal links are at least bidirectional. Similarly, although we have focused on excessive parental control, other studies have documented the negative consequences of too little parental control at this age (see Dornbusch, Carlsmith, Bushwall, Ritter, Leiderman, Hastorf, & Gross, 1985; Fuligni & Eccles, 1990; Steinberg, 1990). Clearly, these results point out the importance of creating educational and family environments for early adolescents that provide a better match to young people's developing needs and desires. How could the creation of such developing appropriate environments be accomplished?

Towards Developmentally Appropriate School Settings

The current situation in traditional junior high schools seems especially problematic. The existing structure of many junior high schools appears to create a climate that undermines both teacher and student motivation and well-being. The large size of these schools, coupled with departmentalized teaching and large student loads, makes it difficult for teachers and students to form close relationships. In turn, this lack of close relationships, coupled with the generally negative stereotypes about adolescents, could be responsible for the prevalence of low teacher efficacy and high use of controlling motivational strategies in junior high school classrooms.

Turning Points (Carnegie Council on Adolescent Development, 1989) outlines a variety of changes in the structure of middle grades educational institutions (e.g., junior highs, middle schools, and intermediate schools) that would make it easier for both teachers and students to maintain a high sense of self-efficacy and to develop a stronger sense of shared community with each other. One potential strategy for remediating the impersonal quality of traditional junior high schools involves within-school reorganization based on the middle school teaching philosophy. Some characteristics of the middle school philosophy that have been identified as potentially helpful are small house programs, team teaching, and advisory sessions (see Eccles & Midgley, 1989).

Field studies of the more successful middle and junior high schools provide numerous examples of classrooms and schools that have more positive and developmentally appropriate learning environments—for example, higher teacher efficacy, greater opportunity for meaningful student participation in both school and classroom decision making, an academic culture that stresses task mastery and improvement and more positive student-teacher relationships (see Ames, 1992a, 1992b; Bryk et al., 1991; Carnegie Council on Adolescent Development, 1989; Dryfoos, 1990; Eccles & Midgley, 1989; Lipsitz, 1981; Maehr & Midgley, 1991). Young adolescents in these schools do not evidence the same declines in intrinsic motivation and school attachment stereotypically associated with students in junior high schools; they also do not engage in the same amount of school misbehavior as students in more traditional junior high schools. Unfortunately, many junior high schools do not provide such a developmentally appropriate environment (see Eccles & Midgley, 1989). Clearly, future research is needed to determine the impact of various restructuring strategies on adolescent adjustment.

Another type of change that may increase opportunities for adolescents to develop their competencies and intrinsic valuing of learning involves restructuring of the "culture" of the school (e.g. Maehr, 1991; Rutter, 1983). Research has demonstrated that particular constellations of classroom and school-level practices have important consequences on children's and adolescents' competencies. Ability-oriented classrooms, characterized by practices that stress social comparison such as ability grouping, use of competitive activities, salient evaluative feedback that focuses on relative ability, and the use of unidimensional tasks, promote student's comparison of their performance with others, and emphasize the demonstration of ability relative to others as the goal of learning (Eccles et al., 1984; Maehr, 1991; Maehr & Midgley, 1991; Nicholls, 1984). Ability-oriented academic environments, either through teacher's differential expectations for students of different ability levels, or through other practices that make social comparison salient, may have detrimental impacts on the development of competence in students, especially low-achieving students, by implicitly or explicitly creating hierarchies of competent and less competent students. These hierarchies based on relative ability status may lead students to believe that only relatively high achieving students can be successful in school and may be a disincentive for low achieving students to try (Covington, 1984; Maehr et al., 1992; Marshall & Weinstein, 1984; Weinstein, 1989). The negative affect and self-perceptions that may arise from ability-oriented school practices may also have more powerful effects in the middle school grades when ability information is more readily used by students in forming their impressions of competence (Nicholls, 1990). Finally, previous research lends support to the hypothesis that environments that stress comparative competence likely cause increased anxiety and stress (Hill & Wigfield, 1984; Covington, 1992). In these types of environments, students may decrease effort and forego participating in activities, exhibit maladaptive coping strategies, and exhibit depressive symptomatology rather than fail publicly (Cole, 1991; Covington, 1992; Gold & Mann, 1985; Roeser, Urdan, & Midgley, 1994).

Fortunately, field and intervention studies have elaborated a host of strategies that educators working at both the classroom and school level can implement to move an academic culture from a stress on relative ability and competition ("ability-oriented culture") to a stress on self-improvement, effort, and task mastery ("task-oriented culture") (Ames, 1990; Maehr & Anderman, 1993; Meece, 1991; Midgley, 1993). Intervention strategies to do this have evolved around six major dimensions of the learning environment: (1) Tasks, (2) Authority, (3) Recognition, (4) Grouping, (5) Evaluation, and (6) Time (Ames, 1990; Epstein, 1988). Using the acronym TARGET, Ames (1990) and others (e.g., Maehr & Midgley, 1991) have elaborated new methods of constructing tasks, distributing authority in the classroom, recognizing students for improvement and effort rather than relative ability, grouping students in heterogeneous ways, using effort as a criterion of evaluation, and using time flexibly to allow for new task and grouping approaches to the curriculum. Early results of intervention studies suggest that both elementary and middle schools can move more towards a task-oriented academic culture by changing certain dimensions of classrooms and schools, and that a task-oriented school culture is most appropriate to meeting the developmental needs of children in late childhood and early adolescence (e.g., Ames, 1990; Maehr & Anderman, 1993; Maehr & Buck, 1992; Roeser, Urdan et al., 1994).

In addition to the structural changes that would facilitate a more community-oriented, task-focused environment in schools, there are other changes that schools can implement to foster a more positive, developmentally responsive environment. One such change is the promotion of greater parent involvement in schools. The evidence is fairly strong that parent involvement in school is linked to better academic performance and overall psychological competence in children (e.g., Comer, 1980; Comer & Haynes, 1991; Epstein, 1990). A school governance that provides a more integral role for parents in policy and curriculum decision making can result in parents feeling more efficacious for influencing their child's education, which, in turn, can be reflected in the adolescent's own improved competence, both academically and psychologically. Indeed, our MAGICS data indicate that by and large seventh graders think it is a good idea for parents to be involved in their school and education. Such involvement should be implemented in such a way to convey to adolescents that the significant adults in their lives care about what goes on in their lives, yet not in a way that could be perceived by adolescents as intrusive upon their personal space. For example, teachers could encourage parent involvement by assigning tasks in which parents and their adolescents work together on tasks or issues that are relevant to the adolescent, such as occupational exploration or delineation of one's family lineage tree (Eccles & Harold, 1993). Tasks such as these encourage parents to be a resource for the adolescents' own self-development.

The promotion of increased parent involvement in school can also be the gateway to greater parent-teacher communication about the child. Such communication can be used to facilitate the integration of the home and school lives of adolescents, enabling both teachers and parents to have a richer picture of what adolescents' lives are like. This integration of contexts would help foster the type of safety zone Simmons and Blyth (1987) advocated as necessary for healthy development during this period—a zone where adolescents can experiment but where the adults are available to catch the adolescent if she or he starts to get into trouble. Again, such communication should not be used as a venue for strict monitoring of adolescents but rather as a means by which teachers and parents can better understand and be attuned to the experiences of their adolescents.

Better efforts could also be made in school environments to increase the degree to which both teachers and parents are attuned to the psychological needs of adolescents. Increased teacher-parent communication could facilitate this. In addition, focus groups in which adolescents are given a forum to openly discuss the issues most relevant to them could provide an excellent opportunity for teachers and parents to learn more about what is happening in the lives of their adolescents. It is also likely that both policy and practice could be greatly informed if we as adults listened to what adolescents themselves are saying about their lives and their social environments.

Overall, each of these reform efforts could serve to increase the degree to which parents and teachers are attuned to the psychological needs of adolescents. For example, classroom environments that stress self-improvement and task mastery would help focus teachers on individual student progress, rather than comparative performance. In addition, facilitated parent involvement in schools and parent-teacher communication could increase the amount of important information shared concerning the progress and wellness of each adolescent in his or her home and school settings. Furthermore, more individualized techniques such as focus groups in which adolescents are given a forum to openly discuss the issues most relevant to them could provide an excellent arena within which the significant adults in young adolescents' lives could learn more about their contemporary concerns, struggles, and lifestyles.

Towards Developmentally Appropriate Family Settings

There is a similar need for developmentally responsive environments in the family. Existing evidence suggests that there is variability in how families adapt to their children's movement into adolescence and that adolescents fare best in family environments that provide a good fit to their increasing need for autonomy. Adolescents fare more poorly in families that respond to their development either by throwing up their hands and relinquishing control or by cracking down too much. Families, like schools, are confronted with a difficult problem—providing an environment that changes in the right way, and at the right pace, to maintain a good fit with their children's developmentally appropriate needs. Unfortunately, we know less about how to help families achieve this balance than we know about how to design schools that help teachers achieve the right balance. There is a great need for programs that will help parents with this difficult task.

Summary

The contexts in which adolescents develop can either facilitate or undermine an adolescent's pursuit of a unique and coherent personal identity and sense of competence. Adolescents' perceptions of their school and family environments as either satisfying or thwarting their developmental needs for autonomy, feelings of relatedness, and opportunities for competence development contribute in significant ways to adolescents' academic motivation, identity formation, and personal adjustment. In this chapter, we reviewed our findings and other studies that suggest teachers and parents of early adolescents might not be very well attuned to the experiences and needs of children who are making the transition into early adolescence. For teachers, this diminished sensitivity probably reflects the demands of the current structure of many middle level schools. For both teachers and parents, this low attunement to adolescent needs in general may reflect a lack of understanding about what early adolescents are up against during this development time in this historical epoch, as well as a lack of understanding about what adolescents need from the adults in their lives. These findings suggest a need for intervention at different levels of the adolescents' social worlds, including school reform and basic education about what adolescence is about for both parents and teachers. It seems incumbent upon us as professionals in the field to serve as advocates for adolescents and as spokespersons to the public in terms of describing their experiences and needs. As such, part of our task is to disseminate information to both the professional and lay public in order to dispel stereotypes, as well as to be involved in the design and implementation of changes at the level of policy and organizations that serve adolescents. Such advocacy on the part of social scientists and practitioners can assist young adolescents to both develop a sense of belonging in this society and realize their potential as productive and valued members of this society.

REFERENCES

Achenbach, T. M., Howell, C. T., Quay, H. C., & Conners, C. K. (1991). National survey of problems and competencies among four to sixteen year olds. *Monographs for the Society of Research in Child Development, 56*, 3.

Ames, C. (1990, April). *Achievement goals and classroom structure: Developing a learning orientation.* Paper presented at the annual meeting of the American Educational Research Association, Boston.

Ames, C. (1992a). Achievement Goals and the Classroom Motivational Climate. In D. H. Schunk & J. L. Meece (Eds.), *Student Perceptions in the Classroom* (pp. 327–348). Hillsdale, NJ: Erlbaum.

Ames, C. (1992b). Classrooms: Goals, structures, and student motivation. *Journal of Educational Psychology, 84*, 261–271.

Barber, B. L., & Crockett, L. J. (1993). Preventative interventions in early adolescence. In R. M. Lerner (Ed.), *Early Adolescence: Perspectives on research, policy, and intervention* (pp. 311–314). Hillsdale, NJ: Lawrence Erlbaum Associates.

Barker, R., & Gump, P. (1964). *Big school, small school: High school size and student behavior.* Stanford, CA: Stanford University Press.

Baron, R. M. & Graziano, W. G. (1991). *Social psychology.* Chicago: Holt, Rinehart, & Winston.

Baumrind, D. (1971). Current patterns of parental authority. *Developmental Psychology Monograph, 4* (l, Pt. 2).

Beeghly, M., & Cicchetti, D. (1987). An organizational approach to symbolic development in children with Down syndrome. *New Directions for Child Development, 36*, 5–29.

Bloom, B. S. (1976). *Human characteristics and school learning.* New York: McGraw-Hill.

Blyth, D. A., Simmons, R. G., & Carlton-Ford, S. (1983). The adjustment of early adolescents to school transitions. *Journal of Early Adolescence, 3*, 105–120.

Braddock, J. H., & McPartland, J. M. (1992). Education of early adolescents. *Review of Research in Education, 19*, 135–170.

Brookover, W., Beady, C., Flood, P., Schweitzer, J. & Wisenbaker, J. (1979). *School social systems and student achievement: Schools can make a difference.* New York: Praeger.

Brophy, J. E., & Evertson, C. M. (1976). *Learning from teaching: A developmental perspective.* Boston: Allyn and Bacon.

Brophy, J. E., & Evertson, C. M. (1978). Context variables in teaching. *Educational Psychologist, 12*, 310–316.

Bryk, A. S., Lee, V. E., & Smith, J. B. (1989, May). *High school organization and its effects on teachers and students: An interpretative summary of the research.* Paper presented at the invitational conference on Choice and Control in American Education, Robert M. La Follette Institute of Public Affairs, University of Wisconsin, Madison.

Buchanan, C. M., Eccles, J. S., & Becker, J. B. (1992). Are adolescents the victims of raging hormones: Evidence for activational effects of hormones on moods and behavior at adolescence. *Psychological Bulletin, 111*, 62–107.

Carnegie Council on Adolescent Development (1989). *Turning points: Preparing American youth for the 21st century.* New York: Carnegie Corporation.

Casper, R. C., Eckert, E. D., Halmi, K. A., Goldberg, S. C., & Davis, J. M. (1980). Bulimia: Its incidence and clinical importance in patients with anorexia nervosa. *Archives of General Psychiatry, 37*, 1030–1035.

Cicchetti, D., & Toth, S. L. (1993). The role of developmental theory in prevention and intervention. *Development and Psychopathology, 4*, 489–493.

Cohen, P., Cohen, J., Kasen, S., Velez, C. N., Hartmark, C., Johnson, J., Rojas, M., Brook, J., & Streuning, E. L., (1993). An epidemiological study of disorders in late childhood and adolescence. I. Age- and Gender-Specific Prevalence. *Journal of Child Psychology and Psychiatry, 34*, 851–867.

Cole, D. A. (1991). Preliminary support for a competency-based model of depression in children. *Journal of Abnormal Psychology, 100*, 181–190.

Collins, W. A. (1990). Parent-child relationships in the transition to adolescence: Continuity and change in interaction, affect, and cognition. In R. Montemayor, G. Adams, & T. Gullotta (Eds.), *Advances in adolescent development, Vol. 2 From childhood to adolescence: A transitional period?* (pp. 85–106). Newbury Park, CA: Sage.

Comer, J. (1980). *School Power.* New York: Free Press.

Comer, J. P., & Haynes, N. M. (1991). Parent involvement in schools: An ecological approach. *The Elementary School Journal, 91* (3), 271–277.

Covington, M. V. (1984). The self-worth theory of achievement motivation : Findings and implications. *Elementary School Journal, 85*, 5–20.

Covington, M. V. (1992). *Making the grade: A self-worth perspective on motivation and school reform.* New York: Cambridge University Press.

Cowen, E. L. (1991). In pursuit of wellness. *American Psychologist, 46*, 404–408.

deCharms, R. (1980). The origins of competence and achievement motivation in personal causation. In L. J. Fyans, Jr. (Ed.), *Achievement motivation: Recent trends in theory and research* (pp. 22–23). New York: Plenum.

Deci, E. L., & Ryan, R. M. (1985). *Intrinsic motivation and self determination in human behavior.* New York: Plenum Press.

Deci, E. L., & Ryan, R. M. (1987). The support of autonomy and the control of behavior. *Journal of Personality and Social Psychology, 53*, 1024–1037.

Donovan, J. E., & Jessor, R. (1985). Structure of Problem Behavior in Adolescence and Young Adulthood, *Journal of Consulting and Clinical Psychology, 53*, 890–904.

Dornbush, S. M., Carlsmith, J. M., Bushwall, S. J., Ritter, P. L., Leiderman, H., Hastorf, A. H., & Gross, R. T. (1985). Single parents, extended households, and the control of adolescents. *Child Development, 56*, 326–241.

Dryfoos, J. G. (1990). *Adolescents at risk: Prevalence and prevention.* Oxford: Oxford University Press.

Eccles, J. S., & Harold, R. D. (1993). Parent-school involvement during the early adolescent years, *Teachers College Record, 94*, 568–587.

Eccles, J. S., & Lord, S. E. (1993). *Round holes, square pegs, rocky roads, and sore feet: The impact of stage / environment fit on young adolescents' experiences in schools and families.* Paper presented at the Rochester symposium on developmental psychopathology: Adolescence: Opportunities and challenges, Rochester, NY.

Eccles, J. S., McCarthy, K. A., Lord, S. E., Harold, R., Wigfield, A., and Aberbach, A. (1990, April). *The relationship of family factors to self-esteem and teacher-rated adjustment following the transition to junior high school environment*. Paper presented at the meeting of the Society for Research on Adolescence, Atlanta, GA.

Eccles, J. S., & Midgley, C. (1989). Stage/environment fit: Developmentally appropriate classrooms for early adolescents. In R. E. Ames & C. Ames (Eds.), *Research on Motivation in Education: Vol. 3, Goals and Cognitions*.(pp. 13-44). New York: Academic Press.

Eccles, J. S. Midgley, C., & Adler, T. (1984). Grade-related changes in the school environment: Effects on achievement motivation. In J. G. Nicholls (Ed.), *The development of achievement motivation* (pp. 283–331). Greenwich, CT: JAI Press.

Eccles, J. S., Midgley, C., Wigfield, A., Buchanan, C.M., Reuman, D., Flanagan, C., & MacIver, D. (1993). Development during adolescence: The impact of stage-environment fit on adolescents' experiences in schools and families. *American Psychologist, 48*, 90–101.

Eccles, J. S. , & Wigfield, A. (1985). Teacher expectations and student motivation. In J. Dusek (Ed.), *Teacher expectancies* (pp. 185–217). Hillsdale, NJ: Erlbaum.

Eggert, L. L, Seyl, C. D., Nicholas, L. J. (1990). Effects of a school-based prevention program for potential high school dropouts and drug abusers. *The International Journal of Addictions, 25*, 773–801.

Elliott, D. S., Huizinga, D., & Menard, S. (1989). *Multiple problem youth: Delinquency, substance use, and mental health problems*. New York: Springer-Verlag.

Epstein, J. L. (1988). Effective schools or effective students: Dealing with diversity. In R. Haskins & D. MacRae (Eds.), *Policies for America's public schools: Teacher equity indicators*. Norwood, NJ: Ablex.

Epstein, J. L. (1990). School and family connections: Theory, research, and implications for integrating sociologies of education and family. *Marriage and Family Review, 15*, 99–126.

Epstein, J. L., & McPartland, J. M. (1976). The concept and measurement of the quality of school life. *American Educational Research Journal, 13*, 15–30.

Epstein, J. L., & McPartland, J. M. (1977). *The Quality of School Life Scale and administrative and technical manual*. Boston: Houghton Mifflin.

Erikson, E. H. (1959). Identity and the life cycle. *Psychological Issues, 1*, 18–164.

Feldlaufer, H., Midgley, C., & Eccles, J. S. (1988). Student, teacher, and observer perceptions of the classroom environment before and after the transition to junior high school. *Journal of Early Adolescence, 8*, 133–156.

Finger, J. A., & Silverman, M. (1966). Changes in academic performance in the junior high school. *Personnel and Guidance Journal, 45*, 157–164.

Finn, J. (1989). Withdrawing from school. *Review of Educational Research, 59*, 117–142.

Flanagan, C. (1986, April). *Early adolescent needs and family decision making environments: A study of person-environment fit*. Paper presented at the meeting of the American Educational Research Association, San Francisco.

Flanagan, C. (1989, April). *Adolescents' autonomy at home: Effects on self-consciousness and intrinsic motivation at school*. Paper presented at the meeting of the American Educational Research Association, Montreal.

Fraser, B. J., & Fisher, D. L. (1982). Predicting students' outcomes from their perceptions of classroom psychosocial environment. *American Educational Research Journal, 19*, 498–518.

Fuligni, A. J., & Eccles, J. S. (1992, March). *The effects of early adolescent peer orientation on academic achievement and deviant behavior in high school*. Paper presented at the biennial meeting of the Society for Research on Adolescence, Washington, DC.

Garfinkel, P. E., & Garner, D. M. (1982). *Anorexia nervosa: A multidimensional perspective*. New York: Brunner / Mazel.

Garmezy, N. (1983). Stressors of childhood. In N. Garmezy & M. Rutter (Eds.), *Stress, coping and development in children*. (pp. 43–84). New York: McGraw-Hill.

Gold, M., & Mann, D. (1985). *Expelled to a friendlier place: A study of effective alternative schools*. Ann Arbor: University of Michigan Press.

Harter, S. (1981). A new self-report scale of intrinsic versus extrinsic orientation in the classroom: Motivational and informational components. *Developmental Psychology, 17*, 300–312.

Harter, S. (1982). The Perceived Competence Scale for Children. *Child Development, 53*, 87–97.

Harter, S., Whitesell, N. R., & Kowalski, P. (1992). Individual differences in the effects of educational transitions on young adolescents' perceptions of competence and motivational orientations. *American Educational Research Journal, 29*, 777–808.

Hartup, W. W. (1989). Social relationships and their developmental significance. *American Psychologist, 44*, 120–126.

Hauser, S., Powers, S. I., & Noam, G. G. (1991). *Adolescents and their families*. New York: Free Press.

Hawkins, J. D., Catalano, R. F., & Miller, J. Y. (1992). Risk and protective factors for alcohol and other drug problems in adolescence and early adulthood: Implications for substance abuse prevention. *Psychological Bulletin, 112*, 64–105.

Hawton, K. (1982). Attempted suicide in children and adolescents. *Journal of Child Psychology and Psychiatry and Allied Disciplines, 23*, 497–503.

Hesse, P., & Cicchetti, D. (1982). Perspectives on an integrated theory of emotional development. *New Directions for Child Development, 16*, 3–48.

Higgins, E. T., & Parsons, J. E. (1983). Social cognition and the social life of the child: Stages as subcultures. In E. T. Higgins, D. W. Ruble, & W. W. Hartup (Eds.). *Social cognition and social behavior: Developmental issues* (pp. 15–62). NY: Cambridge University Press.

Hill, J. (1988). Adapting to menarche: Familial control and conflict. In M. Gunnar & W. A. Collins (Eds.), *Minnesota Symposia on Child Development: Vol. 21. Development During the Transition to Adolescence* (pp. 43–77). Hillsdale, NJ: Erlbaum.

Hill, K. T. (1980). Motivation, evaluation, and educational test policy. In L. J. Fyans (Ed.), *Achievement motivation: Recent trends in theory and research*. New York: Plenum Press.

Hill, K. T., & Wigfield, A. (1984). Test anxiety: A major educational problem and what can be done about it. *Elementary School Journal, 85*, 105–126.

Hunt, D. E. (1975). Person-environment interaction: A challenge found wanting before it was tried. *Review of Educational Research, 45*, 209–230.

Jessor, R., & Jessor, S. L. (1977). *Problem behavior and psychosocial development: A longitudinal study of youth*. San Diego, CA: Academic Press.

Kagan, D. M. (1990). How schools alienate students at risk: A model for examining proximal classroom variables. *Educational Psychologist, 25*, 105–125.

Kavrell, S. M., & Petersen, A. C. (1984). Patterns of achievement in early adolescence. In M. L. Maehr (Ed.), *Advances in motivation and achievement* (pp. 1–35). Greenwich, CT: JAI Press.

Kazdin, A. E. (1993). Adolescent mental health. Prevention and treatment programs. *American Psychologist, 48*, 127–141.

Kennedy, R. E. (1993). Depression as a disorder of social relationships: Implications for school policy and prevention programs. In R. M. Lerner (Ed.), *Early Adolescence: Perspectives on research, policy, and intervention* (pp. 383–398). Hillsdale, NJ: Erlbaum.

Laupa, M., & Turiel, E., (1986). Children's conceptions of adult and peer authority. *Child Development, 57*, 405–412.

Lee, P., Statuto, C., & Kedar-Voivodas, G. (1983). Elementary school children's perceptions of their actual and ideal school experience: A developmental study. *Journal of Educational Psychology, 75*, 838–847.

Lewin, K. (1935). *A dynamic theory of personality*. New York: McGraw-Hill.

Lewinsohn, P. M., Hops, H., Roberts, R. E., & Seeley, J. R. (1993). Adolescent psychopathology: I. Prevalence and disorders of depression and other *DSM-III-R* disorders in high school students. *Journal of Abnormal Psychology, 102* (1), 133–144.

Lipsitz, J. (1981). Educating the early adolescent: Why four model schools are effective in reaching a difficult age group. *American Education*, 13–17.

Lord, S. E. (1994). *Parenting in impoverished urban contexts: Implications for adolescent adjustment*. Unpublished doctoral dissertation, University of Colorado, Boulder.

Lord, S. E., & Eccles, J. S. (1994). *Pubertal development and stage / environment fit in early adolescence*. Unpublished manuscript; University of Michigan, Ann Arbor.

Lord, S. E., Eccles, J. S., & McCarthy, K. (1994). Surviving the junior high school transition. Family processes and self-perceptions as protective and risk factors. *Journal of Early Adolescence, 14*, 162–199.

Maccoby, E., & Martin, J. (1983). Socialization in the context of the family: Parent-child interaction. In E. M. Hetherington (Ed.), P. H. Mussen (Series Ed.), *Handbook of child psychology: Vol. 4. Socialization, personality, and social development* (pp. 1–101). New York: Wiley.

Mac Iver, D., & Reuman, D. A. (1988, April). *Decision making in the classroom and early adolescents' valuing of mathematics*. Paper presented at the annual meeting of the American Educational Research Association, New Orleans.

Maehr, M. L. (1991). The "psychological environment" of the school: A focus for school leadership. In P. Thurstone & P. Zodhiates (Eds.), *Advances in educational administration* (Vol. 2, pp. 51–81). Greenwich, CT: JAI Press.

Maehr, M. L., & Anderman, E. M. (1993). Reinventing schools for early adolescents: Emphasizing task goals. *The Elementary School Journal, 93*, 593–610.

Maehr, M. L., & Buck, R. M. (1992). Transforming school culture. In M. Sashkin and H. J. Walberg (Eds.), *Educational leadership and school culture*. Berkeley, CA: McCutchan.

Maehr, M. L., & Midgley, C. (1991). Enhancing student motivation: A school-wide approach. *Educational Psychologist, 26*, 399–427.

Maehr, M. L., Midgley, C., Pintrich, P., & De Groot, E. (1992). *Personal and contextual influences on adolescent wellness*. Grant proposal submitted to the National Institute for Mental Health.

Marshall, H. H., & Weinstein, R. S. (1984). Classroom factors affecting students' self-evaluations: An interactional model. *Review of Educational Research, 54*, 301–325.

Masten, A. S. (1988). Toward a developmental psychopathology of early adolescence. In M. D. Levine, & E. R. McAnarney (Eds.), *Early adolescent transitions* (pp. 261–278). Lexington, MA: D. C. Heath.

McGuire, D. P., Mitic, W., & Neumann, M. A. (1987). Perceived stress in adolescents: What normal teenagers worry about. *Canada's Mental Health, 35*, 2–5.

Meece, J. L. (1991). The classroom context and students' motivational goals. In M. L. Maehr and P. Pintrich (Eds.) *Advances in motivation and achievement: Vol. 7. Goals and self-regulatory processes* (pp. 261–285). Greenwich, CT: JAI Press.

Midgley, C. (1993). Motivation and Middle Level Schools. In M. L. Maehr and P. Pintrich (Eds.) *Advances in motivation and achievement: Vol. 8. Motivation and adolescent development* (pp. 217–274). Greenwich, CT: JAI Press.

Midgley, C., Anderman, E., & Hicks, L. (1995). Differences between elementary and middle school teachers and students: A goal theory approach. *Journal of Early Adolescence, 15*, 90-113.

Midgley, C., & Feldlaufer, H. (1987). Students' and teachers' decision making fit before and after the transition to junior high school. *Journal of Early Adolescence, 7*, 225–241.

Midgley, C., Feldlaufer, H., & Eccles, J. S. (1988a). Student / teacher relations and attitudes toward mathematics before and after the transition to junior high school. *Child Development, 60*, 375–395.

Midgley, C., Feldlaufer, H., & Eccles, J. S. (1988b). The transition to junior high school: Beliefs of pre- and post-transition teachers. *Journal of Youth and Adolescence, 17*, 543–562.

Midgley, C., Feldlaufer, H., & Eccles, J. S. (1989). Change in teacher efficacy and student self- and task-related beliefs during the transition to junior high school. *Journal of Educational Psychology, 81*, 247–258.

Miller, C. L. (1986, April). *Puberty and person-environment fit in the classroom.* Paper presented at the meeting of the American Educational Research Association, San Francisco.

Miller, C. L., Eccles, J. S., Flanagan, C., Midgley, C., Feldlaufer, H., & Harold, R. D. (1990). Parents' and teachers' beliefs about adolescents: Effects of sex and experience. *Journal of Youth and Adolescence, 19*, 363–394.

Miller, C. L., & Taylor, R. (1986, March). *Pubertal development, self-concept, and behavior: The role of family decision making practices.* Paper presented at biennial meeting of the Society for Research on Adolescence, Madison, WI.

Montemayor, R. (1983). Parents and adolescents in conflict: All families some of the time and some families most of the time. *Journal of Early Adolescence, 3*, 83–103.

Montemayor, R. (1986). Family variation in parent-adolescent storm and stress. *Journal of Adolescent Research, 1*, 15–31.

Moos, R. H. (1979). *Evaluating educational environments.* San Francisco: Jossey-Bass.

Moos, R. H. (1991). Connections between school, work, and family settings. In B. Fraser & H. J. Walberg (Eds.), *Educational Environments: Evaluation, antecedents and consequences* (pp. 29–53). Oxford: Pergamon Press.

Newcomb, M. D., & Bentler, P. M. (1988). *Consequences of adolescent drug use: Impact on the lives of young adults.* Newbury Park, CA: Sage.

Nicholls, J. G. (1980, June). *Striving to develop and demonstrate ability: An intentional theory of achievement motivation.* Paper presented at Conference on Attributional Approaches to Human Motivation, Center for Interdisciplinary Studies, University of Bielefeld, West Germany.

Nicholls, J. G. (1984). Achievement motivation: Conceptions of ability, subjective experience, task choice, and performance. *Psychological Review, 91*, 328–346.

Nicholls, J. G. (1990). What is ability and why are we mindful of it? A developmental perspective. In R. Sternberg & J. Kolligan (Eds.). *Competence Considered.* New Haven: Yale University Press.

Oakes, J. (1981). Tracking policies and practices. School by school summaries. *A study of schooling: Technical report no. 25.* Los Angeles: University of California Graduate School of Education.

Offer, D., Ostrov, E., & Howard, K. I. (1981). The mental health professional's concept of the normal adolescent. *Archives of General Psychiatry, 38*, 149–153.

Office of Educational Research and Improvement (1988). *Youth Indicators 1988.* Washington DC: U.S. Government Printing Office.

Offord, D. R., & Boyle, M. H. (1988). The epidemiology of antisocial behavior in early adolescents, aged 12 to 14. In M. D. Levine & E. R. McAnarney (Eds.), *Early Adolescent Transitions* (pp. 245–260). Lexington, MA: Lexington Books.

Paikoff, R. L. & Brooks-Gunn, J. (1991). Do parent-child relationships change during puberty? *Psychological Bulletin, 110,* 47–66.

Parsons, J. E. & Ruble, D. N. (1977). The development of achievement-related expectancies. *Child Development, 48,* 1975–1979.

Petersen, A. C., Compas, B. E., Brooks-Gunn, J., Stemmler, M., Ey, S., & Grant, K. E. (1993). Depression in adolescence. *American Psychologist, 48,* 155–168.

Rae-Grant, N., Thomas, H., Offord, D. R., & Boyle, M. H. (1989). Risk, protective factors, and the prevalence of behavioral and emotional disorders in children and adolescents. *Journal of American Academy of Child and Adolescent Psychiatry, 28,* 262–268.

Rholes, W. S., Blackwell, J., Jordan, C., & Walters, C. (1980). A developmental study of learned helplessness. *Developmental Psychology, 16,* 616–624.

Roeser, R. W., Lord, S. E., & Eccles, J. S. (1994, February). *A portrait of academic alienation in early adolescence: Motivation, mental health and family indices.* Paper presented at the Society for Research on Adolescence, San Diego, CA.

Roeser, R. W., Midgley, C. M., & Maehr, M. L. (1994, February). *Unfolding and enfolding youth: A development study of school culture and student well-being.* Paper presented at the Society for Research on Adolescence, San Diego, CA.

Roeser, R. W., Urdan, T., & Midgley, C. M. (1994). *Meaning and motivation: A study in middle school culture.* Paper presented as part of a symposium to the American Educational Research Association, New Orleans, LA.

Rosenbaum, J. E. (1976). *Making inequality: The hidden curriculum of high school tracking.* New York: Wiley.

Rosenholtz, S. J., & Simpson, C. (1984). The formation of ability conceptions: Developmental trend or social construction? *Review of Educational Research, 54,* 301–325.

Rounds, T. S., and Osaki, S. Y. (1982). *The social organization of classrooms: An analysis of sixth- and seventh-grade activity structures* (Report EPSSP–82–5). San Francisco: Far West Laboratory.

Rutter, M. (1981). Stress, coping and development. Some issues and some questions. *Journal of Child Psychology and Psychiatry and Allied Disciplines, 22,* 323–356.

Rutter, M. (1983). School effects on pupil progress: Research findings and policy implications. *Child Development, 54,* 1–29.

Ryan, R., & Grolnick, W. (1986). Origins and pawns in the classroom: Self-report and projective assessments of individual differences in children's perceptions. *Journal of Personality and Social Psychology, 50,* 550–558.

Ryan, R. M. & Lynch, J. H. (1989). Emotional autonomy versus detachment: Revisiting the vicissitudes of adolescence and young adulthood. *Child Development, 60,* 340–356.

Schulenberg, J. E., Asp, C. E., & Petersen, A.C. (1984). School from the young adolescent's perspective: A descriptive report. *Journal of Early Adolescence, 4,* 107–130.

Simmons, R. G., & Blyth, D. A. (1987). *Moving into adolescence: The impact of pubertal change and school context.* Hawthorn, NY: Aldine de Gruyler.

Simmons, R. G., Blyth, D. A., Van Cleave, E. F., & Bush, D. (1979). Entry into early adolescence: The impact of school structure, puberty, and early dating on self-esteem. *American Sociological Review, 44,* 948–967.

Smetana, J. G. (1988a). Adolescents' and parents' conceptions of parental authority. *Child Development, 59,* 321–335.

Smetana, J. G. (1988b). Concepts of self and social convention: Adolescents' and parents' reasoning about hypothetical and actual family conflicts. In M. Gunnar & W. A. Collins (Eds.), *Minnesota Symposia on Child Development: Vol. 21. Development during the transition to adolescence* (pp. 79–122). Hillsdale, NJ: Erlbaum.

Smetana, J. G. (1989). Adolescents' and parents' reasoning about actual family conflict. Child Development, 60, 1052–1067.

Steinberg, L. (1981). Transformation in family relations at puberty. *Developmental Psychology, 17,* 833–840.

Steinberg, L. (1987). The impact of puberty on family relations: Effects of pubertal status and pubertal timing. *Developmental Psychology, 23,* 451–460.

Steinberg, L. (1988). Reciprocal relations between parent-child distance and pubertal maturation. *Developmental Psychology, 24,* 122–128.

Steinberg, L. (1990). Interdependence in the family: Autonomy, conflict, and harmony in the parent-adolescent relationship. In S. S. Feldman & G. R. Elliott (Eds.), *At the threshold: The developing adolescent*. Cambridge, MA: Harvard University Press.

Strahan, D. (1988). Life on the margins: How academically at-risk early adolescents view themselves and school. *Journal of Early Adolescence, 8,* 373–390.

Takanishi, R. (1993). The opportunities of adolescence—Research, Interventions, and Policy: Introduction to the special issue. *American Psychologist, 48,* 85–88.

Tisak, M. S. (1986). Children's conception of parental authority. *Child Development, 57,* 166–176.

Trebilco, G. R., Atkinson, E. P., & Atkinson, J. M. (1977, November). *The transition of students from primary to secondary school*. Paper presented at the annual conference of the Australian Association for Research in Education, Canberra.

Trickett, E. J., & Moos, R. H. (1974). Personal correlates of contrasting environments: Student satisfactions in high school classrooms. *American Journal of Community Psychology, 2,* 1–12.

Ward, B. A., Mergendoller, J. R., Tikunoff, W. J., Rounds, T. S., Dadey, G. J., & Mitman, A. L. (1982). *Junior high school transition study: Executive summary*. San Francisco: Far West Laboratory.

Weinstein, R. (1989). Perceptions of classroom processes and student motivation: Children's views of self-fulfilling prophecies. In C. Ames & R. Ames (Eds.), *Research on motivation in education: Vol. 3. Goals and cognitions* (pp. 13–44). New York: Academic Press.

Yee, D. K. (1986, April). *Family decision making, classroom decision making, and student self- and achievement-related attitudes*. Paper presented at the meeting of the American Educational Research Association, San Francisco.

Yee, D. K. (1987, April). *Participation in family decision making: Parent and child perspectives*. Paper presented at the meeting of the Society for Research in Child Development, Baltimore.

Yee, D. K., & Flanagan, C. (1985). Family environments and self- consciousness in early adolescence. Journal of Early Adolescence, 5, 59–60.

Zaslow, M. J., & Takanishi, R. (1993). Priorities for research on adolescent development. *American Psychologist, 48,* 185–192.

III An Evolutionary Perspective on Psychopathology in Adolescence

Laurence Steinberg & Jay Belsky

The study of psychopathology during adolescence—and, more particularly, the study of externalizing psychopathology in adolescence—has always been somewhat paradoxical. On the one hand, common opinion holds that a certain degree of "acting out" and "risk taking" during adolescence is normative. Young people, at least in contemporary industrialized societies, are expected to challenge authority, rebel against their elders, and experiment with risky and dangerous activities. In some circles, it is even held that the total absence of such behaviors may indicate some kind of developmental immaturity. Consider, for example, Anna Freud's oft-cited statement on "normal" development during adolescence:

> We all know individual children who as late as the ages of fourteen, fifteen, or sixteen show no . . . outer evidence of inner unrest. They remain, as they have been during the latency period, "good" children, wrapped up in their family relationships, considerate sons of their mothers, submissive to their fathers, in accord with the atmosphere, ideas and ideals of their childhood background. Convenient as this may be, it signifies a delay of normal development and is, as such, a sign to be taken seriously. (1958, pp. 264–265)

Freud's stance notwithstanding, common opinion also holds that the problem and risk-taking behaviors in which adolescents engage are, in fact, genuinely problematic—not only to society but to the individuals who engage in them. Delinquency, violence, drug and alcohol abuse, and sexual promiscuity among the young command the constant attention of the popular media. Adolescent risk taking and behavior problems drain an immense share of the economic resources of the legal and mental health systems, not only in the United States, but in most industrialized countries. Even more minor adolescent behavior problems, such as oppositionalism, truancy, precocious sexual activity, experimentation with drugs, and rebellion against parental authority, are frequent sources of worry to parents and among the chief reasons that families with teenagers seek the help of mental health professionals (Steinberg & Levine, 1990).

This tension—between viewing adolescent behavior problems as normative and viewing them as disturbed—has existed for as long as experts have been writing about the period. Nearly 40 years ago, Freud (1958) coined her wonderful oxymoron, "normative disturbance," to describe the paradox. Put most succinctly, the *paradox of normative disturbance* is this: The very behaviors we have come to expect of adolescents are those about which we most often worry and those for which we most often refer young people for treatment. It is this paradox that we use as the jumping-off point for this chapter.

The Paradox of Normative Disturbance

Although numerous writers in both the psychoanalytic and nonpsychoanalytic traditions have grappled with this dilemma (see Adelson & Doehrman, 1980)—how can a set of behaviors be viewed simultaneously as developmentally appropriate and problematic?—we have progressed very little since Freud's statement on the subject. Arguments that adolescent "problem behaviors" are not genuinely problematic defy the logic and try the patience of most parents, practitioners, and policymakers. Yet, arguments that adolescent conduct problems are not normative—the current stance taken by most professionals—fly in the face of evidence that an extraordinarily high proportion of the adolescent population has evidenced one or more of these putative disturbances.

Consider some of the following statistics. According to recent national studies, more than one fourth of U.S. eighth graders report drinking alcohol regularly, and more than 40% of all high school seniors regularly abuse alcohol (Gans, 1990). More than 60% of all adolescents admit to having committed one or more delinquent acts before turning 18, and nearly one in six has committed a serious crime before adulthood (McCord, 1990; Moffitt, 1993). By age 15, nearly one in five adolescents has had sexual intercourse. In some inner-city and rural communities, more than half of all adolescents have intercourse before entering high school (Alexander et al., 1989; Zabin et al., 1986). Although experimentation with drugs, violation of the law, and sexual precocity may be taken as signs of disturbance in some circles—witness Jessor's (Jessor & Jessor, 1977) influential and widely cited problem behavior theory, which holds that these varied types of adolescent risk taking and deviance reflect a common, underlying personality deficiency—by all epidemiologic indicators, problem behaviors surely would seem to be common, if not normative, during adolescence.

At the same time, however, the developmental data on the course of problem behavior indicate quite clearly that the prevalence of various conduct problems and risk-taking behavior declines as individuals move into and through adulthood. Rates of drug and alcohol use, criminal activity, violence and aggression, and sexual promiscuity all peak sometime during the adolescent or young adult years and then decline markedly thereafter (Moffitt, 1993). (Interestingly, this is not true for internalized disorders, such as anxiety or depression, whose prevalence either remains stable or increases after adolescence, Kennedy, 1993.) The vast majority of individuals—even those who exhibited disturbances in conduct during adolescence—do not show externalizing problems once they have entered adulthood. From the perspective of the developmental psychopathologist, then, it is interesting to ask why the incidence of externalizing problems should increase during adolescence and decline thereafter.

We use the paradox of normative disturbance as a conceptual jumping off point in this chapter for several reasons. First, the notion that there may be periods in development during which certain disturbed behaviors are normative is intriguing, to say the least, within the framework of developmental psychopathology. Developmental psychopathologists have emphasized the pathways that link pathology across different periods of development and the trajectories that permit us to project present patterns of behavior into the future (e.g., Sroufe & Rutter, 1984). However, if adolescence is a period of inherently heightened problems of conduct—even among individuals whose development is normative—how are we to incorporate adolescence into the pathway and trajectory models that have become so popular in the field of developmental psychopathology? That is, is problematic behavior in adolescence a developmental aberration that has no systematic ties to what preceded or what will follow the period, or can it in fact be understood in light of antecedents and sequelae?

Second, the notion that adolescence is a period of normative disturbance has important implications for understanding *individual differences* in behavior problems during this developmental era. The normative disturbance view implies that there are certain universals in the adolescent experience that may lead to the development of conduct problems. Theories derived from this view may help explain why adolescents, on average, are more inclined to engage in problem behavior and risk taking than younger or older individuals. But such theories do not easily account for the vast individual differences in conduct problems that are observed within the adolescent population. What are we to make of widely observed individual differences in behavior problems at adolescence in light of theories of adolescence as a time of normative disturbance? In particular, are individuals who exhibit relatively higher rates of conduct problems than their peers more normal, or are they more disturbed?

In this chapter, we argue that an evolutionary perspective on psychopathology in adolescence provides at least some insight into these vexing questions. We believe that it does so in several ways. First, it informs our understanding of the continuities and discontinuities between adolescence and other periods of the life span in the development and display of externalizing psychopathology by focusing our attention on adolescence as a critical period in reproductive development. As far as we know, the links between psychopathology and reproductive behavior have seldom been explored. We shall argue that, by looking at adolescence in this fashion, we can better understand why such disturbances as conduct disorders or risk taking may increase in prevalence during adolescence and why they may recede in early adulthood.

Second, the evolutionary perspective we adopt helps us to understand patterns of behavior problems in a way that takes into account both nomothetic and idiographic aspects of normative disturbance during the adolescent decade. The evolutionary perspective we draw on—more specifically, a blend of behavioral ecology (Krebs & Davies, 1981), sociobiology (Hamilton, 1964), and life-history theory (e.g., Chisolm, 1993)—allows for explanations both of within-species universals and individual differences. Thus, the theory we articulate provides a basis for understanding universal aspects of conduct problems and risk taking in adolescence as well as individual variation in the frequency, intensity, and developmental course of such problems.

Finally, and most important in the present context, our view explicitly addresses the paradox of normative disturbance. In particular, the view we propose attempts to explain how it can be that the very same behaviors we label as disturbed in adolescence can simultaneously been seen as normative.

Toward an Evolutionary Theory of Risk Taking and Externalizing Behavior

Central to any evolutionary analysis is an appreciation of the processes of natural selection. From a Darwinian perspective, behaviors are presumed to have evolved because they once promoted reproductive fitness, that is, the dispersal of an individual's genes in subsequent generations. In order to do so, the behaviors in question had to contribute in some manner, shape, and form to survival and reproduction. From this vantage, then, the primary question one asks is: What functions might the phenomena that we have come to refer to as risk taking or problem behavior served in the ancestral environments in which humans evolved? It is important to note that consideration of this question need not imply that functions once served in the environments of evolutionary adaptation that fostered the inclusion—actually selection—of these behaviors into the human behavioral repertoire remain functional today, although they might. Indeed, in speculating about the biological functions of risk taking and externalizing behavior in this evolutionary analysis, we raise the possibility that behaviors that today look problematical to the mainstream majority culture might at one time have promoted reproductive fitness.

It is easy to imagine two simple, but important, functions that risk taking and externalizing behavior might have served in ancestral human environments—ones that might still operate today to some degree. Conceivably, the willingness to take risks, even life-threatening risks, might well have proved advantageous to our ancestors when refusing to incur such risk was in fact even more dangerous to survival or reproduction. However chancy running through a burning savannah or attempting to cross a swollen stream might have been, not doing so might have been even more risky. To the extent, then, that individuals inclined to take such risks were differentially advantaged when it came to surviving and producing descendants who would themselves survive and reproduce in future generations, natural selection would favor at least some risk-taking behavior.

In addition to promoting survival in inherently risky situations, risk-taking and externalizing behaviors might also confer advantages, especially upon males, by means of dominance displays. Acting aggressively and being willing to take risks might well have been a tactic of achieving and maintaining dominance in social-status hierarchies. Such means of status attainment and maintenance might have been selected for not only because they contributed to obtaining for oneself and one's kin a disproportionate share of physical resources (e.g., food, shelter, clothing), but because they also increased reproductive opportunities by preventing other males from mating. These are two common functions of dominance displays among many species of animals today.

While the preceding functional analysis of externalizing behavior and risk taking suggests why these patterns of functioning might exist in the human behavioral repertoire, it does not indicate why these phenomena are especially prevalent during adolescence, nor does it explain the existence of individual differences in the expression of the behaviors in question. We believe that consideration of the central significance of reproduction—to both evolution and adolescence—may be the key to understanding what might be referred to as "the adolescence question" (i.e., why are behavior problems and risk taking especially common in adolescence?) and also help account for individual differences in risk-taking behavior. After all, not only is the emergence of reproductive capability the defining feature of adolescence—both biologically and socially—but reproduction is central to any understanding of evolution. Fundamentally, evolution is about the differential reproductive success of organisms with varying characteristics and traits that enhance or undermine their capacity to produce offspring capable of surviving and reproducing themselves.

The potential dominance function of risk taking and externalizing behavior begins to address the Why adolescence? question since it is during adolescence that the human organism begins to become specifically oriented toward issues of heterosexual relations and mating. But it seems to us that we might gain even more insight if we consider an evolutionary process that is related to, but not the same as, natural selection: the process of *sexual selection*, which we think may illuminate issues of individual differences in risk taking and externalization.

Sexual selection is an evolutionary process that is used to explain phenomena in the animal kingdom such as the intricate beauty of the peacock's tail or the building of elaborate nests by some species of birds (Cronin, 1992; Diamond, 1992). When first considered, these phenomena appear wasteful, both in terms of the energy expended and risks of predation incurred because of them. How, then, might they enhance survival and reproduction and thus come to be selected for by evolution? Contemporary biologists have arrived at an answer that has to do with the messages these phenomena send to prospective mates. The more embellished a male peacock's tail or the more decorated a male bird's nest, the more likely a female will select that male for mating (Diamond, 1992). Why? The biological "reasoning" behind the process is fascinatingly complex. According to evolutionary theorists, the choosy female reasons that an organism who can waste calories to either build a more attractive body or engage in a behavior that ought to make survival more difficult must be an organism who possesses genes enabling it to successfully counteract these costs. Elaborate—and expensive—adornments have evolved among males as a means of signaling that one's genetic makeup is advantagous with respect to survival, and females have evolved to find such adornment attractive. Because all organisms desire or are motivated to enhance their reproductive fitness, it makes biological sense for males to engage in those behaviors that attract females and for females to choose males most likely to bear offspring with high prospects of surviving and reproducing themselves.

We propose, consistent with arguments advanced by the human sociobiologists Wilson and Daly (1993), that risk taking and associated problem behavior during adolescence, especially among males, may well have evolved via sexual selection as well as natural selection. In aboriginal societies that are studied by anthropologists to gain insight into the conditions under which human behavior evolved (e.g., the Ache in Venezuela, the Yamamano in Brazil, the !Kung in Africa), "young men are constantly being assessed as *prospects* by those who might select them as husbands and lovers . . ." (Wilson & Daly, 1993, p. 99, emphasis in original). Moreoover, "prowess in hunting, warfare, and other dangerious activity is evidently a major determinant of young men's marriageability and probably affects their opportunities for extra-marital reproduction, too (e.g., Chagnon, 1988; Kaplan & Hill, 1985)" (Wilson & Daly, 1993, p. 98). In addition to fostering dominance and, thereby, the acquisition of life-enhancing physical resources and the denial of mating opportunities to male competitors, externalizing behavior and risk taking may have evolved, in part, to attract females. In fact, employing the same evolutionary logic used to explain peacocks' tails and nest adornment, we speculate that engaging in externalizing behavior may have once sent signals to females—who have evolved to read and respond to such signs—that the risk taker possessed hardy genes and, therefore, could afford to take such risks.

We emphasize that from the standpoint of an evolutionary perspective it is not necessary to argue that risk taking is still reproductively strategic, although we contend that it might be in certain contexts. All that must be posited to account for the elevated rates of risk taking among adolescents and young adults (especially males) that we see today is that it was advantageous in the environments of evolutionary adapation, and thus was selected for (via natural or sexual selection or both), and that the proximate psychological mechanisms that maintain it are still operative.

With regard to the second condition, Gardner (1993) has offered a provocative analysis of adolescent decision making, which suggests that risk taking actually may be a rational choice for some youth. Central to Gardner's analysis is the empirically substantiated observation that, relative to older adults, the future is more uncertain for adolescents. As a consequence, deferring immediate gratifications (e.g., matings, status, resources, fun) by failing to incur risks—as most mature adults tend to do—may, given the uncertainty of future payoffs, be more risky than incurring certain risks and realizing the probabilistic gains that may accrue from them. In other words, for the young person, risk taking may be a means of hedging his bets by shooting for an immediate payoff that may actually be more certain than waiting for an unpredictable future.

We find Gardner's (1993) analysis intriguing not only because it posits a proximate psychological mechanism for the adaptive value of youthful risk taking and externalized behavior but because his logic can be extended as well to analyses of individual differences in risk taking. We highlight this application of Gardner's argument because it is essential to distinguish between two fundamentally different yet related questions raised in this chapter: Why is risk taking and problem behavior so characteristic of youth? and Why do some youth engage in more risk taking and problem behavior than others?

When Gardner's perspective on adolescent decision making is applied to the analysis of individual differences in adolescent externalization, the possibility arises that some adolescents engage in more risk taking and problem behavior than others because their futures are more uncertain. For these individuals, the apparent costs of taking risks are actually less than they are for those for whom the future is more predictable and, perhaps, promising. In the following section of this chapter, in which we review the basic tenets of an evolutionary theory of socialization and development that we have advanced (Belsky, Steinberg, & Draper, 1991), we shall elaborate our ideas about the development of expectations of the future. For the time being, we simply highlight the potential adaptive function of risk taking and the role of proximate decision-making processes in promoting risky and problem behavior among youth in contemporary society, even though its ultimate (i.e., fitness enhancing) functions may no longer apply.

When risk taking and problem behavior are considered from an individual difference perspective, one of the most interesting—and yet least understood—facets of adolescent externalization is the fact that some youth grow out of it whereas others do not. Further reflection on evolutionary theory may help to illuminate this conundrum. Earlier in this chapter, we argued that externalizing behavior problems may have evolved not only a means of achieving dominance in intrasexual competition for mates but via processes of sexual selection, whereby risk taking signaled to prospective mates the presence of hardy genes. We speculate that such an evolved mechanism may account for the large number of males who engage in high levels of risk taking and problem behavior in adolescence, but who subsequently grow out of it without enduring adverse life-course consequences. Presumably, once individiuals have aged past the prime developmental period for attracting mates, the signaling function of risk taking is less valuable.

But what about disturbed youth who continue to evince high levels of problem behavior well into adulthood? To account for this developmental trajectory from an evolutionary perspective, we draw not upon the notion of sexually selected risk taking performed in the service of attracting a mate, but upon the role of environmental experiences during the preadolescent years. Specifically, we theorize that this other group of risk takers are responding to evolutionary forces that incline them, in light of particularly troubled and stressful rearing histories, to develop psychological, biological, and behavioral processes that lead them to behave aggressively in childhood and antisocially in adolescence.[1] We turn now to an evolutionary theory of human psychosocial development that might help to explain the conditions under which "true and enduring" as opposed to "stage-specific and transient" externalizing psychopathology might develop. This line of theorizing builds upon ideas presented elsewhere (Belsky & Draper, 1987; Belsky, Steinberg, & Draper, 1991; Draper & Belsky, 1990; Draper & Harpending, 1982).

Basic Tenets of the Evolutionary Theory

The theory of adolescent psychopathology we advance draws heavily upon concepts basic to the fields of behavioral ecology (e.g., Krebs & Davies, 1981) and sociobiology (Hamilton, 1964). From sociobiology, we take the maxim that natural selection tends to favor behavior that increases fitness, that is, the representation of an individual's genes (relative to unrelated individuals) in future generations. From behavioral ecologists, we take the maxim that behavioral strategies that contribute to reproductive success are facultative, that is, contextually conditional. In other words, optimal strategies depend upon the options available to individuals given the physical, economic, and social ecology in which they develop and function (Crawford & Anderson, 1989). A given pattern of behavior is only optimal in the context of a specific environment. From the perspective of modern evolutionary theory, complex, highly social organisms like humans evolved to modulate social behaviors like mating and parenting in response to particular environmental cues (Hinde, 1982). Thus, modern evolutionary theory does not presume biological determinism in any narrow sense. Instead, and consistent with contemporary behavioral ecology, it assumes an evolved flexibility of organismic response to variations in context, in the service of biological goals (i.e., reproductive fitness).

Our theory also underscores the need to distinguish among ultimate, distal, and proximate causes of behavioral development (Tinbergen, 1963). A proximate cause is closely spaced in time to a given phenomenon. A distal cause is further away, yet intricately and causally related to the proximate mechanism. An ultimate cause concerns the evolutionary or biological function of a phenomenon, why a process or phenomenon occurs rather than how it occurs. Ultimate, distal, and proximate explanations, then, are by no means mutually exclusive, nor should they be regarded as providing alternative explanations. Rather, they are linked together in a system of causation.

Consistent with other analyses (e.g., Chisolm, 1993; Draper & Harpending, 1982), we assume that evolution has designed humans to vary their mating and child-rearing behavior in accordance with the contextual conditions in which they

---

[1] An alternative, but not mutually exclusive, account of the differences between lifecourse persistent and adolescence-limited externalization is provided by Moffitt (1993). She argues that individuals who engage in life-course persistent antisocial behavior often evince neuropsychological deficits early in life and show a stable pattern of aggressive, antisocial behavior well before adolescence. In contrast, individuals who engage in antisocial behavior during adolescence but desist, Moffit contends, are merely imitating the behavior of their genuinely delinquent peers for purposes of status enhancement.

develop so as to maximize their reproductive success, or to vary these behaviors in ways that *would have* had this effect in the ancestral environments in which human behavior evolved. This notion that rearing context shapes life history, which is itself systematically related to patterns of pair bonding and parenting in a manner designed to maximize the dispersion of an organism's genes in subsequent generations, has a long history in biobehavioral research on many species of animals (Konishi, Emlen, Rickets, & Wingfield, 1989; Lack, 1947), however surprising it may appear to traditionally trained social scientists. There is good reason to believe that humans, too, have evolved to modify their reproductive behavior (i.e., mating and parenting) in the service of fitness considerations and in accord with social and ecological variables.

An Overview of the Theory

Evolutionary ecologists theorize that in order for any organism to reproduce, effort must be apportioned among three fundamental tasks: (1) growth and development, (2) mating, and (3) parenting (including gamete production). Yet species differ dramatically in how they apportion effort across these tasks. Relative to other species, humans emphasize growth and development, as seen in the prolonged period of juvenile dependence and delayed sexual maturation. And, in contrast to most other mammals, we are unusual for the importance we attach to pair bonds and for the high levels of biparental care required to rear children to maturity. But despite this generalization about the species, there is substantial diversity in the ways in which different populations of humans manage growth and development, mating, and parenting.

A central tenet of our theory is that a principal evolutionary function of early experience—the first 5 to 7 years of life—is to induce in the child an understanding of the availability and predictability of resources (broadly defined) in the environment, of the trustworthiness of others, and of the enduringness of close interpersonal relationships, all of which will affect, although not fully determine, how the developing person apportions reproductive effort (Belsky, Steinberg, & Draper, 1991). Individuals whose experiences in and around their families of origin lead them to perceive others as untrustworthy, relationships as opportunistic and self-serving, and resources as either scarce or unpredictable or both will develop behavior patterns that function to reduce the age of biological maturation (within their range of plasticity) (see Barkow, 1984, fn. 7, for a similar prediction), accelerate sexual activity, and orient them toward short-term, as opposed to long-term, pair bonds. In other words, a disproportionate amount of reproductive effort will be allocated toward the individual's growth and development and mating rather than toward parenting.[2]

Individuals, in contrast, whose experiences lead them to perceive others as trustworthy, relationships as enduring and mutually rewarding, and resources as more or less constantly available from the same key persons, will behave in ways that inhibit (relative to the first type) age of maturation, will defer sexual activity, and will be motivated to establish—and be skilled in maintaining—enduring pair bonds, all of which will serve to enhance investment in child rearing. This second type of person, then, disproportionately invests reproductive energies in parenting effort rather than individual growth and development or mating. In essence, we argue that early experiences and the psychological and biological functioning they induce lead individuals to engage in either a "quantity" or a "quality" pattern of mating and rearing.

In arguing that experience shapes development, we do not mean that all, or even most, of the variance on human behavior is contextually controlled. Behavior-genetic studies strongly indicate that this is by no means the case (Plomin et al., 1985). In fact, an obvious alternative to the theory we advance is genetic polymorphism, which asserts that individuals may be genetically predisposed to one developmental pathway or another (see Rushton, 1985). One possible way of integrating such a theory with the more strongly environmental theory we advance is via the notion of differential susceptibility to environmental experience: Whereas some individuals may be genetically predisposed to respond to contextual stress and insensitive rearing by maturing early and engaging in relatively indiscriminate sexual behavior, others may be genetically predisposed to respond to sensitive rearing by deferring sexual behavior and establishing enduring pair bonds in adulthood. Thus, while nurture may determine the direction development takes, nature may determine the likelihood that, and the extent to which, an individual will be influenced by a particular set of environmental conditions.

Figure 1 outlines the interrelation of the major domains of the theory (context, rearing, psychological development, somatic development, reproductive strategy) and contrasts the two developmental pathways mentioned in the preceding section on reproductive strategy. It is important to reiterate that even though these developmental pathways and related reproductive strategies are discussed as distinct types, it may be best to regard them as prototypes that characterize contrasting ends of a continuum of ecologically sensitive behavioral development rather than as the only two expressions of interpersonal development that are possible. The choice, then, is between what Stearns (1982), following Bradshaw (1965), refers to as discrete versus continuous phenotypic plasticity. In discrete plasticity, there are few if any intermediate phenotypes and all individuals show one or the other of only a small number of discrete phenotypes that are triggered by the environment, whereas

[2] We should not that in the time since first presenting our evolutionary theory of human socialization and reproductive strategy, several thoughtful critics have raised questions about propositions central to our thinking (Maccoby, 1991; Hill, 1993; Hinde, 1991; MacDonald, July 14, 1993, personal communication). Most notable we think are concerns regarding the "packaging" of timing of puberty and the number of offspring produced to generate quantity-versus-quality reproductive strategies. As MacDonald has implied and Hill has explicitly noted, it may be useful to maintain the distinction between two life-history variables central to our theorizing—pubertal timing and quantity of offspring produced—if only because these "component processes" may well be under the control of different environmental forces.

Figure 1. Developmental Pathways of Divergent Reproductive Strategies

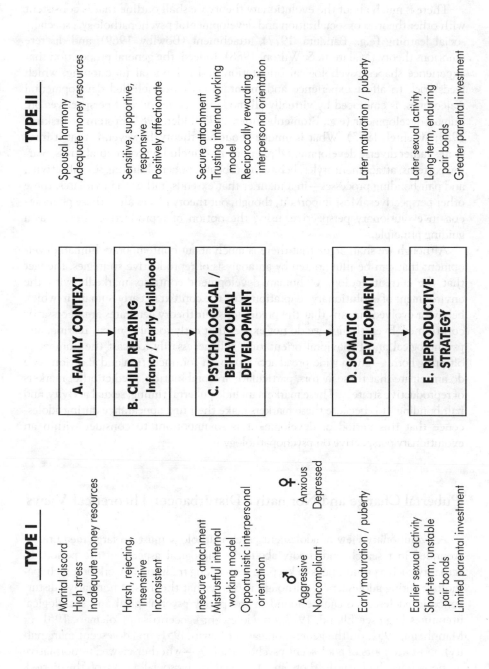

TYPE II

Spousal harmony
Adequate money resources

Sensitive, supportive,
responsive
Positively affectionate

Secure attachment
Trusting internal working
model
Reciprocally rewarding
interpersonal orientation

Later maturation / puberty

Later sexual activity
Long-term, enduring
pair bonds
Greater parental investment

A. FAMILY CONTEXT

B. CHILD REARING
Infancy / Early Childhood

**C. PSYCHOLOGICAL /
BEHAVIOURAL
DEVELOPMENT**

**D. SOMATIC
DEVELOPMENT**

**E. REPRODUCTIVE
STRATEGY**

TYPE I

Marital discord
High stress
Inadequate money resources

Harsh, rejecting,
insensitive
Inconsistent

Insecure attachment
Mistrustful internal
working model
Opportunistic interpersonal
orientation

♂
Aggressive
Noncompliant

♀
Anxious
Depressed

Early maturation / puberty

Earlier sexual activity
Short-term, unstable
pair bonds
Limited parental investment

in continuous plasticity, the phenotypic response is matched or scaled to the environment (Chisolm, 1988).

There is much about the evolutionary theory we shall outline that is consistent with other theories of socialization and developmental psychopathology, especially social learning (e.g., Bandura, 1977), attachment (Bowlby, 1969), and discrete emotions theory (Malatesta & Wilson, 1988). Indeed, the general proposition that experience shapes psychological orientation and behavioral functioning, which feeds back to affect experience and maintain earlier-established developmental trajectories, is embraced by virtually all modern theories of, and perspectives on, human development (e.g., Bronfenbrenner, 1977; Elder, 1981; Epstein & Erskine, 1983; Wachtel, 1973). What is unique about our theory, however, is that it integrates rather diverse developmental phenomena—including contextual stress, rearing patterns, attachment styles, behavior problems, pubertal timing, sexual activity, and pair-bonding processes—in a manner that extends, rather than violates, these other perspectives. Most important, though, our theory places all of these phenomena in evolutionary perspective, using the notion of reproductive strategy as a guiding principle.

Although we shall argue that there is much about contemporary human development that can be illuminated by an analysis of reproductive strategies, the fact that the current ecology of human development contrasts markedly with the environment of evolutionary adaptation (i.e., the context of early humans in which behavior evolved) means that the predictions our theory generates are necessarily constrained. Even though we do not expect that early social context, rearing, and psychological and behavioral orientation will necessarily forecast the number of offspring borne—given widespread access to both contraception and abortion—we do anticipate that these factors, particularly in combination, predict other markers of reproductive strategy. These markers include pubertal timing, sexual activity, and pair bonding. It is because these markers make their first appearance during adolescence that this period of development is so important to consider within an evolutionary perspective on psychopathology.

Pubertal Change and Normative Disturbance: Theoretical Views

A evolutionary view of adolescent psychopathology must, in large measure, be grounded in research and theory about the biological aspects of the period. In actuality, such a grounding is in keeping with a long tradition of work on psychopathology during adolescence. Although it is true that theories of normative disturbance in adolescence can be found in the purely psychological and sociological literatures (e.g., see Elkind, 1978, on adolescent egocentrism; Coleman, 1961, or Mannheim, 1952, on the generation gap; or Lewin, 1951, on adolescent marginality), most accounts of adolescent psychopathology—whether viewed as normative or not—have had sturdy roots in biological or biosocial views of the period.

Specifically, from the work of Hall (1904) on, it has long been held that, in one way or another, the normative biological changes of puberty increase problem behavior. Theorists who have taken this stance generally fall into one of four broad camps.

First, there are those writers who focus on the direct linkages between hormonal that activity and adolescent psychopathology, such as the rise in androgenic activity that is presumed to lead to an increase in aggression (cf. Susmann et al., 1987) or the putative links between hormonal activity in adolescence and sexual behavior (e.g., Udry, 1987). These writers have suggested that both universals and individual differences in externalizing problems can be best understood by examining universals and individual differences in hormonal activity.

A second group of biologically oriented theorists—more specifically, biobehavioral theorists—has suggested that the hormonal changes of puberty do not affect adolescent behavior directly but, rather, increase individuals' susceptibility to environmental stressors, which if present may heighten psychological disturbance (cf. Brooks-Gunn, 1989; Eccles et al., 1988; Petersen, 1985). Thus, for example, it has been suggested that it is the co-occurrence of rapid and variable hormonal change in adolescence with exposure to new environmental demands, such as the transition from elementary to secondary school or the onset of dating pressures, that may trigger behavior problems. These writers acknowledge that the changes of puberty may play a role in the genesis of adolescent conduct problems, but have argued that universalities and idiosyncracies in conduct problems in adolescence can be best understood by looking at similarities and differences in individuals' environments. Essentially one might regard this second perspective on the biological basis of adolescent psychopathology as a sample case of the more general diathesis-stress model of psychopathology. Thus, while hormonal changes make the organism vulnerable to the development of behavior problems (i.e., act as the diathesis), such problems only materialize when the hormonally vulnerable individual encounters stressors like school transitions and dating pressures.

A third group of writers has focused not so much on the internal biological changes of puberty but on the changes in physical appearance that accompany them. Some biologically oriented interpersonal theorists have argued, for example, that changes in the stimulus characteristics of the adolescent precipitate changes in family relationships, which in turn may heighten behavior problems (e.g., Hill et al., 1985a, 1985b; Steinberg, 1987). Increases in the adolescent's physical size, for example, may lead parents to disengage from their children and provide less vigilant supervision, which in turn may lead to increases in antisocial behavior. Others have argued that changes in the adolescent's physical appearance provoke changes in peer relations, which may lead to increases in behavior problems (e.g., Magnusson et al., 1986). Thus, Magnusson has suggested that physical maturation may lead adolescents into peer associations with older individuals, who themselves may introduce younger adolescents to deviant activities.

Finally, psychoanalytically oriented writers have suggested that biology takes its toll on adolescent disturbance through the impact of puberty on intrapsychic functioning. Orthodox analytic theorists, such as Freud (1958), argued that the increase in sex hormones at puberty amplifies libidinal urges and reactivates unconscious Oedipal conflict. Intrapsychic conflict is then acted out through detachment from parents and one or more types of conduct problems. Neoanalytic theorists such as Blos (1979) or Erikson (1968), in contrast, have pointed to the role of pubertal change in disrupting ego functioning, temporarily upsetting the intrapsychic balance established during latency and impelling the young person toward a crisis of "individuation" or "identity." Problems in conduct are then interpreted in light of the adolescent's struggle to establish a more integrated and autonomous sense of self.

As will become clear in subsequent sections of this chapter, the viewpoint we espouse is consonant with many of the basic principles articulated within the endocrinological, biobehavioral, interpersonal, and psychoanalytic views of the connection between biology and psychopathology in adolescence. Specifically, we, too, believe that adolescent psychopathology has roots in the endocrinology of the period but that the effects of the pubertal hormones on the adolescent's behavior are contextually moderated and interpersonally and intrapsychically mediated. Where our view departs from these traditions, however, is in our attempt to place this biology-psychopathology linkage within a broader evolutionary context.

Briefly put, we shall argue that many putative disturbances conduct during adolescence, such as aggression, sexual precocity, and antisocial activity, when viewed within a evolutionary perspective appear to be evolutionarily "sensible" responses to particular environmental circumstances. Specifically, we propose that an overarching pattern of what appears to be externalized psychopathology in adolescence—what Jessor and his colleagues have called "problem behavior syndrome" (Jessor & Jessor, 1977)—may in many cases be a manifestation not simply (or even necessarily) of psychopathology but of a certain reproductive strategy that previously was more functional and adaptive than it appears to be today. This strategy, which we term *opportunistic*, is one that attempts to maximize the number of one's offspring while minimizing the amount of psychological and instrumental investment devoted to forming stable pair bonds or child rearing. It is characterized by sexual precocity and promiscuity, aggressive individualism, and a high degree of risk taking. In a sociocultural context in which stable marriage, interpersonal altruism, and heavy investment in child rearing is valued, this reproductive strategy and its attendant, opportunistic behavior, will appear disturbed. In a context in which stable pair bonding and heavy parental investment is disadvantageous, however, such an orientation will appear to be normative, perhaps even desirable. Fundamentally, we theorize that under the latter conditions this strategy may have been the optimal means of promoting reproductive fitness.

We contrast an opportunistic strategy with one that might be called *mutual reciprocal*. Whereas an opportunistic reproductive strategy is aimed at maximizing the number of one's offspring and minimizing investment in pair bonding and child rearing (what some might call a quantity orientation toward procreation), a mutual-reciprocal one is oriented toward stable pair bonding and maximizing parental investment in a smaller number of offspring (what some might call a quality orientation toward procreation). And, whereas the opportunistic reproductive strategy is associated with higher levels of aggression, risk taking, and sexual precocity and promiscuity, the mutual-reciprocal strategy is associated with low levels of aggression, risk aversion, and sexual restraint. In a sociocultural context in which stable pair bonding, sexual fidelity, and high parental investment is valued, a mutual-reciprocal pattern of behavior will be viewed as normative. In a context in which such traits are disadvantageous, however, an mutual-reciprocal strategy could, in fact, be seen as disturbed.

Although the two social orientations we describe—one geared toward opportunistic advantage taking, the other toward mutual commitment and reciprocal benefit—are often considered in terms of immoral versus moral or improper versus correct behavior, we regard them as being distinctively, yet equally, well-suited for the particular ecological niches in which they have developed. While current Western majority culture may value the mutual-reciprocal over the opportunistic strategy, the evolutionary processes that led to their emergence existed apart from contemporary moral evaluations. Additionally, although the term "strategy" implies a conscious plan to many, we presume no such awareness and use the term in the same sense that behavioral ecologists do when studying the behavior of many species of animals.

One important aspect of the evolutionary perspective we are advancing in this chapter is our assertion that some of the overarching patterns of behavior displayed by adolescents—including some behavior that might be labeled psychopathological—should be viewed within the context of various reproductive strategies. A second, and equally important, component of our model is the association we postulate between reproductive strategy and physical maturation. As we shall make clear, the opportunistic strategy described above includes not only individualistic and ostensibly antisocial behavior but early or accelerated (or both) pubertal maturation as well. In contrast, we assert that the mutual-reciprocal reproductive strategy, characterized by minimal antisocial behavior, stable pair bonding, and heavy investment in parenthood, is associated with either late or slower pubertal maturation or both.

Like many scholars who have come before us, then, we propose a model of adolescent psychopathology that links behavior and biology. The specific nature of the link we propose between enduring externalizing problems and pubertal maturation departs from traditional views in important respects, however. We concur with other writers that there indeed is an association between early physical maturation and disorders of conduct. But whereas traditional scholars of adolescence have portrayed early pubertal maturation as a cause of psychopathology, we argue that the observed association between the two may be due to their shared association with the opportunistic reproductive strategy. That is, we propose that the opportunistic reproductive strategy has its origins prior to adolescence, and that pubertal maturation and externalizing behaviors in some cases may be manifestations of the same underlying phenomenon. Indeed, as we shall make clear, there is just as much evidence that behavioral difficulties in adolescence precede early pubertal maturation as the reverse.

Puberty and Normative Disturbance: Correlational Evidence

Whatever the presumed theoretical linkage between pubertal change and behavioral disturbance during adolescence, there is no shortage of empirical evidence linking the two. Research on the relation between biological change at adolescence and behavioral disturbance falls into two broad categories: studies of the immediate effects of puberty on psychological functioning and behavior (sometimes referred to as research on *pubertal status*); and studies of the differential effects of early versus late maturation (sometimes referred to as research on *pubertal timing*). In the sections that follow, we summarize some of the main findings to emerge in each of these areas of inquiry.

Immediate Effects of Puberty

During the last decade, advances in hormonal assay techniques instigated an upsurge in studies of the relation between endocrinological change at puberty and adolescent psychopathology. To date, research on this topic has proven suggestive, but, for the most part, inconclusive. Although many researchers working on this problem expected to find strong, direct relations between hormones and behavior and that these relations would lend themselves to easy interpretation, the pattern of findings reported in the literature is more complex than had been anticipated and not always internally consistent.

Three tentative conclusions have emerged in recent years that are worthy of note, however. These findings have now been replicated in several investigations conducted in different laboratories. First, there are indeed significant correlations between levels of specific hormones and rates of specific problem behaviors in adolescence. This is seen most clearly in studies of the relation between testosterone levels and sexual activity, especially among males (Udry, 1987); in the relation between levels of adrenal androgens and aggression in both sexes (Susman et al., 1987); in studies of the relation between rapid, fluctuating, or dramatic hormonal activity on the hypothalamic-pituitary axis (HPA) and negative affect in both sexes, as manifested by anger, impatience, behavior problems, and depression (Eccles et al., 1988; Susman et al., 1987); and in studies of the relation between levels of estradiol and progesterone and adolescent susceptibility to negative mood states (Brooks-Gunn, 1989). In general, effects tend to be the strongest in the early stages of puberty, when the system is being turned on (Brooks-Gunn & Reiter, 1990).

Second, hormonal activity at puberty may indirectly affect mood or behavior by altering adolescents' sensitivity to environmental stimuli (Brooks-Gunn, 1989; Eccles et al., 1988). Accordingly, the nature of the impact of hormonal change on mood or behavior depends on the nature of the environmental stimuli. Thus, the negative effect of increases in estradiol or progesterone on mood appears to hold only under conditions of exposure to environmental stress (Brooks-Gunn, 1989). In other words, it appears that estradiol affects "moodiness"—or, more precisely, emotional labiality—but that valence of the mood change depends on the environmental circumstances. Similarly, the impact of testosterone on sexual activity is environmentally moderated, that is, increases in testosterone are correlated with increases in sexual activity only in contexts in which sexual activity is perceived as normative for that age group (Udry, 1987). Thus, the impact of the hormonal change may be more in the form of a predisposition than an actual behavioral change. Interestingly, it appears that the hormone-behavior linkage may be somewhat more environmentally moderated among adolescent females than males, a point to which we shall return later.

Finally, recent studies suggest that there are important individual differences in responses to various hormones. Not all individuals evidence mood or behavior profiles consistent with the observed modal patterns (Eccles et al., 1988). Whether these individual differences reflect individual differences in neuroendocrine organization, which is determined very early in life, or individual differences in environmental circumstances during adolescence itself is not known. Nonetheless, it is important to keep in mind that all adolescents are not affected by the hormonal changes of puberty in identical ways. For instance, there is some suggestion that the adverse effects of hormonal fluctuation during the early stages of puberty are greater among early maturers than late maturers (Brooks-Gunn & Reiter, 1990).

In contrast to research on the immediate impact of the hormonal changes of puberty on adolescent mood and behavior, studies of the immediate impact of the somatic changes of puberty (i.e., the growth spurt, the development of secondary sex characteristics) have not shown comparably dramatic effects. (It is important, however, to differentiate between research on the immediate impact of the somatic changes of puberty, which has not shown strong effects, and research on the differential effects of early versus late somatic maturation, which has.) In general, once the effects on mental health of the hormonal changes are taken into account, the impact attributable specifically to the somatic changes of puberty is negligible (Brooks-Gunn & Reiter, 1990). This conclusion derives from work examining the psychological impact of pubertal change in general as well as the effects of specific pubertal events, such as menarche (first menstruation), spermarche (first ejaculation), the onset of breast growth, and the development of pubic hair.

The one exception to this general tendency has been found in research on the immediate impact of the somatic changes of puberty on family relations. Several studies, employing a diverse set of methodologies, have shown that conflict and distance between parents and children increases at puberty; this effect is more characteristic of females than males and is more commonly observed in mother-child than father-child relations, although findings in this general direction have been reported across studies of all four parent-child dyads (e.g., Hill et al., 1985a, 1985b; Papini & Sebby, 1987; Steinberg, 1987, 1988; Susman et al., 1987). It is not clear from these studies, however, whether the distancing that occurs in the parent-adolescent relationship during the early phases of puberty is more strongly associated with the somatic or the endocrinological changes of puberty. There is also some suggestion in this literature that the adverse impact of pubertal change on parent-adolescent relations is somewhat greater among early maturers than late maturers, especially for females (Hill, 1985b).

Correlates of Early Versus Late Maturation

Research on the differential correlates of early versus late pubertal maturation has a long and distinguished history, dating back to the early investigations of the Berkeley and Oakland Growth Studies (see Livson & Peskin, 1980) and continuing today. As a result, the extensive literature that has been amassed on this topic to date permits us to draw reasonably confident conclusions.

Generally speaking, early pubertal maturation is associated with poorer mental health among both boys and girls, although there are some interesting gender differences in the ways in which this disadvantage is manifested. Among boys, the psychological pattern associated with early maturation is one called pseudo-adult hypermasculinity. For example, early maturing boys are more involved in athletics and, as a result, more confident and more popular with their peers, but they also are more likely to be involved in risk taking and other problem behaviors, including truancy, delinquency, school misconduct, cigarette smoking, drug and alcohol use, and early sexual intercourse (Andersson & Magnusson, 1990; Duncan et al., 1985). Moreover, despite their external appearance of adult maturity, early-maturing boys are more likely to experience "frequent temper tantrums" (Livson & Peskin, 1980, p. 73).

Among girls, early maturation is also associated with increased delinquency and pseudo-adult behavior, such as smoking, involvement with drugs and alcohol, and early sexual intercourse (Aro & Taipale, 1987; Magnusson et al., 1986). Yet, unlike their male counterparts, early-maturing girls also may have self-image and emotional difficulties (including higher rates of anxiety and depressed affect) (Aro & Taipale, 1987; Simmons & Blyth, 1987). Thus, whereas early maturation among boys is associated with various forms of externalized distress (e.g., delinquency, drug use, sexual precocity), among girls early maturation is associated with both externalizing (e.g., delinquency, risk taking) internalizing (e.g., depression) problems.

Interestingly, the impact of early maturation on girls' behavior and, especially, their self-image, is contextually variable. In general, in environments where a higher premium is placed on heterosocial popularity, such as mixed-sex schools or "cliquish" communities, early maturation may be most strongly associated with behavioral and psychological "disturbance" (e.g., Caspi et al., 1993; Richards et al., 1990). Nonetheless, the finding that early-maturing girls exhibit more behavior problems than late-maturing girls has been reported in dozens of studies across a variety of national and cultural contexts. It seems reasonably safe to conclude that early maturation among females, as it is among males, is associated with elevated rates of externalizing problems.

Early Puberty and Externalization: Cause or Effect?

The correlation between early maturation and adolescent externalization historically has been interpreted as evidence that early puberty causes (or at least contributes to) behavior problems, perhaps owing to the psychoanalytic perspective on "raging hormones" and adolescent storm and stress. However, because longitu-dinal studies indicate that many of the behaviors examined as outcomes in research on pubertal timing are stable between childhood and adolescence (Caspi & Bem, 1990; Digman, 1989), it is quite possible that observed personality and behavioral differences between early and late maturers antedate and affect puberty, rather than the reverse. Indeed, however unusual it may seem to students of human development that timing of puberty may be affected by prior and contemporaneous social conditions, this notion is widely accepted among scientists who study reproductive development among nonhuman primates and other mammals. Indeed, there are several lines of evidence derived from both human and animal research that support the hypothesis that the timing of maturation is contextually plastic and socially mediated.

The plasticity of human reproductive functioning in response to social stimuli is well established. For example, studies have documented the synchronization of menstrual cycling among roommates, close friends, and co-workers (Graham, 1991; McClintock, 1980); that sexual activity may induce or quicken ovulation (Trevathan & Burleson, 1989); and that cross-sex contact induces menstrual regularity and shortens the menstrual cycle (Trevathan & Burleson, 1989). Although these studies do not examine pubertal timing specifically, they provide clear evidence of the link between social experience and human endocrinological functioning. Rather than viewing the hormone-behavior relation as unidirectional, recent developments in socioendocrinology demonstrate that the link between social experience and hor-monal activity is reciprocal. As Worthman (1990) notes, there are reasons to expect that this reciprocal relation should be especially potent in the hormonal regulation of sexual behavior and reproduction, including the regulation of pubertal maturation.

Most of the evidence on social influences on pubertal maturation comes from research on family relations in nonhuman primates. Among most nonhuman primates living in the wild, pubertal maturation is associated with increased dis-tance in the parent-child relationship, either in terms of increased physical distance or heightened aggression and, ultimately, forced migration (Caine, 1986; Steinberg, 1989). When voluntary or forced migration is prevented among animals in captivity, however, reproductive maturation is inhibited, especially in the case of females. This effect has been documented in a variety of mammals, including several species of monkeys as well as hamsters, wolves, and wild dogs (Evans & Hodges, 1984; Levin & Johnston, 1986; Tardif, 1984). Similarly, the sexual development of males housed in captivity in large groups is inhibited by the presence of dominant males and stimulated (or disinhibited) by the removal of more dominant members from the groups (Goy, March 13, 1986, personal communication). When separation of juvenile from parent occurs after an inhibitory period, however, reproductive maturity is rapidly attained (Evans & Hodges, 1984; Tardif, 1984). Taken together, these studies suggest that reproductive maturation may be inhibited by physical closeness to parents and accelerated by distance from them. Sociobiological expla-nations of this phenomenon make good sense since postpubertal distance in the parent-child relationship would minimize inbreeding and thereby increase repro-ductive fitness (Steinberg, 1989).

Needless to say, an adequate test of the proposition that social experience may affect human pubertal timing requires the longitudinal assessment of preadolescent psychological functioning, pubertal timing, and behavior during adolescence. In the past few years, several research teams, using data from animal as well as human studies, have tested the hypothesis that certain social and behavioral experiences accelerate pubertal maturation. The results of these investigations suggest that, during preadolescence, one form of externalized distress in particular—parent-adolescent conflict—may provoke early maturation.

Specifically, several studies of human adolescents and their parents indicate that parent-child distance may accelerate pubertal onset, at least in homes of girls, and that female adolescents who have more strained family relations mature earlier than their peers whose family relations are closer. In the first of these investigations, Steinberg (1988) found that preadolescent and early adolescent girls who reported more troubled relations with their parents (and especially, with their mother) matured faster physically over the next 12 months than did age-mates who began the year at an equivalent stage of puberty but reported initially closer family ties. Ellis (1991) similarly reported that girls' physical maturation was more advanced in households characterized by more adolescent autonomy and higher parent-adolescent conflict.

The constrained time period of the Steinberg study leaves open the possibility that there were preassessment differences in youngsters' rate of maturation (although not in their actual pubertal status at the initial assessment) that caused the observed differences in parent-child relations. For this reason, a recent study of Moffitt et al. (1992) is especially interesting. In their sample of New Zealand girls, family conflict at age seven and age nine was significantly, albeit modestly, predictive of earlier menarche, even after the effects of parental divorce (which may also accelerate puberty—see below) and weight (which is known to be associated with earlier maturation; Frisch & McArthur 1974) were taken into account.

Several recent studies corroborate these results and extend them to families of adolescent males. Graber and Brooks-Gunn (1992), for example, report that menarche occurs earlier among girls reporting more negative family relationships, as indexed by greater parent-child conflict and less parental approval. In this study, the researchers controlled for mother's age at menarche, which had been posited by other investigators as a possible genetic confound (cf. Campbell & Udry, 1992). Mekos, Hetherington, & Clingempeel (1992) report similar effects in families of boys, but, interestingly, not girls. Boys who reported more distant relations with their mother prior to adolescence matured faster than their counterparts whose family relations were relatively closer. Finally, a report by Kracke and Silbereisen (1992) suggests that autonomy from parents may speed pubertal maturation in both males and females.

Further evidence that preadolescent family stress is associated with accelerated maturation among girls is found in research on the developmental consequences of father absence. Campbell and Udry (1992), Jones et al., (1972), Mekos et al., (1992), Moffit et al. (1992), and Surbey (1990) each report that girls reared in father-absent households attain menarche at an earlier age than their counterparts in intact homes;

moreover, early father absence (Jones et al., 1972; Mekos et al., 1992; Moffit et al., 1992) and years of father absence (Mekos et al., 1992; Surbey, 1990) are especially predictive of early puberty in girls. Surbey (1990) also reports that higher levels of stress are associated with earlier menarche in families in which fathers were present, although Mekos et al. (1992) and Moffit et al. (1992) report that the father-absence finding holds even after accounting for variations in family conflict.

One explanation for the accelerating effect of father absence on female pubertal development concerns the increased exposure of girls from divorced homes to unrelated adult males, a phenomenon seen most clearly in the case of stepfamilies. There is strong evidence in the animal literature that exposure to nonrelated adult males induces precocious puberty in juvenile females (Levin & Johnston, 1986). Exposure to unrelated adult males is not limited to adolescents growing up in stepfamilies, of course; this is also the case in families with single mothers who may be dating.

Thus, despite the powerful role that genetic factors play in influencing individual differences in pubertal and sexual maturation, a number of studies suggest that adolescents, and especially, girls, who have been exposed to conflictual conditions in the family both before and around the time of puberty may attain menarche earlier and mature at a faster rate. We suggest that family stress may affect pubertal timing by making the individual more biologically reactive to social conditions as she approaches the age of pubertal onset. This suggestion is consistent with findings that the effects of early maturation on girls' behavior and adjustment may be greater in settings in which contact with the opposite-sex is permitted and encouraged.

It is beyond the scope and purpose of this paper to discuss the possible neuroendocrinological processes by which social experiences may be related to the timing of puberty. Suffice it to say that it is widely accepted that the timing of puberty is regulated by a neuroendocrine subsystem that is intertwined with other endocrine systems, and that this system is both dynamic and responsive to environmental stimuli (Brooks-Gunn & Reiter, 1990).

In essence, then, we suggest that the conventional view that early pubertal maturation is a cause of externalizing psychopathology in adolescence may have confused causation with correlation. In our model, early maturation and various "disorders" of conduct in adolescence may be, among some youth, different manifestations of a single, underlying phenomenon. As we shall make clear in the next section, this phenomenon is not, as some have suggested, only a syndrome of problem behavior or personality deficiency, but rather, a reflection of an op- portunistic reproductive strategy. To better understand this argument, we need to turn now to the interrelations of pubertal maturation, externalization, and sexual activity.

Pubertal Timing. Externalization. and Sexual Activity

We have argued thus far that early pubertal maturation is part of a reproductive strategy that has evolved to foster early mating and limited parental investment.

Moreover, we suggest that in the environment of evolutionary adaptation, early maturation fostered the early onset of sexual activity in the service of procreation. Although one would no longer predict, given current cultural conditions, contraceptive accessibility, and the widespread availability of abortion, that early maturation necessarily leads to early childbearing, there is good reason to expect that it is related to earlier sexual activity.

In point of fact, the onset of sexual activity in adolescence is strongly linked to biological development: Boys and girls who mature early initiate intercourse at a younger age than their peers (Aro & Taipane, 1987; Magnusson et al., 1986; Smith et al., 1985). The hormonal changes of puberty—especially the surge in testosterone—increase the adolescent's sex drive, interest in sex, and level of arousal when exposed to sexual stimuli (Smith et al., 1985; Udry et al., 1986; Udry, 1987).

Because not all adolescents of equivalent pubertal status are equally likely to engage in sexual activity, studies of nonbiological factors that appear to augment the direct impact of puberty on sexual behavior are especially interesting. Consistent with our perspective, adolescents who come from father-absent households or conflicted family environments, or who evince signs of delinquency and substance abuse, are more likely than their same-aged, comparably developed, peers to be sexually active (Jessor et al., 1983; Newcomer & Udry, 1987; Thorton & Camburn, 1987).

At first glance, the link between sexual activity and various problem behaviors and environments is easily explained within conventional theories of socialization. But upon further reflection, however, one is forced to wonder about the underlying mechanism. In a society that no longer labels nonmarital sexual activity as "deviant," one must ask why this behavior covaries with such censured behaviors as delinquency and substance abuse. As we noted earlier, Jessor (Jessor et al., 1983; Jessor & Jessor, 1977) has suggested that an underlying personality constellation composed of impulsivity, independence striving, and unconventionality accounts for the clustering of problem behaviors in adolescence, including substance abuse, delinquency, and precocious sexual activity. According to our theory, these behaviors *should* cluster together, but not simply because they are problematic. Rather, we predict that these behaviors sometimes covary because they all manifest the same opportunistic, risk-taking reproductive strategy. By the same token, we believe that delayed pubertal maturation and the attendant diminution of externalized disorder are manifestations of a more cautious, mutual-reciprocal reproductive orientation.

Sex Differences in Rates of Externalizing Psychopathology

Thus far, we have argued that externalizing psychopathology—delinquency, drug and alcohol use, sexual precocity, risk taking, and disorders of conduct—should be associated with each other and with earlier physical maturation in certain cases because externalization and accelerated puberty may both be manifestations of the same underlying opportunistic reproductive strategy. Given the fact that both men and women can be opportunistic in their orientation, however, it is important to ask why rates of externalized disorders are so much higher among adolescent males than females. We believe that three explanations are plausible.

First, it is clear from research on socialization that the behaviors associated with externalizing psychopathology are much more socially accepted among males than females. Thus, even if the biological underpinnings of an opportunistic reproductive strategy predisposed a young woman toward various externalized disorders, she would face many more social constraints on the actual expression of this predisposition than would a male with similar biological inclinations. Consistent with this view is our earlier observation that the links between biology and behavior are more contextually moderated among females than males. Among males, for example, high testosterone levels predict sexual activity, more or less independent of contextual factors. Among females, however, high testosterone levels predict sexual activity only when adolescent sexual activity is socially accepted (Udry, 1987).

Second, although it is conceivable that both males and females can adopt an opportunistic outlook toward reproduction, it is clear that the potential costs of reproductive opportunism are considerably greater for females than males. Pregnancy and childbirth are substantial mortality risks for women and were certainly even more so in the environments of evolutionary adaptation. Thus, while we would expect to find variability among women along the quantity-quality reproductive continuum, we would expect to find a more restricted range among females than males on this dimension and far fewer high-quantity women than comparably high-quantity men. As a consequence, we would expect to find more extreme forms of the behavioral correlates of reproductive opportunism within the adolescent male population than within the adolescent female population.

Finally, as Wilson and Daly (1985, 1993) have pointed out in their evolutionary analysis of homicidal conflict, risky and confrontational competition—certainly an aspect of the antisocial behavior that comprises many externalizing disorders—is more frequent when within-sex fitness (i.e., reproductive) variance is large. Because fitness variance is greater among males than females, competition among them for females is more intense than it is among females for males. For this reason, aggressive, competitive behavior in general is more common among males than females, and most common of all among young adult males in the late adolescent and early adult years, that is, the period during which we would expect competition for mates to be most intense. This third explanation of sex differences in externalizing psychopathology, then, is consistent with our earlier analysis of risk taking. Because it is males who may benefit more from dominance in the social hierarchy, as well as because it is males who seek to demonstrate their genetic hardiness (i.e., via sexual selection processes), it is males who are more likely to be risk takers and evince problematic behavior.

Adolescent Psychopathology in Life-Span Perspective

How should adolescent problem behavior be viewed within the course of the life span? As we have argued elsewhere (Belsky et al., 1991), the reproductive strategies under consideration do not arise de novo at adolescence, but develop over time, and especially during the early portion of the life span. Following Draper and Harpending (1982), we suggest here that humans have evolved to be sensitive and responsive to the context of early rearing and, as a consequence, that they develop certain behavioral patterns and psychological orientations, which subsequently guide their reproductive functioning. Fundamentally, evolution has primed humans to learn particular lessons during the first 5 to 7 years of life that will shape their subsequent pair bonding and child-rearing behavior. Many of the behaviors exhibited in adolescence—including those we label as psychopathological—have their origins in early experience.

Consistent with traditional cultural anthropology (Whiting & Whiting, 1975), we presume that patterns of child-rearing reflect and are derivative of the general ecology in which families reside and that, implicitly (if not explicitly), rearing strategies represent attempts by parents to prepare their children for the world which they "expect" their offspring to encounter. Where we differ from the traditional view is in theorizing that a critical feature of this (often unconscious) interpretation of the future involves expectations regarding reproductive success, especially the chances of establishing an enduring pair bond in which a man and a women reciprocally exchange resources for purposes of facilitating the rearing of their progeny (Draper & Belsky, 1990).

In our view and that of others (e.g., Burgess & Draper, 1989), contextual stressors—among them, marital discord, single parenthood, and unstable employment—will foster more insensitive, harsh, rejecting, inconsistent, and / or unpredictable parenting behavior. This, in turn, will induce in the child what students of attachment theory refer to as an insecure or mistrustful internal working model (particularly of the insecure-avoidant variety) of the self, others, and relationships (Ainsworth et al., 1978; Bowlby, 1969; Sroufe, 1979). That is, when contextual stress and correlated rearing patterns are extant, children will be relatively more likely to come to view themselves as unlovable and as unworthy of love; others as undependable and uncaring; and, as a consequence, close interpersonal bonds as ephemeral and undependable.

In contrast to this scenario of insensitive rearing and insecurity, the child growing up in an environment in which resources are (or at least are perceived to be) either relatively abundant or predictable or both—including care and nurturance—develops a distinctly different working model of the interpersonal world. When parents trust and count on others and rear their children to engage in reciprocally rewarding and enduring interpersonal bonds and to expect paternal investment, children develop predispositions to selectively attend to, encode, and anticipate experiences that are strikingly different from those of children who experience insensitive, rejecting, or neglectful care. Whereas rejected or neglected children will home in on slights, offenses, and mistreatments—because they have learned to be vigilant about being taken advantage of—secure children focus upon empathic and considerate social encounters and, in so doing, foster more of them. For all children then, subsequent experience builds upon past experience and, importantly, expectations and understandings set earlier in life are reinforced via self-constructing interpersonal processes.

This linkage between context, rearing, and psychological orientation to the interpersonal world fosters, as we have implied, correlated patterns of interpersonal behavior during childhood and adolescence. Under conditions of family stress, insensitive rearing, and insecure feelings, behavior problems are more likely to develop—high levels of aggression, impulsivity, or noncompliance with adults and socialization norms—the classic components of externalizing symptoms. In the case of individuals following an opportunistic reproductive strategy, these behaviors are theorized to develop prior to adolescence and continue thereafter, unless effective interventions to remediate them are implemented.

For many social scientists today, these psychological and behavioral patterns are regarded as problems because they appear dysfunctional within mainstream society—and often are. We speculate, however, that these behavior patterns appear functional when viewed from the standpoint of evolutionary biology (see Chisholm, 1988, 1993). This is because they may serve to mediate and, indeed, contribute to the influence that early experience exerts upon somatic development. In particular, many of the externalizing behaviors commonly observed in adolescence may have evolved as proximal biobehavioral mechanisms that accelerate the timing of puberty. As noted earlier, risk taking and externalizing behavior may also be—or at least once were—means of securing physical resources, attracting mates, and denying mating opportunities to competitors.

Early maturation is an important component of an opportunistic reproductive strategy among both males and females because it makes earlier procreation possible. In addition, early sexual activity among females may enhance long-term reproductive success—or at least would have in the ancestral environments in which processes of human development evolved. In the human female, there is an extended period of subfertility between menarche and genuine fertility (Lancaster, 1986; Short, 1976). Moreover, there is some suggestive evidence that both sexual behavior and associating with nonrelated males shortens and regulates the menstrual cycle (e.g., Cutler, Garcia, & Krieger, 1979; Trevathan & Burleson, 1989). A shorter and more regular cycle affords, over a life time, more fertile periods and, thus, more opportunity to conceive children (Metral, 1981; Trevathan & Burleson, 1989). Thus, we speculate that early puberty and accelerated sexual activity might actually function to shorten the period of post-pubertal subfertility and, in so doing, organize the menstrual cycle to permit more conceptions across the life span.

We have also suggested that aggressive, disobedient, and oppositional behavior stimulates earlier maturation among boys and may attract—or at least once served to attract—mates. Aggressive, noncompliant behavior, which seems to be inherently opportunistic vis-à-vis others, should also foster indiscriminant and opportunistic sexuality and, in concert with earlier puberty, increase the likelihood of such males becoming fathers before other men. It seems likely that earlier maturing, opportunistic males would be the least likely to use contraception. Moreover, if such males come from maritally discordant or single-parent homes, it seems unlikely that they will anticipate and seek enduring pair bonds and thus anticipate investing time, energy, and resources in their offspring (i.e., they will evidence low paternal investment).

To many readers it will no doubt seem counterintuitive to assert, as we and others do (MacDonald, 1988), that a developmental trajectory characterized by attachment insecurity, behavior problems, early maturation, precocious sexuality, unstable pair bonds, limited parental investment, and high total fertility makes any biological sense at all. In Western society, this reproductive strategy appears clearly dysfunctional and disadvantageous, and the associated behavioral patterns disturbed. Why, then, would it have evolved?

To answer this question, it is important to keep in mind that past environments of evolutionary adaptation have been highly variable. In some contexts, such as those affording rapid expansion into favorable and previously unexploited niches, there may have been an advantage to rapid reproduction and no penalty (in terms of offspring survival) to transient bonds between sexual partners and loose attachment between mothers and offspring, provided only that other group members could be relied upon to provide food and protection to juveniles (Draper & Harpending, 1982).

What evolutionary mechanism, though, would underlie an accelerated reproductive schedule in the context of unfavorable environments? We point again to the maxim that organisms have been selected to reproduce themselves and to attend to important environmental cues in the process. In the absence of indications that delayed maturation and reproduction can have benefits, early sexual activity and high fertility has much to recommend it. This strategy may be associated with higher offspring mortality, but from the point of view of fitness, individuals living in such adverse circumstances who delay reproducing may well be selected against (i.e., leave few or no offspring). In such an environment, a man who invests disproportionately in one woman and in children (who may not be his own), will leave relatively few of his own offspring behind. Likewise, a young woman who waits for the right man to help rear her children may lose valuable reproductive opportunities at a time when her health and physical capability are at their peak and when her mother and senior female kin are young enough to be effective surrogates. In such circumstances, nonbonded and relatively indiscriminate sexuality, as well as high fertility, can be positively selected. It is in this sense, then, that we assert that both of the reproductive strategies that we detailed make biological sense, in that they are optimal given the contexts in which they develop—and for which they were selected. And it is in this sense that the evolutionary view of adolescent psychopathology we have proposed illuminates the paradox of normative disturbance.

Evolutionary Views and the Paradox of Normative Disturbance

We began this chapter by drawing attention to a paradox about psychopathology in adolescence that has puzzled social scientists for several decades. As we wrote earlier, this paradox inheres in the observation that many of the very behaviors we label as normative during adolescence are those which we also label as disturbed.

The evolutionary view we have espoused helps explain how a certain externalized disorders can be simultaneously normative and disturbed. Risk taking, impulsivity, aggression, delinquency, and sexual precocity—classic forms of externalizing psychopathology in adolescence—may be manifestations of an underlying reproductive strategy that has logical developmental antecedents, clear contextual correlates, and evolutionarily consistent biological manifestations. In view of the likely developmental history and environmental experiences of the opportunistic individual, many of the behaviors we have come to label "deviant" are anything but this. Rather than viewing the constellation of behaviors as forming a problem behavior syndrome, then, it may be more helpful to examine them in terms of their adaptive significance within the context in which the individual has been reared. Is the individual adolescent with a high level of externalization more normative or more disturbed? The answer depends on the context in which the individual has developed. Behaviors that may be disturbed when viewed against a backdrop of one set of environmental conditions may, in fact, be normative when viewed against another.

We also believe that the evolutionary perspective also informs discussions about nomothetic and idiographic aspects of adolescent psychopathology. With regard to nomothetic concerns, rather than looking at psychopathology within adolescence as disjunctive or discontinuous with psychopathology in other periods, we suggest that the field might profit by looking at adolescence as a window on the nature and organization of psychopathology over the entire life span. The pattern of behavior displayed during adolescence may reveal important information about reproductive strategy over the life course and, as such, may say a great deal about the individual's past and future.

With regard to idiographic concerns, we suggest that individual differences in externalizing problems should be magnified during developmental periods like adolescence, when reproductive imperatives are strong and salient. To the extent that externalizing problems are linked to biological maturation and reproductive inclinations, our understanding of the origins and course of externalized psychopathology is particularly well-informed by focusing on adolescence, when biological change is especially dramatic and reproductive inclinations are at their most intense.

This is not to say that because a pattern of behavior can be seen as evolutionarily adaptive, it is something that is "good," or desirable from society's standpoint. Many evolved patterns of behavior that are deeply ingrained in human nature no longer serve their original purpose, and many have obvious disadvantages within the current context in which the majority of individuals come of age. Our point, however, is that an evolutionary view on adolescent psychopathology can help put some putatively disturbed behaviors in perspective. Perhaps in so doing we can look at them with more understanding and greater compassion.

REFERENCES

Adelson, J., & Doehrman, M. (1980). The psychodynamic approach to adolescence. In J. Adelson (Ed.), *Handbook of adolescent psychology* (pp. 99-116). New York: Wiley.

Ainsworth, M., Blehar, M., Waters, E., & Wall, S. (1978). *Patterns of attachment*. Hillsdale, NJ: Erlbaum.

Alexander, C., Ensminger, M., Kim, Y., Smith, B. J., Johnson, K., & Dolan, L. (1989). Early sexual activity among adolescents in small towns and rural areas: Race and gender patterns. *Family Planning Perspectives, 21*, 261–266.

Andersson, T., & Magnusson, D. (1990). Biological maturation in adolescence and the development of drinking habits and alcohol abuse among young males: A prospective longitudinal study. *Journal of Youth and Adolescence, 19*, 33–42.

Aro, H., & Taipale, V. (1987). The impact of timing of puberty on psychosomatic symptoms among fourteen- to sixteen-year-old Finnish girls. *Child Development, 58*, 261–268.

Bandura, A. (1977). *Social learning theory*. Englewood Cliffs, NJ: Prentice-Hall.

Barkow, J. (1984). The distance between genes and culture. *Journal of Anthropological Research, 40*, 9–14.

Belsky, J., & Draper, P. (1987). Reproductive strategies and radical solutions. *Transactions / Society, 24*, 20–24.

Belsky, J., Steinberg, L., & Draper, P. (1991). Childhood experience, interpersonal development, and reproductive strategy: An evolutionary theory of socialization. *Child Development, 62*, 647–670.

Blos, P. (1979). *The adolescent passage*. New York: International Universities Press.

Bowlby, J. (1969). *Attachment and loss: Vol. 1. Attachment*. New York: Basic Books.

Bradshaw, A. (1965). Evolutionary significance of phenotypic plasticity in plants. *Advances in Genetics, 13*, 115–155.

Bronfenbrenner, U. (1977). Toward an experimental ecology of human development. *American Psychologist, 32*, 513–531.

Brooks-Gunn, J. (1989). Pubertal processes and the early adolescent transition. In W. Damon (Ed.), *Child development today and tomorrow* (pp. 155–176). San Francisco: Jossey-Bass.

Brooks-Gunn, J., & Reiter, E. (1990). Pubertal development. In S. Feldman & G. Elliott (Eds.), *At the threshold: The developing adolescent* (pp. 16-53). Cambridge, MA: Harvard University Press.

Burgess, R., & Draper, P. (1989). The explanation of family violence: The role of biological behavioral and cultural selection. In L. Ohlin & M. Tonry (Eds.), *Family violence* (pp. 59-116). Chicago: University of Chicago Press.

Caine, N. (1986). Behavior during puberty and adolescence. In G. Mitchell & J. Erwin (Eds.), *Comparative primate biology: Vol. 2A. Behavior, conservation, and ecology* (pp. 327–361). New York: Alan R. Liss.

Campbell, B., & Udry, J. R. (1992, March). *Mother's age at menarche, not stress, accounts for daughter's age at menarche*. Paper presented at the biennial meeting of the Society for Research on Adolescence, Washington, D.C.

Caspi, A., & Bem, D. (1990). Personality continuity and change across the life course. In L. A. Pervin (Ed.), *Handbook of personality theory and research* (pp. 549–575). New York: Guilford.

Caspi, A., Lynam, D., Moffitt, T., & Silva, P. (1993). Unravelling girls' delinquency: Biological, dispositional, and contextual contributions to adolescent misbehavior. *Developmental Psychology, 29,* 19–30.

Chisholm, J. (1988). Toward a developmental evolutionary ecology. In K. MacDonald (Ed.), *Sociobiological perspectives on human development* (pp. 78–102). New York: Springer-Verlag.

Chisolm, J. (1993). Death, hope, and sex. *Current Anthropology, 34,* 1–24.

Coleman, J. S. (1961). *The adolescent society.* New York: Free Press.

Crawford, C., & Anderson, J. (1989). Sociobiology: An environmentalist discipline? *American Psychologist, 44,* 1449–1460.

Cronin, H. (1992). *The ant and the peacock: Altruism and sexual selection from Darwin to today.* New York: Cambridge University Press.

Cutler, W. B., Garcia, C. R., & Krieger, A. B. (1979). Sexual behavior frequency and menstrual cycle length in mature premenopausal women. *Psychoneuroendocrinology, 4,* 297–309.

Diamond, J. (1992). *The third chimpanzee: The evolution and future of the human animal.* New York: HarperCollins.

Digman, J. (1989). Five robust traits and dimensions: Development, stability, and utility. *Journal of Personality, 57,* 195–214.

Draper, P., & Belsky, J. (1990). The relevance of evolutionary thinking for issues in personality development. *Journal of Personality, 58,* 141–162.

Draper, P., & Harpending, H. (1982). Father absence and reproductive strategy: An evolutionary perspective. *Journal of Anthropological Research, 38,* 255–273.

Duncan, P., Ritter, P., Dornbusch, S., Gross, R., & Carlsmith, J. (1985). The effects of pubertal timing on body image, school behavior, and deviance. *Journal of Youth and Adolescence, 14,* 227–236.

Eccles, J., Miller, C., Tucker, M., Becker, J., Schramm, W., Midgley, R., Holmes, W., Pasch, L., & Miller, M. (1988, March). *Hormones and affect at early adolescence.* Paper presented at the meetings of the Society for Research on Adolescence, Alexandria, VA.

Elder, G. (1981). History and the life course. In D. Bertaux (Ed.), *Biography and society: The life history approach in the social sciences* (pp. 77–115). Beverly Hills, CA: Sage.

Elkind, D. (1978). Understanding the young adolescent. *Adolescence, 13,* 127–134.

Ellis, N. (1991). Pubertal maturation and parent-adolescent relations: A test of the Steinberg accelerating hypothesis. *Journal of Early Adolescence, 11,* 221–235.

Epstein, E., & Erskine, N. (1983). The development of personal theories of reality from an interactional perspective. In M. Magnusson & V. Allen (Eds.), *Human development: An interactional perspective* (pp. 133–147). New York: Academic Press.

Erikson, E. (1968). *Identity: Youth and crisis.* New York: Norton.

Evans, S., & Hodges, J. (1984). Reproductive status of adult daughters in family groups of common marmosets (*Callithrix jacchus jacchus*). *Folia Primatologica, 42,* 127–133.

Freud, A. (1958). Adolescence. *Psychoanalytic Study of the Child, 13,* 255–278.

Frisch, R., & McArthur, J. (1974). Menstrual cycles: Fatness as a determinant of minimum weight necessary for their maintenance or onset. *Science, 185,* 949–951.

Gans, J. (1990). *America's adolescents: How healthy are they?* Chicago: American Medical Association.

Gardner, W. (1993). A life-span rational choice theory of risk-taking. In N. Bell and R. Bell (Eds.), *Adolescent risk-taking* (pp. 66–83). Newberry Park, CA: Sage.

Graber, J., & Brooks-Gunn, J. (1992, March). *The antecedents of menarcheal age: heredity, family environment, and stressful life events*. Paper presented at the biennial meeting of the Society for Research on Adolescence, Washington, DC.

Graham, C. (1991). Menstrual synchrony: An update and review. *Human Nature, 2*, 293–311.

Hall, G. S. (1904). *Adolescence*. New York: Appleton.

Hamilton, W. (1964). The genetical theory of social behavior. *Journal of Theoretical Biology, 7*, 1–52.

Hill, E., Young, J., & Nord, J. (1993). *Rearing environment, attachment security and adult relationships: A preliminary study*. Unpublished manuscript, Department of Psychiatry, University of Michigan, Ann Arbor.

Hill, J., Holmbeck, G., Marlow, L., Green, T., & Lynch, M. (1985a). Pubertal status and parent-child relations in families of seventh-grade boys. *Journal of Early Adolescence, 5*, 31–44.

Hill, J., Holmbeck, G., Marlow, L., Green, T., & Lynch, M. (1985b). Menarcheal status and parent-child relations in families of seventh-grade girls. *Journal of Youth and Adolescence, 14*, 301–316.

Hinde, R. (1982). Attachment: Some conceptual and biological issues. In C. M. Parkes & J. Stevenson-Hinde (Eds.), *The place of attachment in human development* (pp. 60–76). London: Tavistock Press.

Jessor, R., Costa, F., Jessor, L., & Donovan, J. (1983). Time of first intercourse: A prospective study. *Journal of Personality and Social Psychology, 44*, 608–626.

Jessor, R., & Jessor, S. (1977). *Problem behavior and psychosocial development: A longitudinal study of youth*. New York: Academic Press.

Jones, B., Leeton, J., McLeod, I., & Wood, C. (1972). Factors influencing the age of menarche in a lower socio-economic group in Melbourne. *Medical Journal of Australia, 21*, 533–535.

Kennedy, R. (1993). Depression as a disorder of social relationships: Implications for school policy and prevention programs. In R. Lerner (Ed.), *Early adolescence: Perspectives on research, policy, and intervention* (pp. 383–398). Hillsdale, NJ: Erlbaum.

Konishi, M., Emlen, S., Ricklefs, R., & Wingfield, J. (1989). Contributions of bird studies to biology. *Science, 46*, 465–472.

Kracke, B., & Silbereisen, R. (1992, March). *Behavioral autonomy and pubertal maturation*. Paper presented at the biennial meeting of the Society for Research on Adolescence, Washington, DC.

Krebs, J., & Davies, N. (1981). *An introduction to behavioral ecology*. Sunderland, MA: Sinauer.

Lack, D. (1947). *Darwin's finches*. New York: Cambridge University Press.

Lancaster, J. (1986). Human adolescence and reproduction. In J. Lancaster & B. Hamburg (Eds.), *School-age pregnancy and parenthood* (pp. 17–38). New York: Aldine de Gruyter.

Levin, R., & Johnston, R. (1986). Social mediation of puberty: An adaptive female strategy? *Behavioral and Neural Biology, 46*, 308–324.

Lewin, K. (1951). *Field theory and social science*. New York: Harper & Row.

Livson, N., & Peskin, H. (1980). Perspectives on adolescence from longitudinal research. In J. Adelson (Ed.), *Handbook of adolescent psychology* (pp. 47–98). New York: Wiley.

MacDonald, K. (Ed.). (1988). *Sociobiological perspectives on human development*. New York: Springer-Verlag.

Magnusson, D., Stattin, H., & Allen, V. (1986). Differential maturation among girls and its relation to social adjustment in a longitudinal perspective. In P. Baltes, D. Featherman, & R. Lerner (Eds.), *Life span development and behavior* (Vol. 7, pp. 135–172). Hillsdale, NJ: Erlbaum.

Malatesta, C., & Wilson, A. (1988). Emotion / cognition interaction in personality development: A discrete emotions, functionalist analysis. *British Journal of Social Psychology, 27*, 91–112.

Mannheim, K. (1952). The problem of generations. In K. Mannheim (Ed.), *Essays on the sociology of knowledge* (pp. 276–322). London: Routledge & Kegan Paul.

McClintock, M. K. (1980). Major gaps in menstural cycle research: Behavioral and physiological controls in a biological context. In P. Komenich, M. McSweeney, J. Novack, & N. Elder (Eds.), *The menstural cycle: Vol. 2* (pp. 7–23). New York: Springer-Verlag.

McCord, J. (1990). Problem behaviors. In S. Feldman and G. Elliot (Eds.), *At the threshold: The developing adolescent* (pp. 414–430). Cambridge, MA: Harvard University Press.

Mekos, D., & Hetherington, E. M. (March, 1992). *Family antecedents of pubertal timing.* Paper presented at the biennial meeting of the Society for Research on Adolescence, Washington, DC.

Metral, G. (1981). The action of natural selection on the human menstrual cycle: A simulation study. *Journal of Biosocial Science, 13,* 337–343.

Moffitt, T. (1993). Adolescence-limited and life-course persistent antisocial behavior: A developmental taxonomy. *Psychological Review, 100,* 674–701.

Moffitt, T., Caspi, A., & Belsky, J. (1992). Family context, girls' behavior, and the onset of puberty: A test of a sociobiological model. *Child Development, 63,* 47–58.

Newcomer, S., & Udry, J. (1987). Parental marital status effects on adolescent sexual behavior. *Journal of Marriage and the Family, 49,* 235–240.

Papini, D., & Sebby, R. (1987). Adolescent pubertal status and affective family relationships: A multivariate assessment. *Journal of Youth and Adolescence, 16,* 1–15.

Petersen, A. (1985). Pubertal development as a cause of disturbance: Myths, realities, and unanswered questions. *Genetic, Social, and General Psychology Monographs, 111,* 205–232.

Plomin, R., Loehlin, J. C., & DeFries, J. C. (1985). Genctic and environmental components of environmental influences. *Developmental Psychology, 21,* 391–401.

Richards, M., Boxer, A., Pctersen, A., & Albrecht, R. (1990). Relation of weight to body image in pubertal girls and boys from two communities. *Developmental Psychology, 26,* 313–321.

Rushton, J. (1985). Differential K theory: The sociobiology of individual and group differences. *Personality and Individual Differences, 6,* 441–452.

Short, R. V. (1976). The evolution of human reproduction. *Proceedings of the Royal Society of London, 195,* 3–24.

Simmons, R., & Blyth, D. (1987). *Moving into adolescence.* New York: Aldine de Gruyter.

Smith, E., Udry, J., & Morris, N. (1985). Pubertal development and friends: A biosocial explanation of adolescent sexual behavior. *Journal of Health and Social Behavior, 26,* 183–192.

Sroufe, L. A. (1979). The coherence of individual development. *American Psychologist, 34,* 834–841.

Sroufe, L. A., & Rutter, M. (1984). The domain of developmental psychopathology. *Child Development, 55,* 17–29.

Stearns, S. (1982). The role of development in the evolution of life histories. In J. T. Bonner (Ed.), *Evolution and development* (pp. 47-73). New York: Springer-Verlag.

Steinberg, L. (1987). The impact of puberty on family relations: Effects of pubertal status and timing. *Developmental Psychology, 23,* 451–460.

Steinberg, L. (1988). Reciprocal relation between parent-child distance and pubertal maturation. *Developmental Psychology, 24,* 122–128.

Steinberg, L. (1989). Pubertal maturation and parent-adolescent distance: An evolutionary perspective. In G. Adams, R. Montemayor, & T. Gullotta (Eds.), *Advances in adolescent development, Vol. 1* (pp. 71–97). Beverly Hills, CA: Sage.

Steinberg, L., & Levine, A. (1990). *You and your adolescent: A parent's guide for ages 10 to 20.* New York: Harper & Row.

Surbey, M. (1990). Family composition, stress, and human menarche. In F. Bercovitch & T. Zeigler (Eds.), *The socioendocrinology of primate reproduction* (pp. 1–25). New York: Alan R. Liss.

Susman, E., Inhoff-Germain, G., Nottlemann, E., Loriaux, D., Cutler, G., Fr., & Chrousos, G. (1987). Hormones, emotional dispositions, and aggressive attributes in young adolescents. *Child Development, 58,* 1114–1134.

Tardif, S. (1984). Social influences on sexual maturation of female *Saquinas oedipus oedipus*. *American Journal of Primatology, 6,* 199–209.

Thorton, A., & Camburn, E. (1987). The influence of the family on premarital sexual attitudes and behavior. *Demography, 24,* 323–340.

Tinbergen, N. (1963). On the aims and methods of ethology. *Zeitschfrift fur Tierpsycologie, 20,* 410–433.

Trevathan, W., & Burleson, M. (1989, August). *Human pair bonding and non-ovulatory sexual activity in females: New hypotheses.* Paper presented at the first annual meeting of the Human Behavior and Evolution Society, Northwestern University, Chicago.

Udry, J. (1987). Hormonal and social determinants of adolescent sexual initiation. In J. Bancroft (Ed.), *Adolescence and puberty* (pp. 70–87). New York: Oxford University Press.

Udry, J., Talbert, L., & Morris, N. (1986). Biosocial foundations for adolescent female sexuality. *Demography, 23,* 217–230.

Wachtel, P. (1973). Psychodynamics, behavior therapy, and the implacable experiments. *Journal of Abnormal Psychology, 82,* 324–334.

Whiting, B., & Whiting, J. (1975). *Children of six cultures.* Cambridge, MA: Harvard University Press.

Wilson, M., & Daly, M. (1985). Competitiveness, risk-taking, and violence: The young male syndrome. *Ethology and Sociobiology, 6,* 59–73.

Wilson, M., & Daly, M. (1993). Lethal confrontational violence among young men. In N. Bell and R. Bell (Eds.), *Adolescent risk-taking* (pp. 84–106). Newberry Park, CA: Sage.

Worthman, C. (1990). Socioendocrinology: Key to a fundamental synergy. In T. Ziegler & F. Bercovitch (Eds.), *The socioendocrinology of primate reproduction* (pp. 187–212). New York: A. R. Liss.

Youngblade, L., & Belsky, J. (1989). Child maltreatment, infant-peer attachment security, and dysfunctional peer relationships in toddlerhood. *Topics in Early Childhood Education, 9,* 1–15.

Zabin, L., et al. (1986). Ages of physical maturation and first intercourse. *Demography, 23,* 595–605.

IV   Neither "Safe Sex" Nor "Abstinence" May Work—Now What?:
Toward a Third Norm for Youthful Sexuality

Robert Kegan

July 18, 1993

My 15-year-old patient with pelvic pain lay quietly on the gurney, as I asked her the standard questions. "Are you sexually active?" "Yes." "Are you using any form of birth control?" "No." Her answers didn't surprise me. . . . Condoms are not being used. Many studies confirm this, including one survey among college women—a group we might presume to be as well-informed as any on the risks of herpes, genital warts, cervical cancer, and AIDS. In 1989, only 41 percent insisted on condom use during sexual intercourse! If educated women cannot remember to use condoms, how can we expect teenagers or the uninformed to do so?

Boston Globe

The sexual behavior of America's adolescents at the close of the 20th century understandably arouses the concerns of parents, educators, and helping profession-als. This chapter frames the alarming dimensions of teenage sexual behavior in the context of *risk* (rather than *psychopathology*) and—through the lens of constructive-developmental theory—offers a fresh perspective on our cultural and professional efforts to get teenagers to reduce these risks by changing their behavior.

How widespread is teenage sexual behavior? Among unmarried adolescents between the ages of 15 and 19, 51% of females and 60% of males have had sexual intercourse (Guttmacher Institute, 1992). More teenagers are having sex for the first time at younger ages: for example, 25% of unmarried females between the ages of 15 and 19 had had intercourse in 1970, a figure that has now doubled (Forrest and Singh, 1990). The largest increase in intercourse rates was found among 15-year-old females (Center for Disease Control [CDC], 1991). Among females, 27% of 15 year olds and 75% of 19 year olds have had intercourse; among males, 33% of the 15 year olds and 86% of the 19 year olds have had intercourse (Guttmacher Institute, 1992). In a 1990 study, one in four 12th graders reported four or more sexual partners (CDC, 1992).

What risks are associated with this widespread behavior? There is a veritable cascade of risk for sexually active teenagers, burdening teens themselves and the culture at large with enormous psychological and social costs. This cascade begins with the fact that large percentages of sexually active teens do not regularly use contraceptives. This risky behavior leads to unintended pregnancies and sexually transmitted diseases, each of which creates its own set of associated risks.

In a 1988 study (Forrest et al., 1990), 52.6% of females aged 15–19 did not use a condom at first intercourse. In a separate analysis of the same data (Mosher, 1990), 67% of females 15–19 reported they did not use condoms when they had intercourse. In a 1990 study, 59% of high school students with multiple partners reported they did not use condoms at last intercourse, and 55% of all sexually active students reported they did not (Center for Disease Control, 1992).

According to Zabin et al. (1979), two thirds of teens who are having unprotected intercourse will become pregnant within the first two years after initiating intercourse, and one quarter will become pregnant within the first month. More than 40% of all adolescent women in the United States will become pregnant before they get out of their teens (Hayes, 1987), producing one of the highest pregnancy rates in the industrialized Western world (Henshaw and Van Vort, 1989). Eighty-four percent of all pregnancies to teens are unintended (Jones and Lincoln, 1986).

What are the documentable risks of unintended teen pregnancy? Teen mothers run risks to their babies' health and to their health; teen parents run risks to their ability to further or complete their educations, to their employment opportunities, and to their marital stability (Kirby, Short, Collins, Rugg, Kolbe, Howard, Miller, Sonen-Stein, and Zabin, 1994; Nord, Moore, Morrison, Brown, and Myers, 1992). Forty-six percent of teenage mothers receive no prenatal care during their first trimester; an additional 9% receive no care during their second trimester, and an additional 4% receive no prenatal care at all (NCHS, 1989). Women who receive no prenatal care are 40 times more likely to have their baby die during the neonatal stage than women who receive adequate prenatal care (Children's Defense Fund, 1988). Even with adequate prenatal attention, teen mothers are at higher risk for medical difficulties including anemia, pregnancy-induced hypertension, cervical trauma, and premature delivery (National Commission to Prevent Infant Mortality, 1988).

Pregnancy is the number one reason cited by adolescent females for dropping out of school (Walker and Vilella-Velez, 1992). Of females who give birth at or before age 17, only half will graduate from high school by age 30 (Guttmacher Institute, 1992). Only 39% of teen fathers receive high school certification by age 20, compared to 86% of male teens who are not fathers (Marsiglio, 1989), and teen fathers are only half as likely to finish college (Guttmacher Institute, 1981).

According to the Alan Guttmacher Institute (1992), teen mothers earn about half the lifetime income of women who first give birth in their 20s. In 1987, 70% of single females under the age of 25 with children were living below the poverty level, and teen mothers were 50% more likely than nonparents to be receiving welfare (Hayes, 1987).

Over half of all AFDC payments are made to families begun when the mother was a teenager (Burt, 1986). In 1988, AFDC, Medicaid and food stamps payments to families where the mother first gave birth as a teenager were $19.83 billion (Center for Population Options, 1989). By 1989, the U.S. government spent over $21 billion for social, health, and welfare services to families begun by teen mothers (Guttmacher Institute, 1992).

Staggering as are the costs of unintended teen pregnancy, the risks associated with sexually transmitted diseases (STD) must be cast in the even more stark terms of life and death. Teens have the highest rates of STD of any age group (DHHS, 1990); each year, between 2.5 and 3 million adolescents (about 1 of every 6) are infected with an STD (CDC, 1990). Between 1960 and 1988, most age groups experienced a decline in rates of gonorrhea, but the rates among teens aged 15–19 increased by 170% (Conference on the Health of America's Youth, 1991). By 1991, the rate of gonorrhea among 15–19-year-old females was 22 times higher than the rate for women age 30 or older (Webster and Berman, 1993). Most frightening of all, of course, are the statistics concerning HIV infection and the spread of AIDS. AIDS is now the sixth leading cause of death among 15–24 year olds in the United States (CDC, 1992). Between December 1990 and December 1992, the number of cases of AIDS among 13–24 year olds increased by 43% (CDC, 1992, 1993). Between 1987 and 1990, more than 1 in 300 teens aged 16–21 who were applying for entrance into the U.S. Job Corps tested HIV positive; among black and Hispanic entrants at age 21, the infection rate was nearly 1 in 80 (St. Louis, Conway, Hayman, Miller, Petersen, and Dondero, 1991). A survey conducted in 1990 on 16,861 students at 19 universities found that 1 in 500 tested positive for HIV (Gayle, Keeling, Garcia-Tunon, Kolbourne, Narkunas, Ingram, Rogers, and Currean, 1990).

Thus the concerns we have about the sexual conduct of our teenagers have never been more legitimately alarming. Despite greater support for the use of contraceptives and more widespread sex education, the risks they run wreak havoc on present lives and human futures. If Americans are really at their best when things are at their worst, we might finally be ready to face what I will suggest here are a few truths we have been long avoiding.

Truth #1. In contrast to the foreboding picture these statistics paint, the world of sexuality is a world of irresistible allure to most teenagers. However newly arrived and however ambivalent about their new world, adolescents enter a human realm whose appeal we ourselves find unabating. Our finest writers and poets do not tire of trying to name its pleasure, power, warmth, heat, sweetness, stimulation, delight, and satisfaction. The more our worries increase about teen sexuality, the more the whole subject becomes a "social problem" to us, further and further removed from an understanding of what teen sexuality is to teens themselves. Sexuality to most teens may not be a problem in need of solution. Our anxieties about their sexuality (organized around themes of danger and risk) may not be at all the same as their anxieties (more commonly organized around themes of their own acceptability and competence). Sex to teenagers is a world of extraordinary attraction, unexplored possibility, scary but thrilling companionable adventure. Sex is what God gives to teenagers when the appeal of Disneyland starts to fade.

Truth#2. Teens are not ordinarily well-socialized, responsible, or future-oriented when they develop sexual interests and capacities. In fact, as I will discuss, sound developmental theory and an array of careful research into the cognitive and interpersonal meaning making of adolescents make clear that the teen years are distinguished by the quite gradual transformation from a concrete, short-term, and self-interested orientation to the world toward a more abstract, future-oriented, and socialized orientation. The willingness to subject one's immediate goals to considerations of their future costs or implications and the ability to subordinate self-interests to relationships of trust, loyalty, or mutual obligation develops only gradually. The lack of a future time sense and the lack of identification with the values and expectations of one's social surround is widespread during much or most of one's teens. This is true across social positions and genders. In one place it may express itself in crime against one's community, while in another in self-endangering risk taking, and in a third, the practiced deception of one's parents. The social costs and level of personal pathology may differ, but the level of psychological complexity may be the same.

Truth #3. Neither of the current norms for teen sexuality that we are trying contentiously as a culture to advance—abstinence and safer sex—actually squares well with the developmental realities of most adolescents' consciousness constraints. While each norm is an authentic expression of concern for the young, an effort to get teens to modify their risky behavior, neither is proving particularly successful (as the statistical profile at the start of this chapter makes clear). Teens of the present and future are ill-served by the limits of a cultural battle between insufficiently worthy contenders.

The "abstinence" norm's biggest failing is that it denies how irresistible sexual experience is. Now I never met a denial I couldn't feel sympathy for. There is always a very good reason why it would be more pleasant to assume the denied reality is not real. The denial of teen's interest in sexuality is *not* the province alone of obviously out-of-touch, repressed, or parochial people. It is my privilege to consult to high schools and private schools staffed by sophisticated, urbane, progressive professionals. They frequently are alarmed by crises that require them to face a picture of teen sexual activity far more widespread than the one they have been carrying around in their heads. The expectation of abstinence is unrealistic because it asks adolescents to put on hold, disable, or disengage a qualitatively new medium for connecting to and experiencing the world. It is one thing to ask people to forego a feature of their diets, for example, to stop eating high-cholesterol oils or high-fat meats; it is quite another thing to ask people to stop eating altogether.

The "safe sex" norm's biggest failing is that it assumes, unwarrantedly, an order of consciousness capable of such responsibility, farsightedness, and future-mindedness. Given how infrequently even college-age youth make use of the condoms and dental dams passed out during freshman orientation week, is it not wildly unrealistic to believe that sexually active 12, 13, 14, 15, 16, and 17 year olds are going to have the presence of mind to regularly involve these devices in their sexual activity, when these devices are completely unnecessary to, and unenhancing of, the only real goals and interests they have in mind at the moment? This may be as big a denial on the part of safe-sex proponents as is the denial of sexuality's irresistibility on the part of abstinence proponents. As I said, I have never met a denial I couldn't sympathize with. It is frightening, but I think necessary, to consider that not only is a great proportion of teens' sexual behavior and judgment compromised by alcohol, spontaneity, susceptibility to the partner's pressure, and silencing embarrassment about the technical details of sex, but even when that judgment is completely free to do its best work, it is constrained by an order of consciousness that considers the future as the present that hasn't happened yet rather than something real, right now, in the present, and commanding of our attention. Why do I say this?

Adolescent Development in the Context of Social Expectation: The Constructive-Developmental View

All teenagers in America are under a rather constant barrage of expectation from home, school, the community, and even from some of their friends. Sometimes in public discourse we hear these expectations proclaimed by the schools, the Department of Labor, or politicians. Most of the time, they are present but private, particular, subtle, and unspoken in the intimate arenas of family and neighborhood.

If we survey these expectations (Kegan, 1994), we will find that we want teenagers to be employable, good citizens, critical thinkers, trustworthy and dependable in their relationships, emotionally self-reflective, possessed of common sense, and committed to meaningful ideals. Regarding their sexual conduct in particular, we want teens to express their sexuality respectfully and responsibly, with concern for the welfare of the other and oneself and with regard to the future implications of present acts. This is a lot to want. It grows out of people's concerns for themselves, their concerns for others who live with adolescents, and their concerns for adolescents themselves. But will adolescents be up to all our expectations? To answer we first have to ask a different, perhaps bigger, question (and one that developmental psychology is particularly well suited to answer). What kinds of expectations are these really? I suggest that although these look like expectations for adolescents' outer behavior they are really about more than how adolescents should behave. Although these look like expectations about adolescents' inner feelings or attitudes, they are about even more than what and how they should feel because our feelings and attitudes come from *how we know*. More than expectations about how we want teenagers to behave, how we want them to feel, what we want them to know, they are expectations about *how* we want them to know. They are claims on adolescents' minds.

Although we don't realize it, we have some shared expectations about what the mind of a teenager should be like. Whatever definition of adolescence we might cull from a textbook, the one that is operating most powerfully on the human being who happens to be going through adolescence is the hidden definition derived from the cultures' claims or expectations about *how* an adolescent should *know*.

Over the last 20 years my colleagues and I have been studying the mental structures by which people organize reality and the evolution of these structures throughout the life span (Henderson & Kegan, 1989; Kegan, 1976, 1982, 1985, 1986, 1994; Kegan, Broderick, & Popp, 1992; Kegan & Lahey, 1983; Kegan, Noam, & Rogers, 1982; Noam & Kegan, 1982; Souvaine, Lahey, & Kegan, 1990). This work has included the development of a reliable interview-assessment procedure that identifies the level of complexity of an individual's prevailing "epistemological structure." The subject-object interview (Lahey, Souvaine, Kegan, Goodman, and Felix, 1988) clarifies what aspects of meaning organizing one has control over, can make use of, or can reflect on (what is object in one's meaning making) and what aspects control one or one is captive of, identified with (what one is subject to in one's meaning making). Longitudinal study shows that if people change the underlying structure of their meaning making, they do so in a developmental direction, in other words, they *differentiate* from structures to which they were subject, thus making those structures into objects, and *integrate* these structures into a more complex organizational principle to which they are newly subject (Kegan, 1994).

What are the expectable changes of subject and object during adolescence? Consider, for example, some widely agreed upon distinctions in cognitive capacity (Piaget, 1952), as summarized in Table 1. Each capacity can do what the preceding one cannot. Each capacity contains the abilities of the preceding ones, organized at a greater level of complexity. Each capacity is organized by a more complex or inclusive *principle of organization*. The very principle of a prior capacity becomes an element or object, which can itself be organized by the new capacity's principle of organization. What this means is that the relation between the principles is essentially developmental (each one is included in the next) and that the organizational principles themselves are essentially epistemological: their structure can be described in terms of what is taken as subject (the principle of organization) and what is taken as object (the element that can be organized). The capacities, thus, are not so much about *what* the mind knows, but about *how* it knows, with each capacity describing a more complex organization of mind in the logical-cognitive domain.

We could add to these descriptions of qualitatively different orders of complexity in one's *cognitive* organizing, descriptions of different orders of complexity in other domains of meaning making. For example, such a description of distinctions in our social-cognitive domain (Selman, 1980; Kohlberg, 1984) is summarized in Table 2. Here, too, we can see that what is being traced is a development in which the prior capacity's principle of knowing becomes an element in the next principle.

Consider, for example, that, using the second-order capacity, people are able to distinguish their own point of view from another's. They can see that others' actions issue from their own purposes and intentions, which may not be the same as one's own purposes and intentions. Consider the story about which Piaget often questioned children (Piaget, 1948): There are two children. One is told by his mother not to touch a collection of cups; he deliberately picks one up and drops it and it breaks. The other child is told nothing but sees that his mother needs a tray of cups taken into the kitchen. While carrying them, he slips and all the cups break. Now which child is naughtier? Piaget asked. Those who make use of the second capacity see that the first child is naughtier because they orient to the child's point of view

Table 1. Three Capacities in Logical-Cognitive Organizing

First Principle Roughly 2 to 6 years	Second Principle Roughly 6 years to teens	Third Principle Teenage years and beyond
Can: recognize that objects exist independent of own sensing of them ("object permanence")	*Can:* grant to objects their own properties irrespective of one's perceptions; reason consequentially, that is, according to cause and effect; construct a narrative sequence of events; relate one point in time to another; construct fixed categories and classes into which things can be mentally placed	*Can:* reason abstractly, that is, reason about reasoning; think hypothetically and deductively; form negative classes (for example, the class of all not-crows); see relations as simultaneously reciprocal
Cannot: distinguish own perception of an object from the actual properties of the objects; construct a logical relation between cause and effect.	*Cannot:* reason abstractly; subordinate concrete actuality to possibility; make generalizations; discern overall patterns; form hypotheses; construct ideals	*Cannot:* systematically produce all possible combinations of relations; systematically isolate variables to test hypotheses

or intention. But those who use the prior capacity orient to the result, do not distinguish between points of view and think the second child naughtier. Yet what those using the second capacity cannot do is subordinate the category of point of view to an even broader context that will relate points of view together. Selman (1980), for example, asked children who correctly recited the Golden Rule what it told them to do if someone hit them. "Hit 'em back!" is a common answer among the children. They seem to understand the Golden Rule as "Do unto others like they do unto you." Only by using the third-order capacity, where one can hold multiple points of view simultaneously, can one come to a different understanding of the Golden Rule.

We can thus describe a sequence of qualitative developments in the way one knows in the organizing of the *cognitive* and *social spheres*. And we can do the same in the *personal sphere*, as summarized in Table 3.

The question naturally now arises: Are we describing an evolution of three independent sequences, or are these the expression, in three different domains, of a single development? This is actually a complicated and controversial question just now for the field of mind-oriented developmental psychology, especially when it appears in this form: Should we conceive of individual mental development as consisting in the gradual evolution of a single process of increasing complexity, or as consisting in the gradual evolution of a number of relatively independent

Table 2. Three Capacities in Social-Cognitive Organizing

First Principle Roughly 2 to 6 years	Second Principle Roughly 6 years to teens	Third Principle Teenage years and beyond
Can: recognize that persons exist separate from oneself	*Can:* construct own point of view and grant to others their distinct point of view; take the role of another person; manipulate others on behalf of own goals; make deals, plans, and strategies	*Can:* be aware of shared feelings, agreements, and expectations that take primacy over individuals interests
Cannot: recognize that other persons have their own purposes independent of oneself: take another person's point of view as distinct from one's own	*Cannot:* take own point of view and another's simultaneously; construct obligations and expectations to maintain mutual interpersonal relationships	*Cannot:* construct a generalized system regulative of interpersonal relationships and relationships between relationships

processes of increasing complexity? At some level this becomes a wave / particle kind of question, in which light can be reasonably looked at either way. Both views have merit. At this time, however, the field of developmental psychology is in far greater danger of losing the wisdom in the *common mental enterprise* view than the *multiple mental adventures* view. But if there is a wisdom to be found in the former view, it must address a prior question: How does it make sense even conceptually to speak of a single order of mind or order of consciousness that is organizing a person's experience across such an array of domains? That is, if the sequences of increasing complexity in one's cognitive, interpersonal, and intrapersonal capacities are not to be seen as reflective of three different consciousnesses but one common consciousness, then what is the single underlying form reflective across the various, apparently unrelated, capacities?

Consider the second rung of capacities in each of the domains, now summarized together in Table 4. In all these capacities a common order of consciousness is being expressed, in the apparently disparate capacities to construct a concrete world, see that others have a point of view distinct from one's own, conceive of the self as consisting of durable (rather than moment-to-moment) preferences, abilities, intentions. What is that common order of mind?

In each of the three domains, what is being demonstrated is the ability to construct a mental *set, class, or category*, the ability to order the things of one's experience—whether physical objects, other people, oneself or one's desires—as *property-containing phenomena*. Consider these not intuitively comparable discoveries: (1) the quantity of liquid is not changed by its being poured into a smaller glass;

Table 3. Three Capacities in Intrapersonal-Affective Organizing

First Principle Roughly 2 to 6 years	Second Principle Roughly 6 years to teens	Third Principle Teenage years and beyond
Can: distinguish between inner sensation and outside stimulation	*Can:* drive, regulate, or organize impulses to produce enduring dispositions, needs, goals; dealy immediate gratification; identify enduring qualities of self according to outer social or behavioral manifestations (abilities—"fast runner"; preferences—"hate liver"; habits—"always oversleep")	*Can:.* internalize another's point of view in what becomes the co-construction of personal experience, thus creating new capacity for empathy and sharing at an internal rather than merely transactive level; coordinate more than one point of view internally, thus creating emotions experienced as internal subjective states rather than social transactions
Cannot: distinguish one's impulses from oneself, that is, the self is embedded in or driven by one's impulses	*Cannot:* internaly coordinate more than one point of view or need organization; distinguish one's needs from oneself; identify enduring qualities of the self according to inner psychological manifestations (inner motivations—"feel conflicted"; self attributions—"I have low self-esteem"; biographic sources—"My mother's worrying has influenced the way I parent")	*Cannot:* organize own state or internal parts of self into systematic whole; distinguish self from one's relationship; see the self as the author (rather than merely the theater) of one's inner psychological life

(2) a person who could have no way of knowing that you would be made unhappy by his actions cannot be said to be mean; and (3) when I tell you "I don't like spinach" or think to myself "I'm a Catholic girl," I mean that these are things that are ongoing about me, not just how I feel or think *now*—these are how I *am* or tend to *be*. Now as different as these three discoveries are (they are about one's understanding of the physical, social, and personal world), it is a same, single epistemological principle or way of knowing that makes these discoveries possible.

In each case, the discovery arises out of the same, single ability to see that the phenomenon being considered (things, others, the self) has its own properties,

Table 4. Second-Order Capacities in Three Domains

Logical-cognitive domain	*Can:* grant to objects their own properties irrespective of one's perceptinos; reason consequentially, that is, cause and effect, construct a narrative sequence of events, relate one point in time to another; construct fixed categorie and classes into which things can be mentally placed
	Cannot: rason abstractly; subordinate concrete actuality to possibility; make generalizations; iscern overall patterns; form hypotheses; construct ideals
Social-cognitive domain	*Can:* construct own point of view and grant to others their distinct point of view; take the role of another person; manipulate others on behalf of one's own goals; make decisions
	Cannot: take own piont of view and another's simultaneously; construct obligations and expectations to maintain mutual interpersonal relationships
Intrapersonal-affective domain	*Can:* drive, regulate, or organize impulses to produce enduring dispositions, needs, goals, delay immediate gratification; identify enduring qualities of self according to their outer social or behavioral manifestations (abilities—"Fast runner"; preferences—"hate liver"; habits—"always oversleep"
	Cannot: internally coordinate more than one point of view or need-organization; distinguish its needs from itself; identify endurig qualities of the self according to their inner psychological manifestations (inner motivations—"feel conflicted"; self-attributions—"low self-esteem"; biographic sources—"I have my mother's problems with anger"

which are elements of a class to which these properties belong, and that the phenomenon (things, others, the self) is itself known as this class (which like all classes has durable, ongoing rules that create the idea of class membership and regulate that membership). "Liquid" is a class that has as a member the property of *quantity*, and that property is not regulated by my perception; "other person" is a class that has as a member the property of *intention*, and that property is not determined by my wishes; "self" is a class that has as members the properties of *preference*, *habit*, and *ability*, and—the self being a class, something that has properties—these are things about me in some ongoing way, as opposed to just what I want to eat now, for example. Hence, new ways of knowing in such disparate domains as the inanimate, the social, and the introspective may all be occasioned by a single transformation of mind.

The essence of the transformation of consciousness from second- to third-order capacities is the subordination and integration of the earlier form—durable categories—into a new form that is capable of simultaneously relating one durable category to another. The form of consciousness reflected in all the capacities of the third order of mind is this *trans-categorical* or *cross-categorical* construction. The capacity to think abstractly involves the subordination of the concrete to a higher-order structure that relates multiple concretes simultaneously. The same cross-categorical structure allows for the ability to hold multiple points of view simultaneously and so construct mutual interpersonal relationships.

To illustrate, if the adolescent were knowing the world through the second order, she could certainly understand, for example, her parents' point of view on a nightly curfew. She could see it as distinct from her own, provide her parents with the accurate sense that she understands their point of view, even take on this point of view when it costs her own point of view nothing. She could thereby confuse her parents into thinking she is actually identified with their point of view, in other words, that she not only understands their sense of its importance but shares that sense. She could do all that from the second order. But all that is not their expectation. In order for her to actually hold their point of view in a way in which she identifies with it, she would have to give up an ultimate or absolute relationship to her own point of view. In order to subordinate her own point of view to some bigger way of knowing to which she would be loyal, in order to subordinate it to some integration or co-relation between her own and her parents' point of view, in order for her sense of herself to be more about the preservation and operation of this co-relation than about the preservation and operation of her own independent point of view— for all of that to happen this teenager would have to take her own point of view as object rather than subject. There is no way she can do this unless she is operating out of an order of mind at least as complex as the third. The parents' demand, in other words, is an unrecognized epistemological claim that their daughter will operate out of a way of knowing at least as complex as that described by the third order.

And it is not just these more social relating-type expectations that require the third order. All the expectations do. For this same young woman to think reflectively, inferentially, connotatively, thematically—all these require that the concrete become an object of her knowing rather than a subject of her knowing. "Definition" is minimally a third-order way of knowing because it takes the concrete example as an instance or element of a bigger category of knowing that includes all the concrete examples. Examples must therefore be object, not subject. "Inference" is a minimally third-order way of knowing because it takes the datum as an instance or element. Data must therefore be object, not subject. *Reflective* thinking requires a mental "place" to stand apart from, or outside of, a durably created idea, thought, fact, or description. The idea, thought, fact, or description is made subordinate, figure, or object to a *super*ordinate, ground, or subject that is now capable of bending back (the literal meaning of "reflective") its attention on its own products. Each of these expectations about thinking is really an expectation for yet another expression of what it means to think abstractly. They require abstract relations to phenomena, a capacity that itself requires at least the third order of consciousness.

The expectation that adolescents experience their emotions as inner psychological states is also a demand for the subordinating or integrating of the simpler, categorical self ("I'm mad at my sister; I like BLT sandwiches; I don't like it when my father cooks my eggs too runny") into a more complex context that relates to the categorical self ("I'm much more confident; I used to be just super insecure, very self-conscious"). Thus, the expectations that adolescents be able to identify inner motivations, hold onto emotional conflict internally, be psychologically self-reflective, have a capacity for insight all implicate the third-order capacity to experience the self as a relation to a given set or category rather than the set or category itself.

The construction of values, ideals, and broad beliefs also requires minimally a third order of mind. Knowing that adolescents are operating out of such a consciousness tells us nothing about what their values or ideals will be, or what they will set their heart on. But in order for them to construct any kind of generalizable value or ideal, they must subordinate the factual and actual to the bigger array of the possible or currently contrary-to-fact.

The very idea of *the future as something one lives with as real in the present* rather than the future as *the-present-that-hasn't-happened-yet* requires the same epistemological emancipation from actual / factual / present reality. The most common kind of lack of common sense we find in teenagers is often mistakenly referred to as "poor impulse control," an imprecise characterization paying too great a respect to the "raging hormones" view of adolescence. But the second order of mind is enough to handle impulse control. What is being asked here of adolescents is more complex because it is rarely unmediated impulses actually that lead adolescents into the more foolish risks they are willing to run. Much more often it is an embeddedness in the short-term, immediate present—a present lacking a live relation to the longer term future.

What we want of adolescents, I am suggesting, is not just a new set of behaviors, nor even a new collection of disparate mental abilities. What we want is a single thing, a qualitatively new way of thinking or reasoning. What is the common, single organizational principle at work in every instance of third-order capacity? What is it that makes this collection of capacities a single constellation or "order of mind"?

In this becoming, a new form of consciousness becomes evident. It is a consciousness that includes the still-existing categories of persons (self and others) as containing their properties, wants, or preferences. But the new consciousness sub-ordinates and links these categories to a cross-categorical construction, which brings a whole new phenomenon into being, the person as superordinately relational. "I used to worry that I would mess up," as our newly cross-categorical adolescent told us, "and that others would make me pay for it. Now I worry that if I mess up, others will worry." Both the "I" and the "others" have moved from categorical phenomena (defined by their own properties of intention and need) to cross-categorical phenomena (defined by the link between the categories). Figure 1 illustrates this link.

That there may exist an unrecognized claim on teenagers—by employers, parents, teachers, neighbors, psychotherapists, fellow citizens and even by other teens—for

Figure 1. The Transformatioon from Durable Categories to Cross-Categorical Meaning-Making (and its products in the cognitive, social-cognitive, and intrapersonal-affective domains)

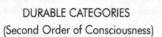

DURABLE CATEGORIES
(Second Order of Consciousness)

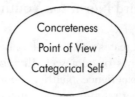

Concreteness

Point of View

Categorical Self

CROSS-CATEGORICAL MEANING MAKING
(Third Order of Consciousness)

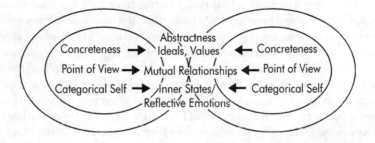

Abstractness
Concreteness → Ideals, Values ← Concreteness

Point of View → Mutual Relationships ← Point of View

Categorical Self → Inner States/ ← Categorical Self
Reflective Emotions

a particular level of consciousness, what we call the third order of mind, is only half of the untold story about adolescence in contemporary culture. The other half is this: *Adolescents do not generally construct the various realms of their experiencing at this order of complexity the moment we think adolescence begins,* whether that is when they become thirteen, reach puberty, or start fighting with their parents. In fact, if a wide range of mind-oriented developmental studies are to be believed, it makes more sense to conceive of the period between ages 12 and 20 as a time when normal mental development consists in the gradual transformation of mind from the second to the third order (Gold, 1980; Hudgins and Prentice, 1993; Jurkovic and Prentice, 1977; Kohlberg and Freundlich, 1973; Villegas, 1988). This means that it would be *normal* for persons during much of their adolescence to be unable to meet the expectations their culture holds out for them, including the expectation to "express one's sexuality respectfully and responsibly, with concern for the welfare of the other and oneself, and with regard to the future implications of present acts."

Does this mean it is *unreasonable* to ask adolescents to act responsibly and foresightfully in their sexual expression? Absolutely not. But it does mean we must realize every adolescent in America will have trouble meeting this expectation for some portion of his or her teen years. It means we cannot expect people to be where they are not. It means we will do less good pointing fingers of blame and more good extending a hand to support adolescents to develop the mental capacities that would enable them to meet these challenging, but reasonable expectations.

Toward a Third Norm for Youthful Sexuality

Every wholesome culture creates safe, provisional environments for its youth in all the most frequented arenas of adolescent living. For most adolescents this includes the arena of sexual experience. Neither of our currently contending norms for youthful sexuality—neither safe sex nor abstinence—successfully creates such an environment. The norm of abstinence creates no provisional environment at all. The norm of safe sex creates a wholesome provisional arena, but it also creates an epistemological barrier to entry that is beyond the mental reach of practically all adolescents during at least some portion of their teen years. Sex education programs on behalf of either (or more often, both) norms have had less than spectacular results. A 1989 review of programs concludes that "the available evidence indicates that traditional sex education programs in junior and senior high school have little or no effect positively or negatively on altering the age of onset or frequency of adolescent sexual activity, on increasing contraceptive use, or on preventing un-planned pregnancy" (Stout and Kirby, 1993, p. 121). A 1994 review similarly con-cludes that no good evidence exists that these programs decrease frequency, pregnancy, births to teen mothers, or STD or HIV rates (Kirby et al., 1994). The best that can be claimed for such programs is that there is evidence they do not hasten the onset of intercourse (Kirby et al., 1994), which epitomizes just how unspectacular the results have been. Is there a norm that would constitute a better fit between society's hopes and concerns and adolescents' desires and mental capacities?

In our culture, for whatever reason, we have come to equate sexuality with the act of genital intercourse. The vast variety of sexually pleasuring acts that do not involve the penis inside the vagina are referred to professionally as "foreplay," as if they are mere appetizers and unacceptable as the main course. The meaning of "going all the way" suggests that the only complete realization of a sexual experi-ence is intercourse. When adolescents today, and many adults as well, use the expression "having sex," they mean that the penis went inside the vagina. The term "sexually active teenager" does not mean that the teens fondle each other's genitals or bring each other to climax with their hands or mouth. Apparently these acts do not constitute being sexually active. Only if the penis enters the vagina is a heterosexual couple said to be sexually active. There is nothing ordained by biology or divinity that this is what sexuality should be. It is wholly a cultural invention. And it is even an odd one for this day and age when sexuality is no longer fused with procreation. The reduction of sexuality to genital intercourse has come home to haunt us as we see it reflected in the sexual behavior of our adolescents.

All norms draw a line. In contrast to a norm that draws a line between behaviors that are or are not associated with sexual pleasure (abstinence) or a norm that draws the line between protected intercourse and unprotected intercourse (safe sex), I suggest a norm that permits as wide a range of sexual pleasure as adolescents feel comfortable sharing, but that draws the line between sexuality with intercourse and sexuality without it.

In contrast to the stance of abstinence ("limit sexual activity to what you can do with your clothes on, hands above the waist, buttons buttoned, zippers zipped") and that of safe sex ("if you are going to have intercourse make sure the young man is wearing a condom"), I would invite us to consider the merits of a stance more like this: Adolescents have become sexual people. It's only natural that this powerful new way of experiencing and expressing themselves is going to be an important addition to the ways they relate to others and to themselves. Their sexuality might naturally become one of the ways they explore a variety of important personal interests, needs, concerns, and issues, including their developing sense of themselves as a man or woman, their wish to feel attractive, accepted or loved by another of their own age, their need to express their fondness, affection, attraction to, or love for another of their own age, and, of course, their desire to experience and share the physical pleasures of sexuality. Their sexuality is an understandable and natural way of expressing or pursuing any and all of these. Some percentage, in every culture at every time in history, will naturally feel drawn sexually to those of their same sex; most will be attracted to those of the opposite sex.

How adolescents will express this new capacity will be shaped to a large extent by interaction between the "curriculum" of their culture and the way they understand that curriculum. As makers of that curriculum, we must be aware that the wish to exercise this new capacity is irresistible and the ways adolescents will understand our curriculum will vary. If our curricular aims are somewhat over the heads of the entering "student" (e.g., that people express their sexuality respectfully and responsibly, with concern for the other's feelings and with regard to the future implications of present acts), then we must build a transitional or bridging context for younger sexuality that is both *meaningful* to those who will not yet understand that curriculum and *facilitative* of a transformation of mind that will come to understand that curriculum. We cannot simply stand on our favored side of the bridge and worry or fume about so many who have not yet passed over. A bridge must be well-anchored on both sides, with as much respect for where it begins as where it ends.

A bridging context might somehow convey a message such as this one concerning the particulars of how sexuality among teens is expressed:

So what, you may ask, do we think is okay and not okay for you to actually do? What does your culture tell you? First, your culture tells you that you should not do anything more than you want to or feel comfortable doing, and obviously that may involve some trial and error for you to figure out. But within the limits of your comfort, and the comfort of your partner, your culture says what you actually do is almost entirely up to you. As you will see, there is a continuum of sexual arousal and satisfaction that begins with the slightest degree of excitation and ends, for young men and women, with the unusually intense satisfaction and release of orgasm. Obviously your own comfort and imagination, and those of your partner, will suggest a variety of ways you can touch (and even talk to) each other that will lead to you and your partner becoming more and more aroused. If the question, How far can we go?, refers to this continuum of arousal, your culture's answer is, You may go just as far as the comfort of you and your partner permits. If "all the way" means experiencing the climactic end of this continuum—you or your partner, or both of you, having an orgasm—then your culture is saying, When you are ready to go all the way you can; this is not prohibited behavior. Nor does your culture seek to impose a limit on the number of partners you may have, whether the relationship is casual or committed, whether you are in only one sexual relationship at a time or have multiple partners. That is all your choice. Of course, your culture prohibits lying to your partners about your true feelings for them or your sexual conduct with others, but that is a particular instance of the more general cultural norm against lying and deceit, which applies to, but is not peculiar to, norms of sexual behavior. The only line your culture wishes to draw regarding your sexual behavior is that until such time as you are ready to enter into a serious relationship, you do not engage in intercourse. Bluntly put, the penis does not enter the vagina or anus. A serious relationship is marked by a different intention toward each other (e.g., to remain together, or to decline other sexual relationships, or to be held by one's community as a couple, or to create a family) and by a different order of responsibility for each other (including responsibility for one's sexual relationship by the practice of protected intercourse). Your culture makes a distinction between sexual relationships marked by this seriousness of commitment and those that are not. It values both. Each may have its proper time. The culture does not reserve a greater degree of sexual pleasure for one kind of relationship over the other. Both kinds may engage in fully pleasurable sexual expression. It simply reserves one form of sexual expression, intercourse, as an exclusive marker of the more serious relationship.

Now before we consider the psychological plausibility of such a norm's actually being adopted by adolescents at the dawn of the 21st century, let us quickly make note of the array of concerns that would be significantly impacted if just this single norm were made viable: (1) *The incidence of sexually transmitted diseases*—especially the most frightening of these, contracting of the virus that currently leads inexorably to AIDS—while not eliminated, would be dramatically reduced, and without the intermediation of prophylactic devices. Is oral sex a risk free sexual practice? According to experts and the best evidence on HIV transmission it is not absolutely risk free. It is certainly a higher risk than protected intercourse. But is this really the relevant comparison? Given the unlikelihood of universal use of condoms by 12 to 18 year olds, the realistic comparison is between the safety of oral sex and *un*protected intercourse. Oral sex is a greatly less risky sexual practice than unprotected intercourse. Faced with our concern about a deadly virus, if we could choose between adolescents practicing only safe sex or only oral sex, of course, we would choose the former. But if the former is not realistic, the latter is enormously preferable to the current circumstances of widespread unprotected intercourse.

(2) *The incidence of unplanned pregnancy* would be greatly reduced, and without the need for contraceptive devices, because the single prohibition not only impacts those who would have intercourse without regard to pregnancy but those countless teens who now employ ineffective prevention methods such as "withdrawal" or having intercourse during "safe" times of the month.

(3) *Shifting the focus, purpose, conclusion, or even the very meaning of shared sexual intimacy* away from genital penetration *toward the feelings* and, if desired, sexual satisfaction that each partner is giving and receiving, may eventually promote the mutuality, reciprocity, sensitivity, thoughtfulness, and even technical skillfulness of adolescent sexual activity. There is a much greater likelihood of youthful sexuality being an arena in which one comes to understand the other's sexuality; of each partner's keeping the other, as well as oneself, in mind; of making the other's experience, as well as one's own, central to what shared sexual activity is about.

(4) The new norm *redraws a distinction, which many value and are concerned has been lost, between premarital and marital intimacy*, or, at least, between casual and committed relationship; and, in its own fashion, it reasserts the sanctity of the family, which many feel has been profaned. It fosters the idea that one conducts oneself differently in committed relationships such as those that bear or care for children.

Thus by drawing this single new line, the culture addresses its own concerns about disease, unwanted pregnancy, emotional insensitivity, and the decay of the institutions of marriage and family. It addresses issues dear to a variety of positions on the political spectrum. It speaks to the concerns of advocates of both the current contesting norms, incorporating features of each (although I am sure it will please neither since it is neither as safe nor as abstinent as either would want).

But is it plausible that a new norm of fully pleasurable sex without intercourse could actually be internalized by teens of the future? (I am thinking of tomorrow's teens, of people who are today young children.) My answer is that the barrier to its being internalized is greater in the minds of adults who themselves equate sexuality with intercourse than in the minds and bodies of teenagers whose motives and interests in sexuality carry no necessary demand for the particular act of genital penetration. Nothing that calls adolescents to sexual expression requires genital penetration for its satisfaction. Not the motive to feel loved or lovable, to be taken care of or care for, to give or receive pleasure, to win or confer a valuable prize, to make another person jealous, to hold a boyfriend or girlfriend, to establish one's identity as a sexual person, to satisfy curiosity about another's body, to fill the gaps in the conversation, to become more separate from one's parents, to experience physical tenderness, to have something to do on Saturday night, to have an orgasm, to have an orgasm with another person, to create and sustain an intimate relationship—not one of these *requires* the act of intercourse for its satisfaction. (In fact, the only motive I can think of that does is the motive to have a baby, which does exist among some adolescents, usually disadvantaged young girls who consider having a child not as a further burden but as the increase of a preferable social role and an object to love. But this is more an instance of a claim to readiness for a "serious relationship," challengeable or not, rather than needing intercourse to be a part of one's noncommitted sexuality.)

Fully pleasurable sex without intercourse, implausible though it may sound in our current cultural climate, is actually quite responsive both to the irresistibility of sexuality and to the inevitable variation in the consciousness capacities of adolescents. More than this, it represents a way to foster the arena of sexual behavior as yet another supportive context for the gradual evolution from the second to the third order of mind. How so?

Consider that efforts to persuade teens—for example, to abstain from drug use, by just saying no or to stay in school by appealing to the importance down the road of a high school diploma—are not likely to be compelling to adolescents organizing reality at the second order of mind. Adolescents of categorical consciousness do not say yes to drugs primarily as a result of peer pressure but out of their own interest in the experience they derive from getting high. If I construct the world out of the short-term, quid-pro-quo logic of categorical consciousness, you are proposing that I give up something I value in return for exactly nothing. Why would I want to do that? There is no short-term reward for giving up getting high. You want me to give up something I like and you leave me with nothing but the internally or externally depressing condition I was seeking to escape in the first place. Stay in school? Why? If I don't like school, am humiliated, or bored there, what is the short-term payoff for staying in school? There is none. Appeals to the long-term consequences of drug abuse or of dropping out of school are not compelling when the future is constructed as the present that hasn't happened yet.

Now consider that the norm of fully pleasurable sex without intercourse does not require cross-categorical knowing to be immediately meaningful and is not first of all about privation. In sanctioning a provisional or transitional arena for younger sexuality, the culture would *not* be setting as the price of admission the ability immediately to understand and identify with a set of prosocial values. Nor would values, such as "Act respectfully and responsibly, with concern for the other's feelings and with regard to the future implications of present acts," even be what teens see first when they enter. What they would see first is the alluring opportunity for self-enhancement and the pursuit of their own ends. That's what they would perceive their culture seemed to be saying sex was all about at the moment, and that precisely is what it would seem to be about to them. The fact that there is a rule involved would not be fatal to the match of meaningfulness between the student and the activity. It is a small enough price to pay, and it doesn't reduce the number of goodies available. As I have said, not a single motive that draws an adolescent into sexual experience requires the act of intercourse. That it sounds strange to say the norm does not reduce the number of available goodies might only be a sign of how embedded we are in the equation of sexuality with intercourse.

"But the very act of excluding something," you may object, "makes it that much more alluring. Adolescents are going to want to have intercourse all the more for your trying to reserve it to some later time in one's life." Actually, it is untrue that reserving something until later in life automatically makes it more appealing.

Adolescents do not want the long-term, day-in-and-day-out responsibilities of child care, for example (but they do want to earn money babysitting); they do not want to carry a monthly mortgage (but they do want a room of their own); they do not want to have to sustain themselves economically (but they would like an allowance or a part-time job). There are many activities already reserved for later in life that are not made one bit more alluring by being so. What adolescents of categorical consciousness want is not actually a function of whether adults reserve it for later; it's a function of whether it meets immediate needs. The culture reserves the commitment to full-time employment until later in life, but in doing so, it doesn't make this activity more alluring. Teens may want the culture to provide more part-time jobs so they can generate more spendable cash, but they have no heightened need for full-time employment. Aside from the money, they may have a need to try on the experience of work, but they have no real need to take on the full trappings of employment. A teen may want to have sexual experience, may feel the *need* for sexual experience, but, odd as it sounds, they may have no real need for genital penetration unless the culture builds a dazzling shrine to it. If we continue to believe that the climax of sex is genital penetration, rather than that the climax of sex is climax, then it is true that in trying to get adolescents to forego intercourse until they are more responsible we will simply make intercourse even more alluring. But the reason will not be that the act of reserving something until later in life automatically makes it more desirable. The reason will be that we are being dishonest and disingenuous and adolescents can tell.

The practice of unprotected sexual intercourse among teens does not create a safe, wholesome arena for provisionally experiencing and expressing one's sexuality. As I have said, the norm of safer sex creates a wholesome provisional arena, but it also creates a barrier to entry that is beyond the mental reach of practically all adolescents during at least some portion of their teen years. The norm of abstinence creates no provisional environment at all. Only the norm of fully pleasurable sex without intercourse creates a safer environment with an entry fee whose cost the adults in the culture can reduce to affordability. Given our own initiation to sexuality in an intercourse-centric culture, it takes an act of imagination to consider what happens to sexuality when this center is shifted. If we placed orgasm rather than intercourse at the end of the sexual continuum, adolescents would correctly see that the arena of provisional sexuality the third norm creates is one that offers them a full range of expression. We would not be saying (or meaning), "You can go almost 'all the way,' but not quite."

The visual and spoken explicitness that now characterizes our popular media, especially film, could be turned to good advantage on behalf of this new norm. What would be the effect on adolescent viewers of watching appealing young characters in compelling stories living out intensely passionate and exciting sexual relationships that were mutually pleasuring, emotionally and physically fully satisfying, but not focused on genital penetration?

I do not pretend that adolescents who are closer to the second order of consciousness than the third will suddenly be less exploitative or more responsible because they practice their sexuality within this provisional context. I fully expect them to be as captive of their short-term interests as ever. That's one reason a provisional environment *is* provisional; people can make poor choices and be mistaken at less cost. Teens of both sexes will continue to make use of each other for their own ends. Some will still look to "score." There is no way to abstain from categorical consciousness. But there is a way to make its expression less dangerous. Exploitation of one sort or another will continue. But it could continue with less actual intercourse, less HIV transmission, less unplanned pregnancy.

And it could continue in a context that is actually more conducive to overturning categorical consciousness. A well-schooled culture is a tricky culture. It creates environments that are not only intensely meaningful to the current way its members construct their experience, but it increases the likelihood that interacting with this environment will disturb this very way of constructing reality and promote its transformation. When the goal or aim or end of sex is shifted from the purely bodily cooperation of fitting one's genitals together to a goal aimed at pleasure, sensation, and satisfaction, the form of cooperation necessarily requires fitting our minds together as well as our bodies. I am called to think about you, what feels good to you, what you need, at the same time I am thinking about me, what feels good to me, what I need. This is just the sort of activity that comes to relativize categorical consciousness and promote the cross-categorical consciousness that makes more meaningful the values of mutual respect and responsibility.

At this time, the adolescent is ready to pass out of the provisional environment into a new realm of sexual practice, a realm that not only sanctions intercourse but warrants that those who have intercourse—be they more mature teenagers or adults—are responsible enough to do so if they wish this to be a part of their sexual expression, in a way that abjures the risks of disease, unplanned pregnancy, and exploitation. The idea is not that genital penetration is somehow itself the acme of mature sexual expression or that this is what people should do when they are more mature (e.g., women whose partners are women would be the first to attest to this), but only that if intercourse is part of one's sexual expression, it should occur in relationships where partners have the capacity to love each other responsibly.

The promoting of a new cultural norm is very hard work, to be sure. How can we get people to change their behavior? But it is not impossible. Were I to have told you fifty years ago that Americans as a group would radically reduce their smoking or significantly alter their diets to reduce fat intake, you might have been as incredulous about those suggestions then as you are about this one now. Not a single food company in America sat down and said, "Gee, our customers don't know it, but our products aren't really that healthy for them. Even though these consumers are making us rich, let's see if we can sour them on our existing products. Let's promote an altogether new health consciousness in them and then design a new set of more wholesome products for them to buy." The great corporate engines didn't throw themselves into reverse. They were thrown by the one market force more powerful than the conglomerate producer, namely the shifting cultural value of the consumers themselves.

Why did our smoking habits and food choices change? It takes a compelling, simple, and clear signal to effect a change like this one, preferably a signal of alarm. More people stopped smoking and eating so much fat because they came to believe if they didn't it would kill them. The threat of death is a compelling signal. And it is the threat of death, above all, that fuels our culturewide concern about adolescent sexuality.

Who can be moved by this threat of death? Not the adolescent of categorical consciousness, for whom the future is not a part of the present. Why spend time trying to create fear where there is none? I would spend it trying to mobilize fear where it is already present or latent—in the minds and hearts of parents capable of assuming the responsibilities that are ours as keepers of the school. It is we adults, not the adolescents, who are failing to create wholesome provisional experimental spaces that adolescence needs to make mistakes and safely learn. I admire those relatively few adults who by actively promoting the norms of safe sex or abstinence are without doubt seeking to assume some responsibility for this important aspect of our culture-as-school.

The promoting of a new cultural norm is very hard work: daunting, slow, incremental, fraught with failure amidst only gradual success. But if we are to take up the work of promoting the internalization of a new norm, then let us elect one, unlike abstinence or safe sex, that reflects the *realities* as well as our *concerns*. Let us elect one that reflects the realities of the irresistibility of sex and the consciousness constraints of adolescence. And let us elect one that makes of this important and highly frequented arena of adolescent activity a hospitable environment for the mental growth our culture requires of people of this age. When it comes to sexuality and adolescents it may be that the way we will get them to change their behavior tomorrow will depend on our changing the way we think today.

REFERENCES

Burt, M. R. (1986). *Estimates of public costs for teenage childbearing.* Washington, DC: Center for Population Options.

Center for Disease Control. (1991, June). *HIV/AIDS Surveillance:* Center for Disease Control.

Center for Disease Control. (1992, Jan.). *HIV/AIDS Surveillance, Year-End Edition:* Center for Disease Control.

Center for Disease Control. (1992, Oct.). *HIV/AIDS Prevention Newsletter:* Center for Disease Control.

Center for Disease Control. (1993, Feb.). *HIV/AIDS Surveillance, Year-End Edition:* Center for Disease Control.

Center for Population Options. (1989). *Teenage pregnancy and too-early childbearing: Public costs, personal consequences. Washington, DC: Center for Population Options.*

Children's Defense Fund. (1988). Adolescent and young adult fathers: problems and solutions. Washington, DC: Chidren's Defense Fund.

Conference on the Health of America's Youth. (1991, March). *Current trends in health status and utilization of health services*. University of California at San Francisco.

Forrest, J. P. and Shusheela Singh (1990). The sexual and reproductive behavior of American women, 1982–1988. *Family Planning Perspectives, 22*(5), Sept./Oct.

Gayle, H. D., Keeling, R. P., Garcia-Tunon, M., Kilbourne, B. W., Narkunas, J. P.; Ingram, F. R., Rogers, M. R. and Curra, J. W. (1990). Prevalence of the Human Immunodeficiency Virus among university students. *New England Journal of Medicine, 323*, 1538.

Gold, S. (1980). Relations between level and ego development and adjustment patterns in adolescence. *Journal of Personality Assessment, 44*(6), 630–638.

The Alan Guttmacher Institute. (1981). Teenage pregnancy: The problem that hasn't gone away.

The Alan Guttmacher Institute. (1992). Facts in brief: Teenage sexual and reproductive behavior in the U.S.

Hayes, C. D. (1987). *Risking the future: Adolescent sexuality, pregnancy, and childbearing*. Washington, DC: National Academy Press.

Henderson, A. F., & Kegan, R. (1989). Learning, knowing, and the self. In K. Field, B. Cohler, and G. Wool (Eds.), *Motive and meaning: Psychoanalytic perspectives on learning and education*. New York: International Universities Press.

Henshaw, S. K., & Van Vort, J. (1989). Teenage abortion, birth and pregnancy statistics. *Family Planning Perspectives, 21*, 85.

Hudgins, W., & Prentice, N. (1973). Moral judgment in delinquent and nondelinquent adolescents and their mothers. *Journal of Abnormal Psychology, 82*, 145–152.

Jones, E. F. and Lincoln, R., eds. (1986). *Teenage pregnancy in industrialized counries*. New Haven: Yale University Press.

Jurkovic, G., & Prentice, N. (1977). Relation of moral and cognitive development to dimensions of juvenile delinquency. *Journal of Abnormal Psychology, 86*, 414–420.

Kegan, R. (1976). Ego and truth. Doctoral dissertation, Harvard University.

Kegan, R. (1982). *The evolving self*. Cambridge, MA: Harvard University Press.

Kegan, R. (1985). The loss of Pete's dragon: Transformation in the development of the self during the years five to seven. In R. Leahy (Ed.), *The development of the self*. New York: Academic Press.

Kegan, R. (1986). The child behind the mask: sociopathy as developmental delay. In W. H. Reid, J. W. Bonner III, D. Dorr, J. I. Walker (Eds.), *Unmasking the psychopath*. New York: Norton.

Kegan, R. (1994). *In over our heads*. Cambridge, MA: Harvard University Press.

Kegan, R., Broderick, M., & Popp, N. (1992). A developmental framework for assessing youth in prorammatic interventions. Unpublished report for U.S. Dep. of Labor and Public Private Ventures, Philadelphia.

Kegan, R., & Lahey, L. (1983). Adult leadership and adult development. In B. Kellerman (Ed.), *Leadership: multidisciplinary perspectives*. New York: Prentice-Hall.

Kegan, R., Noam, G., & Rogers, L. (1982). The psychologic of emotions. In D. Cicchetti and P. Pogge-Hesse (Eds.), *Emotional development*. San Francisco: Jossey Bass.

Kirby, D., Short, L., Collins, J. Rugg, D., Kolbe, L., Howard, M., Miller, B. Sonenstein, F. and Zabin, L., (1994). School-based programs to reduce sexual risk-taking behaviors: Sexuality and HIV/AIDS education, health clinics and condom availability programs. Santa Cruz, CA: Education, Training, and Research Associates, March 21, 1994.

Kohlberg, L. (1984). *The psychology of moral development*. New York: Harper & Row.

Kohlberg, L., and Freundlich, D. (1973). Moral reasoning and delinquency. Unpublished paper, Harvard University, Lab of Human Development.

Lahey, L., Souvaine, E., Kegan, R., Goodman, R., & Felix, S. (1988). A guide to the subject-object interview: its administration and iterpretation. Unpublished manual, Harvard University, Lab of Human Development.

Marsiglio, W. (1989). Adolescent fathers in the U.S.: Their initial living arrangement, marital experience and educational outcomes. *Family Planning Perspectives, July/Aug.*

Mosher, W. D. (1990). Contraceptive practice in the United States, 1982–1988. *Family Planning Perspectives, 22*(5), Sept./Oct.

NCHS. (1989, June). *Advance report of final natality, 1987.*

National Commission to Prevent Infant Mortality. (1988). *Death before life: The tragedy of infant mortality.* Washington, DC: National Commission to Prevent Infant Mortality.

Noam, G., & Kegan, R. (1982). Social cognition and psychodynamics: Toward a clinical-developmental psychology. In W. Edelstein and M. Keller (Eds.), *Perspektivat und interpretation.* Frankfurt: Suhrkamp Verlag.

Nord, C. W., Moore, A., Morrison, R., Brown, B. Myers, D. E. (1992). Consequences of tenage parenting. *Journal of School Health, 62,* 7.

Piaget, J. (1948). *The moral judgment of the child.* Glencoe: Free Press.

Piaget, J. (1952). *The origins of intelligence in children.* New York: International Universities Press.

Rogers, L., & Kegan, R. (1990). Mental growth and mental health as distinct concepts in the study of developmental psychopathology. In H. Rosen and D. Keating (Eds.), *Constructivist approaches to psychopathology.* Hillsdale, NJ: Erlbaum.

Selman, R. (1980). *The growth of interpersonal understanding.* New York: Academic Press.

Souvaine, E., Lahey, L., & Kegan, R. (1990). Life after formal operations. In C. N. Alexander and E. J. Langer (Eds.), *Higher stages of human development.* New York: Oxford University Press.

St. Louis, M. E., Conway, G. A., Hayman, C. R., Miller, C., Petersen, L. R. and Dondero, T. J. (1991). Human Immunodeficiency Virus infection in disadvantaged adolescents: findings from the U.S. Job Corps. *Journal of the American Medical Association, 266,* 2378–2391.

Stout, J. and Kirby, D. (1993). The effects of sexuality education on adolescent sexual activity. *Pediatric Annals, 22*(2): 120–126.

U.S. Department of Health and Human Services (DHHS). (1990). *Healthy people 2000: National health promotion and disease prevention objectives.* Washington, DC: U.S. Department of Health and Human Services.

Walker, G., & Vilella-Velez, F. (1992). Anatomy of a demonstration. Philadelphia: Public-Private Ventures.

Webster, L. A., & Berman, S. M. (1993). Surveillance for gonorrhea and primary and secondary syphilis among adolescents, U.S., 1981–1991. *Morbidity and Mortality Weekly Report, 42,*. 55–3, 1–11.

Zabin, L. S., Kantner, J. S. and Velnik, M. (1979). The risk of adolescent pregnancy in the first months of intercourse. *Family Planning Perspectives, July/Aug.*

V Language in High-Functioning Adolescents with Autism: Questions About Deviance and Delay

Catherine Lord

Autism is a developmental disorder that the general public usually associates with childhood. Until the production of *Rainman*, many people had never heard of an adult with autism; it is still common for students and professionals to ask what happens to autistic children when they grow up? The answer seems on the surface rather simple; in almost all cases, children with autism grow up to be adults with autism. Unfortunately, the amount of research about autism beyond childhood is slim. Even for the period of adolescence, which is the topic of this chapter, there is relatively little empirical information. The purpose of this chapter is to put together several lines of research in order to identify factors in language development that are important for high-functioning adolescents with autism. The goal is to use this information to exemplify developmental questions having to do with differences between deviance and delay and with issues related to the usefulness of developmental models in understanding psychopathology.

Autism is a disorder that is typically recognized in early childhood. Symptoms generally develop in the first years of life. It is characterized by deficits in three areas: reciprocal social interaction, communication, and restricted and repetitive behaviors (American Psychiatric Association, 1994; World Health Organization, 1992). It is most commonly associated with mental retardation, but there is a substantial minority of individuals with verbal and nonverbal skills in the average range of intelligence (Bryson, Clark, & Smith, 1988; Lotter, 1966). The body of evidence suggests that autism arises from central nervous system dysfunction, probably occurring prenatally (Minshew, 1991). There is also a strong genetic component (Bolton & Rutter, 1990), though the exact pattern(s) of transmission is unknown.

Autism is a disorder that is interesting for its own sake because of the contrast it represents between relatively intact abilities to operate in a visual-spatial world, coupled with extraordinary deficits in social-communicative behaviors that are taken for granted by most people. It also results in tremendous costs to society and to families because of its lifelong course and because of the behavioral difficulties that limit the independence and productivity of many persons with autism. Even most high-functioning people with autism, people of normal intelligence, remain very limited socially and continue to need support and assistance as adults in order to find and maintain employment and to live independently (Rumsey, Rapoport, & Sceery, 1985; Venter, Lord, & Schopler, 1992).

Autism as a Developmental Disorder

Autism is a particularly intriguing example of a developmental disorder because the manifestation of the syndrome both affects development and is affected by development in fairly obvious ways. For example, parents of autistic children spend more time worrying about them than parents of children with other disabilities, but less time interacting with them (Bebko, Konstantareas, & Springer, 1987). Autistic children are less likely to go places other than school (Bartak, 1978) and spend less time interacting with other children, whether in categorical (Attwood, Frith, & Hermelin, 1988) or integrated settings (Lord & Magill, 1989). Deficits in basic social behaviors such as imitation, the timing of visual attention, and use of facial expression to communicate (Dawson & Adams, 1984; Klin, Volkmar, & Sparrow, 1992; Rogers & Pennington, 1991; Sigman, Mundy, Sherman, & Ungerer, 1986) all seem likely to affect the kind and amount of information taken in by autistic children from very early on in development.

From the opposite perspective, development also affects how autism is manifested. Many symptoms cited as classic examples of autism, such as odd intonation or excessive need for sameness, are common in mildly to moderately retarded autistic school-age children, but rare in younger or more developmentally delayed children and in higher functioning or older adolescents or adults (Volkmar, 1994). Particular tasks, such as those measuring first-order theory of mind, that effectively discriminate autistic children with mental ages of four and five years from nonautistic mentally handicapped children, may be much less effective at older ages or with more advanced children or adolescents (Ozonoff, Pennington, & Rogers, 1991; Prior, Dahlstrom, & Squires, 1990). As they become older, autistic children may discover compensatory strategies for accomplishing tasks in which nonautistic children are competent almost automatically (Hobson, 1993). Thus, behaviors that best characterize moderately retarded eight-year-old children with autism may be quite different than those that describe the unique features of severely retarded autistic three-year-old children or high-functioning adolescents with autism (Lord et al., in press). Fitting into this puzzle is the fact that comparison groups for autistic individuals, such as nonautistic, mentally handicapped adolescents, or developmentally normal toddlers, are also developing, often at a faster pace than the autistic children under study. Thus, increasingly obvious deficits in the behavior of autistic children, such as the failure to develop flexible, imaginative play or receptive understanding of complex syntax, are often defined by the acquisition of these skills by nonautistic children rather than by the loss or increased deviance of these behaviors in the children with autism.

For young children with autism, the effects of the language deficit in particular can be overwhelming. Preschool studies suggest that many autistic children have virtually no comprehension even of single words outside of familiar contexts up until age three (DiLavore, Lord, & Opsahl, 1994). Other studies of comprehension suggest that few mildly to moderately retarded autistic children of elementary school age understand simple instructions as well as normally developing two year olds (Lord, 1985). Comprehension, particularly of connected discourse, is one structural aspect of language measurable outside of social contexts that continues into adolescence to differentiate high-functioning autistic children from children with severe receptive-expressive language disorders (Cantwell, Baker, & Rutter, 1978). Language measures, ranging from scores on formal tests of receptive vocabulary or comprehension of stories to verbal IQ to parent reports of communicative competence, are all good predictors of outcome in adolescence and adulthood, specifically of adaptive competence and academic achievement in high-functioning individuals with autism (Venter et al., 1992).

On the other hand, the order of acquisition of skills within narrowly defined linguistic systems, such as syntax or phonology, seems similar in children and adolescents with autism and other youngsters (Tager-Flusberg, 1981). Yet when we look more closely at the relationships among these systems and at the natural history of language development in autism, the severity and pattern are unlike any other disorder. The focus of this chapter is the study of the language of verbally fluent, high-functioning adolescents with autism with the goal of understanding the relationships among linguistic, social, and cognitive aspects of development. We can ask when do asynchronies in these lines of development begin to represent deviance as opposed to separate developmental delays? And what do these findings mean in terms of our conceptualizations of developmental psychopathology?

A focus on high-functioning adolescents allows us the opportunity to observe a range of fairly well-developed skills and their relationships to each other. For more able youngsters with autism, adolescence is a time when verbal fluency and language comprehension are sufficiently established such that one can study the occurrence of low-frequency phenomena without reifying very low numbers. In addition, adolescence is a time when simple behavioral and social conventions have presumably already been established and there are increasing demands and expectations for independence. Adolescents are also faced with acquiring social mores increasingly defined by peers rather than adults. All of these factors make adolescence an interesting developmental period in which to study language in high-functioning youngsters with autism.

Language Development in Adolescents with Autism

The absence of speech by two years or other indicators of delayed language is the primary reason that parents of young autistic children seek help (Siegel, Pliner, Eschler, & Elliott, 1988). Autism is just one of several developmental disorders associated with language delay. Yet, when we look closely, the severity and pattern of language deficits even in very young children with autism are unlike those seen in any other developmental disorder. In one study, two year olds judged likely to be autistic consistently scored below the 6-month level in estimated age equivalence for receptive language (DiLavore et al., 1994), despite mental ages of more than 15 months on standard developmental scales such as the Bayley Scales of Mental Development (Bayley, 1969). What does receptive language below the 6-month level mean? It means little or no response to the child's name, limited response to gestures such as pointing, some response to familiar sounds (such as television or candy wrappers), but little or no understanding of object or person names out of context (such as "Where's Mommy?" or "Show me your shoe").

Eventually, by school age or adolescence, at least half of autistic children are estimated to acquire truly functional language (Lotter, 1966). In a follow-up study of high-functioning 10-year-old boys with autism, compared to boys with severe receptive-expressive language disorders of the same age who had had equivalent nonverbal skills at age seven, autistic children had higher nonverbal IQs and equaled the language-impaired children in many measures of expressive language, but still remained behind in receptive language and pragmatic skills (Cantwell et al., 1978). Other studies have shown that many autistic children with verbal mental ages above 6 years can pass first-order theory of mind tasks (Prior et al., 1990) and begin to show some success on measures of executive functioning (Ozonoff et al., 1991). As full-scale IQ approaches 100, an increasing number of autistic adolescents show higher VIQs than PIQs (Rumsey et al., 1985) and many become much more interested in peers (Lord & Venter, 1992).

In a study of 58 high-functioning autistic adolescents and adults seen first at age 6 to 8 years and followed up between 12 and 25, early nonverbal IQ predicted about 30 percent of the variance in adaptive scores and 41 percent of the variance in academic achievement scores in adolescence / young adulthood (Venter et al., 1992). When early receptive vocabulary scores were available, they predicted 65 percent of the variance in later adaptive functioning and 54 percent of the variance in later achievement. Current vocabulary scores predicted 44 percent of the variance in adaptive functioning; current verbal IQ predicted 78 percent of the variance in academic achievement. Academic achievement was important because it was the best predictor of employment in the autistic individuals 20 years old or over, even though few of those employed used academic skills in their jobs. On the whole, language measures, including receptive vocabulary, verbal IQ and measures of more complex language skill were virtually interchangeable in their power to predict functioning in other areas.

Of the 23 adolescents over age 19 who had been out of high school two or more years, 8 were or had been employed (only one found a job without special help), 4 lived independently, 2 had acquaintance relationships with a peer, and none had ever had a truly reciprocal friendship. In the last five years, 2 had had "psychotic" episodes that resulted in hospitalization, 5 had been hospitalized because of aggressive behavior, and 5 others had been treated for affective disorder. In the four years since these data were collected, life situations have changed, with several more adults finding employment and beginning to live independently and several more having periods of very difficult behavior.

Methods of Present Language Studies

Once a high-functioning adolescent with autism has reasonable control of communicative language, what happens then? The following studies are all based on transcriptions of the speech of adolescents and young adults taken from various parts of the Autism Diagnostic Observation Schedule (ADOS), a standardized protocol of communicative and social contexts intended to elicit a range of language and social behaviors from verbal individuals suspected of having autism (Lord et al., 1989). The ADOS takes about 30 minutes to administer and is usually videotaped so that detailed codings and transcriptions can be made later. Codings are also made as the interview progresses. The examiner follows a set protocol that may vary in order, but contains activities designed to create a "press" (Murray, 1938) for turn taking, asking for help, imaginative use of representational materials, helping, and conversation. The examiner follows a set hierarchy of ways to structure the situation, initially placing the onus for initiating conversation on the subject, but providing increasing help and prompts if the adolescent does not participate otherwise. In addition, there are several specific series of questions about socioemotional topics, such as emotions and friendship, which also follow a set hierarchy of prompts, from open-ended questions to examples and specific yes / no questions.

These studies were carried out on approximately 80 subjects (i.e., depending on the requirements of the task and the particular matching variable, from 2 to 10 different subjects were excluded in two of the studies), who comprised four groups of equal chronological age (Ms ranged from 12.03 to 13.24 years) and distribution of gender (12 male, 8 female). The four groups were equated in pairs for verbal mental age (computed from age equivalents from Wechsler tests): a high-functioning autistic group with a mean verbal mental age of 12.4 years (M VIQ = 95.1; range 80–116) and a normal group with a mean chronological age of 12.03, for whom IQ was not tested; and a lower functioning autistic group with a mean verbal mental age of 7.8 years (M VIQ = 60.5, range 40–76) equated to a nonautistic mentally handicapped group with a mean verbal mental age of 7.8 years (M VIQ = 59.7, range 40–71). It should be noted that the lower functioning autistic group would actually be considered "high functioning" by some standards because its mean nonverbal IQ was 80, significantly higher than that of the nonautistic mentally handicapped adolescents, whose mean nonverbal IQ was 66.4. The high-functioning autistic group was exceptional (e.g., as compared to the sample in the follow-up study above) in including only individuals who had verbal IQs above 80. Subjects were part of a larger study on sex differences in autism, which is why there was a relatively high number of females. All individuals with autism met formal diagnostic criteria for autism on the basis of medical records at age five (using *DSM III* criteria; American Psychiatric Association, 1980) and currently (using data from current and retrospective items) on the Autism Diagnostic Interview (Le Couteur et al., 1989).

For these studies, the ADOS was videotaped and transcribed by trained ob-servers who were blind to the purposes of the studies and to the diagnoses of the speakers. Systematic Analysis of Language Transcripts (Miller & Chapman, 1983) was used to mark terms and structures of interest and to compute various linguistic measures.

Language as a Specific Developmental Delay in Autism

The first study looked at the use of "mental" verbs (Tager-Flusberg, 1992) in these adolescents, omitting any subject whose mean length of utterance (MLU: Brown, 1973) was under 3.0 morphemes or who produced fewer than 75 spontane-ous utterances during the socioemotional questions and open-ended conversation portions of the ADOS. Pilot testing had indicated that below the levels specified by these cutoffs, the behavior of interest was too rare in our relatively brief samples to be measured reliably. Mental verbs were defined as terms that referred to the mental state of the speaker, listener, or a third party, coded as to the function that they served. Mental verbs referring to true mental states (such as thinking or wondering) were contrasted to mental verbs used to modulate assertions ("I think it's time to go") or to direct behavior ("I wonder if you would mind closing the door") and idioms, such as "I don't know."

This study grew out of an increasingly large literature suggesting that individuals with autism have a specific deficit in the understanding of "mind" or of the idea that thoughts are separate from behavior (Baron-Cohen, Tager-Flusberg, & Cohen, 1993). An earlier study of young children with autism, for whom extensive tran-scripts of natural interaction with their mothers were available, had suggested that, even in preschool and early school-age years, nonautistic mentally handicapped youngsters exceed autistic children of equal skill at generating sentences, in their use of mental verbs to describe mental states (Tager-Flusberg, 1992). Such differ-ences were not found for terms used to describe emotional states or mental verbs used for other social functions. As more data have emerged about "theory of mind" in both normally developing and autistic populations, increasingly divergent opin-ions have appeared as to whether theory of mind is a modular function that emerges as a unit, or one aspect of a more general developmental phenomenon.

Because even the nonautistic children studied previously were producing very low rates of mental verbs, our ADOS transcripts of generally fluent, higher func-tioning adolescents provided us with an opportunity to test the specificity of the earlier hypothesis with another relevant population (Storoschuk, Lord, & Jaedicke, submitted). In addition, differences between autistic adolescents and all other adolescents were expected in the use of mental verbs for both mental state and social functions, with the most extreme differences predicted for mental state terms because of the very specific deficit in metarepresentation that has now been well-documented in numerous other studies. Use of mental verbs to describe mental states was expected to be extremely rare or even absent in most autistic adolescents, regardless of language skill. Differences across diagnostic groups in social functions that had not been apparent for the younger children were expected for the adolescents because of increased verbal and social sophistication. These social skills were expected to develop at a much slower rate in the autistic adolescents. Thus, use of mental verbs for social functions was expected to be correlated to language skill within diagnoses, but also to differ significantly by diagnosis, such that the adolescents with autism were always the least competent. Results were surprisingly straightforward. Autistic adolescents, across developmental level, consistently used fewer mental *state* terms than MLU- matched mentally handicapped or normally developing subjects, but did not differ in social uses of the same terms. However, verbal mental age and linguistic competence (e.g., MLU) were consistently related to the use of mental state terms *within diagnostic groups* and to the social use of mental verbs (i.e., directing behavior, modulating an assertion) *across diagnostic groups.*

As shown in Figure 1, regression lines for MLU (or verbal mental age) and use of mental state verbs were nearly identical in slope for the high-functioning autistic and two groups of nonautistic subjects, but differed in intersection. That is, though the trajectory of increased use for the high-functioning autistic adolescents was very similar to that of the other groups, the autistic subjects always used fewer mental state terms at any particular point in development (as measured by verbal mental age or MLU). There were no significant correlations with use of complex embedded sentences, scores on the Vineland Adaptive Behavior Scale (Sparrow, Balla, & Cicchetti, 1984), nor several specific Wechsler subtests (WAIS-R: Wechsler, 1981; WISC-R, Wechsler, 1991) that might allow links between various types of memory and metarepresentation to be evaluated. These findings provide an example in which the autistic adolescents showed a clear pattern of specific developmental delay, as opposed to more categorical differences, or deviance. There were also clear developmental trajectories in the acquisition of mental state terms and social use of mental verbs, that differed primarily in starting points across diagnostic groups.

Language as an Example of
Developmental Asynchronies in Autism

For many years now, clinicians and researchers have commented on specific phenomena such as pronoun reversal and delayed echolalia that are frequently associated with autism (Bartak, Rutter, & Cox, 1977). Neologisms, that is, use of made-up non-words as if they were words, has also frequently been listed as a characteristic of autistic language. We had been struck in our work on developing diagnostic instruments by odd words and phrases that did not seem to fit narrow definitions of neologisms or stereotypic speech, but that were used by a significant proportion of verbally fluent individuals with autism. We decided to ask graduate

Figure 1. Relationship of Mean Length of Utterance and Mean Number of Mental Verbs Referring to Mental States per 100 Utterances.

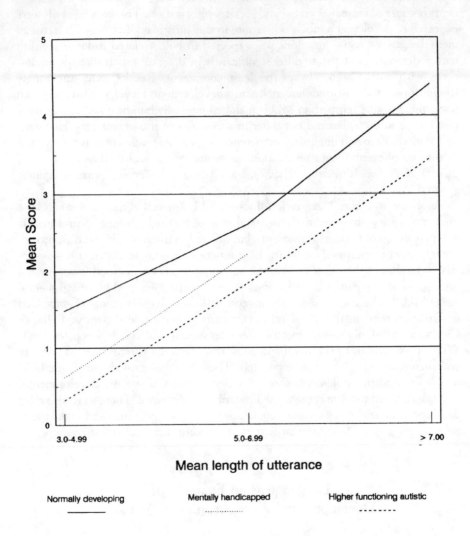

Mean length of utterance

Normally developing   Mentally handicapped   Higher functioning autistic

students in linguistics who were blind to the nature of the disorder we were studying, to read ADOS transcripts and to tag any word or phrase or sentence that sounded "odd" to them. This study used the same data set (slightly varied because subjects did not have to meet the criteria for MLU employed in the study of mental terms) involving the MLU- and verbal IQ-matched individuals (Volden & Lord, 1991). Raters were asked to ignore developmental errors, that is, anything that sounded like it would have been normal for a much younger child, but to mark anything that sounded unusual.

A second set of graduate students was then given these fagged words and phrases in context and asked to code them in terms of the type of error or unusual usage (single word vs. phrase or more; syntactic or semantic; and within syntax, developmental or nondevelopmental). Purely pragmatic errors were not coded because of the reliance on written transcripts and the need for contextual information that was not always available. Three social features were coded: social risk (whether the adolescent indicated uncertainty about using the term), joint reference (whether joint reference was established for the topic and, if so, by whom?), and whether the term applied to a conventional English category (whether a synonym existed in English). The coders also attempted to construct a paraphrase of what they thought the subject intended to say, based on linguistic and physical contexts.

Mentally handicapped subjects made the greatest proportion of developmental syntax errors, followed by the low- and high-autistic groups, and then the non-handicapped group. However, patterns for nondevelopmental syntax and semantic errors were quite different, with the high-autistic students making the largest proportion of these errors, followed by the lower functioning autistic group, followed by the mentally handicapped and normally developing youngsters. For the autistic groups, there were positive relationships between nonverbal IQ and receptive vocabulary and the percent of semantic and nondevelopmental syntax errors. For the mentally handicapped children, these relationships were negative. As autistic children became more fluent speakers, they had more frequent unusual use of language. Mentally handicapped children made fewer errors as they became more linguistically competent.

Out of 20 subjects in each of the four groups, 15 higher autistic, 7 lower autistic, 7 mentally handicapped, and 1 normally developing child were flagged for at least five nondevelopmental errors in usage. These errors were coded as to whether they had any relationship to an English word and whether this relationship was phonological or semantic or both. The largest proportion (over 55%) of errors was semantically related to the English paraphrase; 18% were phonologically related to the English paraphrase, and 11% to both. As exemplified in Table 1, for all groups, most of the neologisms and unusual uses of words had a clear relationship, either semantic or phonological, with a conventional English word or phrase. However, more of the higher and lower autistic subjects had at least one example of an unusual word or phrase whose origins were not discernable to the coders. Of the students who had at least one flagged utterance, more of the lower functioning autistic students used true neologisms than in any of the other groups. There were no differences in social risk or joint reference, probably due to the context (i.e., a diagnostic interview) from which these language samples were taken.

Ungerer (1989) suggested that the search for a specific deficit underlying the social dysfunction of autism may have failed because the source(s) of the dysfunction may be different at different times. The findings in our neologism study also suggest that conceptualizing the language or, more fairly, the communication deficit in autism as one entity does not accurately reflect the dynamic relationship between different aspects of language, not only over time, but also at any one point in time.

Table 1. Interpretation and Codes Assigned to Sample
Flagged Utterances

Flagged utterance	Linguistic categories		
	Develop-mental	Non-developmental	Semantic
High-functioning autistic			
"It makes me want to go as deep as *economical* with it" interpreted to mean "withdraw as much as possible"			X
"They're having a meal and then they're finishing and *siding* the table" interpreted to mean "clearing the table"			X
"And so he's seriously wounded like *cutses and bloosers*" interpreted to mean "cuts and bruises"		X	X
"If they even take it *true enough*" interpreted to mean "If they take it seriously enough"			X
Lower-functioning autistic			
"It *bells* the school" interpreted to mean "The bell rings in the school"		X	
"*turken*," to denote turkeys"			X
"I had a *racket* when I'm a little baby" interpreted to mean "I had a rattle when I was a little baby"	X		X
"It's *ready to come and ready to go*" interpreted to mean "Easy come, easy go"			X
"She's *bawcet*" interpreted to mean "She's bossy"			X
"That woman is *wiping* her hair" interpreted to mean "drying her hair"			X
Mentally handicapped			
"I'm going to *stump* the guy" interpreted to mean "He's going to stick (this fork) into the guy"			X
"Although if the bus is off *sequence* I don't know if I'm on time or not" interpreted to mean "off schedule"			X
Normal			
"She was very *bald*" interpreted to mean "She was naked"			X
"The bee's going to *cut* him" interpreted to mean "sting him"			X

Several years ago in a review article about language and autism, Tager-Flusberg (1981) indicated that syntactic and phonological and, to some extent, semantic development in autism proceeded along a normal but delayed course, in contrast to pragmatic development, which was consistently deviant. One possibility is that development in the former areas may proceed along normal lines but be out of synchrony with each other. No one cognitive or linguistic factor accounted for the majority of ways in which these autistic adolescents and young adults sounded odd or unusual. However, they did more often produce unusual sounding words and phrases than other students of their age and intellectual level. The kinds of errors that the students made were quite similar. The primary difference was that for the autistic students, errors of commission (such as using odd words and phrases) actually increased proportionately with greater linguistic complexity, whereas for the mentally handicapped youngsters, errors of comission decreased with increasing skill. The autistic students were also more likely than other subjects to produce neologisms and odd phrases that did not follow typical error patterns, though these were always the minority.

As shown above, asynchronies in different language skills combine in the autistic students to create unusual patterns that are interpreted as deviance, even though the courses of individual, narrowly defined linguistic skills are merely delayed. Thus, when an older adolescent, commenting that he has not played with action figures for a while, says "I'm just on the crack of the Ninja stage," he is reflecting not only his mismatch with toys, but also the asynchrony between his memory, competence with certain aspects of grammar, and his ambitious but not quite accurate use of idioms.

Language as an Indication of Developmental Devaince in Autism

The third study with the same sample had to do with the ability of verbally fluent individuals to talk about emotions (Jaedike, Storoschuk, & Lord, 1994). In this case, analyses were performed on the content of subjects' responses to questions about emotions, using categories from the developmental studies of Stein and Trabasso (1986). As part of the ADOS, students are asked to identify the source of several common emotions (i.e., What makes you happy? What kinds of things make you sad?) and then to describe how they feel inside during that emotion. Our prediction was that autistic adolescents would be more conventional and less personal in the causes they offered than other students of the same age. It was expected that particularly high-functioning adolescents with autism would respond to questions about emotions by repeating conventional responses that they had learned rather than being able to discuss in a spontaneous way their individual responses, expecially to happiness and sadness. Autistic subjects were expected to make fewer references to social situations in their discussion of major emotions than other groups. It was also predicted that, because of basic deficits in emotions, autistic students would have specific difficulties in describing how they felt.

Only one of these hypotheses was supported. Autistic students consistently made fewer references to social events and relationships as a source of emotions than did other adolescents. In addition, they were significantly less likely than the non-handicapped group, with the mentally handicapped students in between, to relate emotions to either success or failure. There were no group differences in the ability to describe how emotions felt, in part because there was tremendous variability in this measure across all groups. None of the patterns of response were related to chronololgical age, verbal IQ, or measures of language competence.

What is different about these findings, compared to the findings about mental-state terms and neologisms, is that they are not developmental in any way. That is, as a group, nonhandicapped and mentally handicapped subjects ranging in age from childhood to early adulthood, frequently offered social explanations for their feelings (i.e., "I'm happy when I see my friends; I'm sad when someone I like moves away"); autistic students did not. Unlike the previous studies with mental verbs and neologisms, there was no developmental point at which the groups converged. Unlike the previous studies, there was also no relationship, positive or negative, between verbal skills and the tendency to relate emotions to social experiences. Although at first glance, this finding supports recent hypotheses about the importance of deficits in very young autistic children in the ability to share affect (Kasari, Sigman, Mundy, & Yirmiya, 1990), it is also possible that having limited social experience contributed to the autistic subjects' offering social situations less frequently as a source of emotion, as well as, or even rather than, the reverse. The finding of group differences in relating emotions to success and failure was unexpected, but fits in well with theories that emphasize deficits in planning and goal-directedness in autism (Ozonoff et al., 1991). The possible effect of being "handicapped" on experiences of success and failure also cannot be under-estimated.

Clinical Implications for Assessment and Treatment

The preceding studies have a number of implications for assessment and treatment of autism in high-functioning adolescents. The follow-up studies offer the practical suggestion that, in high-functioning adolescents, a variety of different measures provide roughly similar information about verbal skills. Even relatively quick and focused instruments such as the Peabody Picture Vocabulary Test or the verbal subtests of the Wechsler scales are strongly correlated with prognosis. In fact, such a finding should not be surprising. The differences between the social communication of autistic adolescents and other teenagers are generally not particularly subtle, to be ferreted out with complex assessments, but by that age, are large and often obvious. When only more subtle differences in language skill are suspected, then the prognosis is generally better, if not good; in these cases, other factors such as social motivation and interest become particularly important.

Implications about treatment that arise from distinctions among deviance, asynchronies, and delay relate in part to the other factors, outside of language skill, that influence the communication of individuals with autism. Findings from the study of descriptions of affect suggest that in these individuals, social situations and success and failure may not be as important to adolescents with autism as they are to other young people. Interventions that attempt to make autistic adolescents *sound* like other teenagers need to take into account that often the adolescents with autism may not *feel* like other teenagers. The question of what is gained by sounding like age-mates versus accurately communicating what the individual feels will be answered differently depending on whether the goal is to help a teenager with autism build a real, reciprocal friendship (in which case communicating real feelings may take precedence) versus helping the same youngster apply for a part-time job with other teenagers at a fast food restaurant (where sounding "normal" may be highest priority).

The studies discussed here did not address the well-established deficits in receptive language that affect youngsters with autism, particularly in the early preschool years (Cantwell et al., 1978; Lord, 1985). We do not know if the years that many children with autism spend without understanding much of the language that goes on around them has long-term effects on their knowledge of social behavior and on later language skill. Implications of the fact that most autistic youngsters actually speak less often (or at least, less often in a socially directed way) than other children of comparable language skill (Lord, 1985) and so have both less practice talking and fewer opportunities for feedback are not well understood. One wonders if differences in these factors may account for the milder form of autistic symptoms in adults diagnosed with Asperger's syndrome, a form of pervasive developmental disorder that is not associated with language delay (Ghaziuddin, Tsai, & Ghaziuddin, 1992). Similarly, one wonders if the frequent use of neologisms or odd words found in autistic young people might be related to lack of practice in both listening and talking to age-mates.

Data from several of the studies suggest that autism and language skill have related but independent effects on specific behaviors. In designing treatments, particularly those for youngsters already surviving in mainstream educational placements, one must then ask, What is the goal of treatment? To speed up acquisition of particular linguistic forms? While this might be a high priority for a three year old who is moving from single words to simple phrases, with a high-functioning teenager, increasing rate of acquisition of linguistic terms would often seem to merit less attention than pragmatic skills. On the other hand, goals such as helping the individual acquire ways to deal with specific situations, such as entering or leaving a conversation or making a telephone call, may have immediate positive consequences. Using compensatory strategies, which build on strengths that are in fact asynchronous with the adolescent's spontaneous language level such as a very good auditory memory or good visual-spatial skills, may be an effective mode of intervention. For example, helping a teenager overlearn the need to say "Excuse me" or "Fine, how are you?" may allow him or her the option to participate appropriately in interactions in which he would not yet be able to join if more flexibility were required. Interventions teaching language skills at this age seem more likely to result in social improvements than practical linguistic gains, and for good reason. Thus, if one wanted to work with autistic adolescents to improve their "theory of mind," the focus might better be on increasing the frequency with which they note and comment about their own or someone else's mental state or alerting them to typical cues about mental states, rather than working on the definition of mental state at a conceptual level.

Implications for Further Research

The present triad of perspectives on language development and use in high-functioning adolescents with autism leaves many gaps to be filled by further research. Further studies concerning relationships among language skills and other aspects of development in autism are much needed. The notion of asynchronies is an important one that should further our understanding of the behavior patterns associated with autism and have implications for treatment along the way (Tager-Flusberg, 1992). As our understanding of the structure of discourse increases, data on how the talk of an autistic person affects the language directed to him or her and how to break the cycle of poor communication will be important. In addition, a better understanding of the comorbidity between various forms of language disorder and autism will be particularly helpful in determining prognoses, selecting treatments, and understanding the nature of both disorders.

Conclusion

In summary, the preceding studies illustrated numerous relationships between diagnostic differences, particularly those associated with autism, and the development of language. Using the same sample in the same contexts, examples were found of phenomena in autism that were strongly related to developmental factors (i.e., use of mental-state terms and syntactic development), that were related to developmental characteristics in unusual ways (i.e., the positive relationship for the autistic subjects between use of unusual words and phrases and syntactic development), and that did not seem to be related at all in any simple way to developmental factors in language (i.e., the very low frequency of references to social events or relationships made by autistic children and adolescents when asked about reasons behind feelings). These studies provide an example of how a factor, such as language, both provides a window through which we can study other aspects of development (e.g., such as asking questions about emotions) and is a phenomenon worthy of study in its own right. The role of language changes with the purposes of the researcher, but also may vary in much more significant ways in the life of the child or adolescent, depending on the demands of the situation and the point in development under study. In so doing, language development in autism represents a prototype of the relationship between development and psychopathology.

REFERENCES

American Psychiatric Association. (1980). *Diagnostic and statistical manual of mental disorders (DSM-III)* (3rd ed.). Washington, DC: American Psychiatric Association.

American Psychiatric Association. (1994). *Diagnostic and statistical manual of mental disorders (DSM-IV)* (4th ed.). Washington, DC: American Psychiatric Association.

Attwood, A., Frith, J., & Hermelin, B. (1988). The understanding and use of interpersonal gestures by autistic and Down's syndrome children. *Journal of Autism and Developmental Disorders, 18,* 241–258.

Baron-Cohen, S., Tager-Flusberg, H., & Cohen, D. (1993). *Understanding other minds: Perspectives from autism.* Oxford: Oxford University Press.

Bartak, L. (1978). Educational approaches. In M. Rutter & E. Schopler (Eds.), *Autism: A reappraisal of concepts and treatment* (pp. 423–438). New York: Plenum Press.

Bartak, L., Rutter, M., & Cox, A. (1977). A comparative study of infantile autism and specific receptive language disorders. III. Discriminant function analysis. *Journal of Autism and Childhood Schizophrenia, 7,* 383–396.

Bayley, N. (1969). *Manual for the Bayley Scales of Infant Development.* New York: Psychological Corporation.

Bebko, J. M., Konstantareas, M. M., & Springer, J. (1987). Parent and professional evaluations of family stress associated with characteristics of autism. *Journal of Autism and Developmental Disorders, 17,* 565–576.

Bolton, P., & Rutter, M. (1990). Genetic influences in autism. *International Review of Psychiatry, 2,* 67–80.

Brown, R. (1973). *A first language.* Cambridge, MA: Harvard University Press.

Bryson, S. E., Clark, B. S., & Smith, T. M. (1988). First report of a Canadian epidemiological study of autistic syndromes. *Journal of Child Psychology and Psychiatry, 29,* 433–445.

Cantwell, D. P., Baker, L., & Rutter, M. (1978). A comparative study of infantile autism and specific developmental receptive language disorder. IV. Analysis of syntax and language function. *Journal of Child Psychology and Psychiatry, 19,* 351–363.

Dawson, G., & Adams, A. (1984). Imitation and social responsiveness in autistic children. *Journal of Abnormal Child Psychology, 12,* 209–226.

DiLavore, P., Lord, C., & Opsahl, A. (1994, October). *The Early Diagnosis of Autism: Year One.* Paper presented at the annual meeting of the American Academy of Child and Adolescent Psychiatry, New York.

Ghaziuddin, M., Tsai, L., & Ghaziuddin, N. (1992). Brief report: A comparison of the diagnostic criteria for Asperger's syndrome. *Journal of Autism and Developmental Disorders, 22,* 643–649.

Hobson, R. P. (1993). Understanding persons: The role of affect. In S. Baron-Cohen, H. Tager-Flusberg, & D. Cohen (Eds.), *Understanding other minds: Perspectives from autism* (pp. 204–227). Oxford: Oxford University Press.

Jaedicke, S., Storoschuk, S., & Lord, C. (1994). Subjective experience and causes of affect in high-functioning children and adolescents with autism. *Development and Psychopathology*, 6, 273–284.

Kasari, C., Sigman, M., Mundy, P., & Yirmiya, N. (1990). Affective sharing in the context of joint attention interactions of normal, autistic, and mentally retarded children. *Journal of Autism and Developmental Disorders*, 20, 87–100.

Klin, A., Volkmar, F. R., & Sparrow, S. S. (1992). Autistic social dysfunction: Some limitations of the theory of mind hypothesis. *Journal of Child Psychology and Psychiatry*, 33, 861–876.

Le Couteur, A., Rutter, M., Lord, C., Rios, P., Robertson, S., Holdgrafer, M., & McLennan, J. D. (1989). Autism Diagnostic Interview: A semi-structured interview for parents and caregivers of autistic persons. *Journal of Autism and Developmental Disorders*, 19, 363–387.

Lord, C. (1985). Autism and the comprehension of language. In E. Schopler & G. Mesibov (Eds.), *Communication problems in autism* (pp. 257–281). New York: Plenum Press.

Lord, C., & Magill, J. (1989). Methodological and theoretical issues in studying peer-directed behaviour and autism. In G. Dawson (Ed.), *Autism: Nature, diagnosis and treatment* (pp. 326–345). New York: Guilford.

Lord, C., Pickles, A., McLennan, J., Rutter, M., Bregman, J., Folstein, S., Fombonne, E., Leboyer, M., & Minshew, N. (in press). Diagnosing autism: Analyses of data from the Autism Diagnostic Interviews. *Journal of Autism and Developmental Disorders*.

Lord, C., Rutter, M., Goode, S., Heemsbergen, J., Jordan, H., Mawhood, L., & Schopler, E. (1989). Autism Diagnostic Observation Schedule: A standardized observation of communicative and social behavior. *Journal of Autism and Developmental Disorders*, 19, 185–212.

Lord, C., & Venter, A. (1992). Outcome and follow-up studies of high-functioning autistic individuals. In E. Schopler and G. Mesibov (Eds.), *High-functioning individuals with autism* (pp. 187–199). New York: Plenum Press.

Lotter, V. (1966). Epidemiology of autistic conditions in young children. I. Prevalence. *Social Psychiatry*, 1966, 124–137.

Miller, J., & Chapman, R. (1983). *Systematic analysis of language transcripts (SALT)*. Madison: University of Wisconsin, Waisman Center.

Minshew, N. (1991). Indices of neural function in autism: Clinical and biological implications. *Pediatrics*, 87, (Suppl.) 774–780.

Mundy, P., Sigman, M., & Kasari, C. (1990). A longitudinal study of joint attention and language development in autistic children. *Journal of Autism and Developmental Disorders*, 20, 115–128.

Mundy, P., Sigman, M., & Kasari, C. (1993). Theory of mind and joint attention deficits in autism. In S. Baron-Cohen, H. Tager-Flusberg, & D. Cohen (Eds.), *Understanding other minds: Perspectives from autism* (pp. 181–203). Oxford: Oxford University Press.

Murray, H. A. (1938). *Explorations in personality*. New York: Oxford University Press.

Ozonoff, S., Pennington, B. F., & Rogers, S. J., (1991). Executive function deficits in high-functioning children with autism: Relationship to theory of mind. *Journal of Child Psychology and Psychiatry*, 32, 1081–1106.

Prior, M., Dahlstrom, B., & Squires, T. L. (1990). Autistic children's knowledge of thinking and feeling states in other people. *Journal of Child Psychology and Psychiatry*, 31, 587–602.

Rogers, S. J., & Pennington, B. F., (1991). A theoretical approach to the deficits in infantile autism. *Development and Psychopathology*, 3, 137–162.

Rumsey, J. M., Rapoport, M. D., & Sceery, W. R. (1985). Autistic children as adults: Psychiatric, social, and behavioral outcomes. *Journal of the American Academy of Child Psychiatry,* *24*, 465–473.

Siegel, B., Pliner, C., Eschler, J., & Elliott, G. (1988). How children with autism are diagnosed: Difficulties in identification of children with multiple developmental delays. *Developmental and Behavioral Pediatrics, 9,* 199–204.

Sigman, M., Mundy, P., Sherman, T., & Ungerer, J. (1986). Social interactions of autistic, mentally retarded and normal children and their caregivers. *Journal of Child Psychology and Psychiatry, 27,* 647–656.

Sparrow, S., Balla, D., & Cicchetti, D. (1984). *Vineland Adaptive Behavior Scales.* Circle Pines, MN: American Guidance Service.

Stein, N. L., & Trabasso, T. (1986). The search after meaning: Comprehension and comprehension monitoring. In F. J. Morrison, C. Lord, & D. P. Keating (Eds.), *Applied developmental psychology* (Vol. 2, pp. 33–56). New York: Academic Press.

Storoschuk, S., Lord, C., & Jaedicke, S. (1996). *Autism and the use of mental verbs.* Manuscript submitted for publication.

Tager-Flusberg, H. (1981). On the nature of linguistic functioning in early infantile autism. *Journal of Autism and Developmental Disorders, 11,* 45–56.

Tager-Flusberg, H. (1992). Autistic children's talk about psychological states: Deficits in the early acquisition of a theory of mind. *Child Development, 63,* 161–172.

Tager-Flusberg, H., Calkins, S., Nolin, I., Baumberger, T., Anderson, M., & Chadwick-Denis, A. (1990). A longitudinal study of language acquisition in autistic and Down syndrome children. *Journal of Autism and Developmental Disorders, 20,* 1–22.

Ungerer, J. A. (1989). The early development of autistic children. In G. Dawson (Ed.), *Autism: Nature, diagnosis and treatment* (pp. 75–91). New York: Guilford.

Venter, A., Lord, C., & Schopler, E. (1992). A follow-up study of high-functioning autistic children. *Journal of Child Psychology and Psychiatry, 33,* 489–507.

Volden, J., & Lord, C. (1991). Neologisms and idiosyncratic language in autistic speakers. *Journal of Autism and Developmental Disorders, 21,* 109–130.

Volkmar, F. R., Klin, A., Siegel, B., Szatmari, P., Lord, C., Campbell, M., Freeman, B. J., Cicchetti, D. V., Rutter, M., Kline, W., Buitelaar, J., Hattab, T., Fombonne, E., Fuentes, J., Werry, J., Stone, W., Kerbeshian, J., Hoshino, T., Bregman, J., Loveland, K., Szymanski, L., & Towbin, K. (1994). Field trial for autistic disorder in *DSM-IV. American Journal of Psychiatry, 151*(9), 1361–1367.

Wechsler, D. (1991). *Manual for the Wechsler Intelligence Scale for Children (3rd ed.).* San Antonio, TX: Psychological Corporation.

World Health Organization (1992). *ICD-10. Mental, behavioral and developmental disorders.* Geneva: World Health Organization.

VI   A Developmental Perspective on Depressive Symptoms in Adolescence: Gender Differences in Autocentric-Allocentric Modes of Impulse Regulation

Per F. Gjerde & Jack Block

This chapter presents and integrates the results of an ongoing prospective program of research on depressive symptoms in adolescence and young adults (J. Block & Gjerde, 1990, J. Block, Gjerde, & J. H. Block, 1991; Gjerde, 1993, Gjerde, 1995a, 1995b; Gjerde & J. Block, 1991; Gjerde, J. Block, & J. H. Block, 1988; Gjerde, J. Block, & J. H. Block, 1991). Conceptually, this research builds on principles of developmental psychopathology, psychodynamic theories regarding the direction of impulse expression / regulation, and current research on the origins and implications of gender differences in socialization.

A major goal of developmental psychopathology is to identify the life courses that lead to and from psychological disorders (e.g., Bowlby, 1988; Cicchetti, 1989; Masten & Braswell, 1991; Rutter, 1988; Rutter & Garmezy, 1983; Sroufe & Rutter, 1984). Prospective designs are uniquely relevant for examining the origins, trajectories, and outcomes of psychological disorders. Nevertheless, as noted by Wierson and Forehand (1994), longitudinal designs have only infrequently been used in the study of psychopathology in young people. This chapter illustrates the special contribution to the study of one psychological disorder—depressive symptoms—afforded by this method by referencing our particular longitudinal data base (see J. Block, 1993; J. H. Block & J. Block, 1980, for a description of this research). Therein, in addition to investigating the concurrent manifestations of depressive symptoms in adolescents, we also examine—backward in time—the early antecedents of adolescent dysphoria and—forward in time—the psychological sequelae of adolescent depressive symptoms in young adults. Are depressive symptoms developmentally traceable back to earlier years? Do adolescent depressive symptoms merely reflect an ephemeral state of adolescent-specific despondency or do such symptoms anticipate vulnerability and emotional distress later in young adulthood?[1]

From its inception, gender differences in pathways to adolescent depression have been a central concern of this research. There is no a priori reason to expect that the

---

[1] We recognize that the results obtained in our nonclinical sample may not generalize to samples employing psychiatric definitions of clinical depression. Hence, we use the terms "depressive symptoms"

This study was supported by National Institute of Mental Health Grant MH 16080 to Jack Block and by a Social Science Division Grant from the University of California at Santa Cruz to Per F. Gjerde.

origins and development of depressive symptoms—or any other psychological disorder—are necessarily equivalent in males and females. If this possibility is not evaluated, however, acting on the assumption of such equivalence may prevent identification of gender specific patterns in the origins and course of depressive conditions. Our analyses, therefore, have consistently evaluated the sexes separately, examining whether pathways emanating from childhood to depressive symptoms in adolescence are gender specific or gender neutral. We have found many gender differences.

In summarizing the gender-distinct pathways to depressive conditions in young people, we have found it conceptually useful to focus on the *direction of impulse expression* in dysphoric males and females. Dysphoric adolescent males, we anticipated, would be likely to manifest an *allocentric* (or outer-oriented) mode of symptom expression and manifest their internal motivations and despairs by acting on the world, often in an impulsive and angry manner. This expressive pattern is likely to create manifest problems of external adaptation vis-à-vis the external world— problems often referred to as externalizing in nature. In contrast, dysphoric adolescent females were expected to manifest an *autocentric* (or inner-oriented) mode of symptom expression. This expressive pattern is likely to create problems of internal adaptation such as low self-esteem, anxiety, and preoccupation with the adequacy of self—problems often referred to as internalizing in nature. In the social sphere, the female mode of dysphoric symptom expression, which involves heightened attention to one's own thoughts and feelings of distress, is less likely to find clear behavioral expression and, therefore, also less likely to be socially disruptive than the male mode of dysphoric symptom expression. Zahn-Waxler's (1993) differentiation between males as "warriors" and females as "worriers" in her discussion of gender and psychopathology comes to mind here. Nonetheless, problems regarding both internal and external psychological adaptation can be expected to interfere— albeit in different ways—with the individual's successful resolution of important age-related developmental tasks and increase the probability of pathology. Understanding the circumstances that give rise to each mode of adaptation is therefore of interest to developmental psychopathology.

The Need for Longitudinal Designs in the Study of Depressive Symptoms in Young People

Cross-sectional studies have established useful connections between depression and such psychological characteristics as dependency, self-criticism, introversion,

---

and "dysphoria" throughout this chapter when referring to our research. We also register, however, that there is no clear way to determine when depression constitutes a distinct disease entity and therefore becomes a qualitatively different deviation from the dysphoria often observed in normal samples (e.g., Rutter, 1986). Depressive symptoms, even if they do not meet stringent psychiatric definitions, deserve study in their own right since a substantial proportion of individuals manifesting subclinical depression early in life subsequently go on to develop clinical depression (Lewinsohn, Hoberman, & Rosenbaum, 1988; Lewinsohn, Roberts, Seeley, Rohde, Gotlib, & Hops, 1994).

low self-esteem, and negative attributional style (e.g., Abramson, Seligman, & Teasdale, 1978; Akiskal, Hirschfeld, & Yerevanian, 1983; Arieti & Bemporad, 1980; Blatt, 1974; Hirschfeld, Klerman, Chodoff, Korchin, & Barrett, 1976). However, most cross-sectional studies derive either from analyses of retrospective reports provided by already depressed individuals or from comparisons between actually (or formerly) depressed and nondepressed individuals. Both approaches are ultimately insufficient. Looking backward in time, one can almost always identify (or reconstruct) an event that, plausibly, would account for an outcome (Freud, 1955, pp. 167–168; see also Dawes, 1993; Fischoff & Beyth, 1975; Beck, 1967). Retrospective reports about subjective psychological status should be approached with particular care (Henry, Moffitt, Caspi, Langley, & Silva, 1994). Further, differences observed in the character structure of depressed and nondepressed individuals are equivocal in their implication. They may (1) reflect the aftermath of having experienced depressive episodes, (2) represent the concurrent manifestations of a depressive process that may recede with its remission, or (3) truly precede the emergence of the depressive disorder and thus reflect a dispositional vulnerability to depression (e.g., Barnett & Gotlib, 1988; Farmer & Nelson-Gray; 1990). Finally, during depressive states, individuals appear to attribute more negative qualities to themselves than is objectively warranted (e.g., Hirschfeld, Klerman, Clayton, Keller, McDonald-Scott, & Larkin, 1983). Thus, the extensive reliance on self-report data in most cross-sectional research complicates attempts to distinguish antecedent factors from concurrent symptomatic manifestations. In short, the results of cross-sectional studies are indeterminate in their implication.

The aspiration to comprehend the developmental trajectories of psychopathology and to clarify relations between childhood attributes and subsequent patterns of adaptation thus requires longitudinal inquiry. In the study of depressive conditions, long-term prospective designs are particularly required because the manifestations and prevalence of depressive symptoms vary with age, the connections between childhood and adult depression are complex, and the early dispositional basis of depressive symptoms is not well understood (Barnett & Gotlib, 1988; Bemporad & Wilson, 1978; J. Block et al., 1991; Gjerde, 1995a; Lewinsohn, Roberts, Seely, Rohde, Gotlib, & Hops, 1994; Lewinsohn, Steinmetz, Larson, & Franklin, 1981; Rutter, 1986; Sroufe & Rutter, 1984; Zahn-Waxler, Cole, & Barrett, 1991).

Although highly prevalent in young people, depressive symptoms—as many other internalizing disorders—do not appear to manifest the substantial consistency over time characterizing the developmental course of such societally manifest psychological problems as antisocial behavior or conduct disorder (Loeber, 1991a; Loeber & Farrington, 1994; Ollendick & King, 1994; Olweus, 1979).[2] The elusive

---

[2] We recognize that not all types of antisocial behavior may manifest equally impressive consistency over time. Moffitt (1993), for example, has drawn the distinction between "life-course-persistent" and "adolescence-limited" antisocial behavior. Only the former type, it is argued, shows considerable longitudinal consistency over age, whereas the latter type, although perhaps characteristic of a larger group of individuals, evidences less impressive continuity. Hence, there may exist different categories of antisocial behavior, each characterized by separate etiology and developmental course.

trail of depressive antecedents may reflect the complex developmental nature of this existential problem. Interestingly, Zahn-Waxler et al. (1991) have observed that depression-prone persons as children often appear relatively competent, sensitive, and eager to please others. In the search for early indicators of vulnerability to depressive symptoms in adolescents and young adults, therefore, close personality and environmental assessments of the early years and, entirely independently, close personality and environmental assessments of later life are required.

Depressive Symptoms in Adolescence: Gender Differences in Psychological Dynamics

Research on adolescent psychopathology is a vigorous field of study. In particular, adolescence has been identified as a period crucial for subsequently understanding the nature, trajectory, and treatment of depressive disorders, a major psychiatric concern (Petersen, Compas, Brooks-Gunn, Stemmler, Ey, & Grant, 1993). It is now recognized that many individuals pass through adolescence without experiencing significant psychological problems, thus upending the once widely held view that such problems are normative in adolescence. It is also the case, however, that a substantial number of adolescents do experience significant emotional problems and that these difficulties can have lasting implications for adaptation well beyond adolescence (e.g., Fleming, Boyle, & Offord, 1993; Jessor, Donovan, & Costa, 1991; Kandel & Davies, 1986; Petersen et al., 1993).

Epidemiological studies of depressive disorders have found gender differences in levels of depression. Moreover, the nature of these differences appears to vary with developmental level. Prior to adolescence, there is some evidence that boys manifest equal or more depressive symptoms than girls (Anderson, Williams, McGee, & Silva, 1987; Nolen-Hoeksema, Girgus, & Seligman, 1992). This pattern is reversed after puberty when girls begin to show a substantially higher incidence of depression than boys—a pattern maintained throughout adulthood (Nolen-Hoeksema, 1990). This finding of puberty-related increase in female depression raises additional issues such as why females, after puberty, are more vulnerable to dysphoria. Is the underlying developmental dynamic of this disorder importantly different in males and females and may this difference derive from how boys and girls experience the transition to adolescence (e.g., Gjerde, 1995a; Kaplan, 1986; Kavanagh & Hops, 1994; Petersen, Sarigiani, & Kennedy, 1991; Radloff & Rae, 1979, Weissman & Klerman, 1979)?

Differentiating Between Threshold and Directionality of Impulse Expression

In order to conceptualize gender differences in the psychodynamics of depressive symptoms, we focus on the directionality of impulse expression (allocentric, or outer-directed, versus autocentric, or inner-directed) in dysphoric adolescents and young adults. Theories of motivational states and modes of impulse expression must distinguish between two related, yet conceptually separate, ways of thinking about impulse regulation: the threshold for impulse expression versus the direction for impulse expression. The notion of threshold for impulse expression is exemplified by J. H. Block and J. Block's (1980) construct of ego control, which derives from an integration of Fenichel's (1945) psychoanalytic theory of impulse modulation and Lewin's (1935) attempt to formalize the psychological system of an individual (J. Block, 1950; J. H. Block, 1951). Undercontrol, or low threshold for impulse expression, implies insufficient modulation of impulse, the inability to defer gratification, immediate manifestation of motivations and affects, and vulnerability to environmental distractors. From a developmental perspective, the undercontroller has not acquired sufficient ability to tolerate tension. At the opposite end of the continuum, overcontrol, or high threshold for impulse expression, is manifested through excessive restraint of impulse, unduly prolonged delay of gratification, inhibition of action, and isolation from environmental distractors. From a developmental perspective, the overcontroller invokes inhibitory capacities in a preservative way that often is not adaptationally effective.

Once an impulse has gained regnancy or exceeded its threshold, it can be expressed through actions on the external world (allocentric modes) or through internal expressive conduits (autocentric modes, either cognitive or visceral).[3] This dimension of impulse expression, sometimes referred to in the literature as "externalization" and "internalization," has a long history in psychology and has been conceptualized in several, sometimes incompatible, ways. As early as in 1935, Jones described individuals characterized by inverse relations among emotional expressiveness and skin conductance as "externalizers" and "internalizers." (Jones, 1935). Buck, Savin, Miller, and Caul (1972) used the label "externalizers" to describe individuals characterized by a combination of low physiological reactivity and high ability to accurately communicate their inner emotional states and the label "internalizers" to describe individuals characterized by a combination of high physiological reactivity and low ability to accurately communicate their inner emotional states. It is only more recently that a close conceptual link between the externalization-internalization dimension and psychopathology has become prevalent (e.g., Achenbach & Edelbrock, 1978).[4]

---

[3] A conceptually related differentiation is represented by Lewin's (1951) early distinction between the "locomotion of the person" versus the "change of cognitive structure."

[4] Related empirical studies have often interpretively organized their findings in terms of the centrality of two overarching dimensions of psychopathology labeled as "externalization" and "internalization." As often operationalized, these two dimensions prove to be highly correlated, often as high or higher than .50 (e.g., Achenbach, 1991; Achenbach & Edelbrock, 1978), probably because they both embody also an underlying dimension of maladaptation or lack of resilience. In our view, Achenbach and Edelbrock's conceptualization of externalization and internalization can best be understood as representing unresilient undercontrol of impulse and unresilient overcontrol of impulse, respectively.

There is, furthermore, no necessary relation between the threshold of impulse expression and the direction of impulse direction. That is, an undercontroller need not be an externalizer and an overcontroller need not be an internalizer. Independent of motivational pressures, an allocentric mode of impulse expression is associated with action and affect channeled outward on the world. This mode implies an absence of introspective concerns and of self-awareness. By contrast, an autocentric mode of impulse expression is associated with internal routes for impulse expression, such as introspective concerns and frequent self-awareness. Historically, the directionality of impulse expression should be distinguished from the dimension of extroversion-introversion, as proposed by Jung (1923). For Jung, the idea of extroversion implied a specific perceptual orientation, receptivity oriented toward outer stimuli; the idea of introversion implied a perceptual orientation, receptivity oriented toward the inner world of thought.

Because allocentric individuals typically express their impulses through acting on the external world and autocentric individuals typically express their impulses through internal conduits, the two patterns of impulse expression are likely to differ in the degree to which they are associated with clear behavioral markers. Allocentric impulse expression is more likely than autocentric impulse expression to be associated with a direct translation of inner states into overt behaviors. Hence, the degree of convergence across methods, such as observer-data (O-data) versus self-report (S-data), is likely to be higher for allocentric dysphoric males than for autocentric dysphoric females. From an observer's point of view, that is, allocentric individuals are more "judgable" than autocentric individuals.[5] Self-report data, on the other hand, may be relatively more important in clarifying the nature of female than male depression since problems of internal adaptation are likely to involve fewer clear behavioral manifestations. Such conceptually based lack of convergence across methods thus is not necessarily a threat to measurement validity but is, in itself, a source of relevant information. As Ozer comments (1989), the convergence of results across different methods (such as O- and S-data) should be "a theoretical prediction when warranted, not an unvarying methodological imperative" (p. 230). As we show below, discrepancies between results based on S- and O-data can provide insights into gender differences in the dynamics of depressive symptoms.

A caveat: Although we have found the directionality of impulse expression to be useful in summarizing gender differences in the dynamics of depressive symptoms, we do not exclude other aspects of impulse regulation (such as threshold for impulse expression) that also differentiate between dysphoric males and females. For example, in our work, we find that dysphoric males are typically more undercontrolling of impulse than dysphoric females. We also note that although the directionality and the threshold of impulse expression represent conceptually distinct dimensions, differentiation between the behavioral expressions of directionality and threshold can sometimes be difficult.

---

[5] The term "judgability" was coined to reflect the fact that observers were able to judge the personalities of some individuals with greater accuracy than others (Colvin, 1993).

Depression as a Mixture of Emotions: Sadness and Anger Co-occurrence

Our interest in the directionality of impulse expression in dysphoric individuals derives from the complex emotional nature of depressive disorders. Psychodynamically oriented therapists have long observed that, although the primary emotion in depression is one of intense sadness, depression is often accompanied by hostility. Recent empirical research also has indicated that depression often is a mixture of emotions, such as sadness and anger. Although classical psychoanalytic theory held that depression-related hostility was directed inward toward the self (Abraham, 1927; Freud, 1968), more contemporary observations of overt hostility in depressed individuals have cast doubt on this proposition (Friedman, 1970; Gershon, Cromer, & Klerman, 1968; Kahn, Coyne, & Margolin, 1985; Weissman, Klerman, & Paykel, 1971).

It is also the case that depression, however conceived, may co-occur within the same individual with other problems, such as anxiety and conduct disorders. For example, dysphoria and the impulsivity underlying conduct disorders are both to be seen with some frequency in males who are consequently more likely than females to express their depression-related anger in an outer-directed, or extrapunitive manner (Capaldi, 1991, 1992; Gjerde et al., 1988; Harter & Marold, 1994; Ollendick & King, 1994; Puig-Antich, 1982; Rutter, Tizard, & Whitmore, 1970).[6] Recognitions of comorbidity are paralleled in the developmental literature by increasing interest in the experience of mixed emotions in young children.[7] By middle childhood, children apparently are able to experience simultaneously the emotions both of sadness and anger (Harter & Marold, 1994).

To summarize: There is evidence from both the clinical and the developmental literature that depression is characterized by a mixture of emotions, particularly sadness and anger. The *target* of the anger component in depression need not be the self as the early psychoanalytic view contended, but can be others as well. There is also evidence that the direction of impulse expression differs for Dysphoric males and females: dysphoric females tend to be self-focusing and ruminating while dysphoric males are more likely to manifest their internal unhappiness via aggression and conduct disorders (Allgood-Merten, Lewinsohn, & Hops, 1990; Edelbrock & Achenbach, 1980; Ingram, Cruet, Johnson, & Wisnicki, 1988; Kobak, Sudler, & Gamble, 1992; Nolen-Hoeksema, 1990; Ostrov, Offer, & Howard, 1989; Puig-Antich, 1982). Given these observations, the further study of the direction of

---

[6] Zahn-Waxler (1993), however, observes that antisocial behavior in women may be underdetected or redefined and that we therefore may underestimate the degree of overlap between the sexes in extent of conduct problems.

[7] The term, comorbidity as diagnosis, is often thought to imply the simultaneous occurrence within an individual of different feelings. Psychodynamically, however, within the individual, these different affects may be reciprocally present: The presence of the one affect often displaces the presence of the other (as a way of escaping the other affect).

impulse expression—allocentric versus autocentric—as it relates to depressive disorders in males and females was of special interest.

Gender-Differentiated Socialization Patterns and the Directionality of Impulse Expression in Dysphoric Individuals

Our gender-differentiated hypothesis has its earlier origins in previous conceptualizations and research on gender differences in socialization. J. H. Block (1973, 1983) argues that traditional sex-role socialization tends to extend the range of available experiences for boys and to limit that range for girls. Boys are allowed relatively greater freedom than girls to explore and are encouraged to express greater curiosity, independence, competition, and achievement-related behaviors. For girls, in contrast, socialization pressures tend to circumscribe or restrict domains of activity and experience; proprieties are stressed, and girls tend to be subject to closer and more cautionary adult supervision than boys. J. H. Block (1983) suggests that these differences contribute to gender-dependent orientations toward the self. Boys will tend to "develop a premise system about the self that presumes or anticipates having consequences, instrumental competence, and mastery" (p. 1345). Girls, on the other hand, perhaps due to their reduced exposure to situations encouraging awareness of "the evocative role they themselves play in eliciting effects from the environment" (p. 1345), will be less likely to develop a sense of resourcefulness on which later instrumental competence builds. Hence, the self-percepts of girls are less likely to include qualities such as agency, initiative, and a sense of personal enablement.

Socialization may also promote differences between males and females in their tendencies to perceive the world as either contingent or noncontingent, responsive or nonresponsive, to their actions. Dweck, Davidson, Nelson, and Enna (1978), for example, report that girls are more likely than boys to attribute their failures to stable internal characteristics (e.g., lack of ability) and their successes to unstable external causes (e.g., luck). Boys, in contrast, are more likely to perceive their failures as accidental and their successes as due to personal attributes. Thus, failure is likely to produce amplified feelings of helplessness and rumination among girls but impunitiveness and no particular instigation toward rumination among boys. This view finds further support in a study of daily feelings in which dysphoric females were more likely than boys to feel helpless and without control of their environment (Csikszentmihalyi & Larson, 1984).

These findings have implications for the direction and target of impulse expression in dysphoric individuals, especially with respect to the anger component. McCranie (1971) proposed that when feelings of helplessness (and by implication, the experience of the world as nonresponsive, or noncontingent to one's actions) become amplified, an individual may experience depression. This dysphoria may or may not also involve internal feelings of anger. When an instrumental response, such as aggression, is anticipated as unlikely to succeed, overt expressions of anger and aggression are likely to be inhibited. Bemporad (1980) has remarked that "the capacity to feel angry and to use anger as a direct mode of achieving an objective implies a sense of autonomy and independence that ... the depressive lacks" (p. 177). This conjecture, conjoined with the finding that males are more likely to have a personal sense of autonomy and efficacy, suggests that dysphoric adolescent boys, more than dysphoric adolescent girls, will express their despair in an allocentric, extrapunitive, and aggressive manner.

Concurrent Gender Differences in the Allocentric-Autocentric Balance of Impulse Expression in Dysphoric Adolescents

An initial study (Gjerde et al., 1988) evaluated the gender-differentiating hypothesis in our longitudinally followed sample when the subjects were 18-years-old. This study focused on concurrent patterns of impulse expression and depression-related personality attributes. Results supported our hypotheses. First, depressive symptoms in males were associated with many behavioral markers and with convergence between O-data and S-data. Second, according to observer judgments, dysphoric 18-year-old males manifested an allocentric pattern: internal despair was expressed in the form of interpersonal antagonism, lack of behavioral restraint, manifest discontent with self, and unconventionality of thought and behavior. These observer evaluations of the directionality of impulse expression were consistent with the male subjects' self-perceptions: Dysphoric 18-year-old males viewed themselves as aggressive, alienated, highly stress-reactive, and socially distant from others. These young men were angry and alienated, and their anger was more likely to be extrapunitive (directed toward others) than intropunitive (directed toward self). Both internally (in terms of self-evaluations), and externally (in terms of observer-evaluations), these young dysphoric men were allocentric.

The results for dysphoric 18-year-old females were substantially different. First, depressive symptoms in females were associated with very few behavioral markers and with low convergence between O-data and S-data. Although the dysphoric young women perceived themselves as angry, alienated, and lacking in psychological well-being and self-esteem, these experienced feelings referred to problems of internal adaptation that were not discernible in their manifest behavior. Observer evaluations indicated that dysphoric adolescent women withheld behavioral expressions aversive to others; behaviorally, they exhibited an autocentric mode of impulse expression—a tendency to reflect on the adequacy of self, introspectiveness, even self-dislike. They did not express outwardly the feelings of aggression and alienation that, in their self-evaluations, they had acknowledged.

Five years later, at age 23, the personality correlates of depressive tendencies were evaluated once more (Gjerde, 1993). Although dysphoric males were still viewed as undercontrolling of impulse and antagonistic, their self-experienced feelings of lack of psychological well-being was behaviorally more visible in terms of psychological vulnerability (such as ego brittleness) than at age 18. A coherent set of personality characteristics also characterized the correlates of depressive symptoms in 23-year-old females: vulnerability, brittleness, anxiety, and interpersonal difficulties. In contrast to age 18, however, age-23 dysphoric women were more likely to by seen by observers as hostile and distrustful of others, a finding suggesting that the self-experienced feelings of interpersonal antagonism that earlier could not be discerned by observers have found a less mitigated behavioral expression in young adulthood. To summarize, the behavioral expression of self-acknowledged depressive symptoms at age 23 has begun to converge for the sexes, with males more likely now than in adolescence to show behavioral signs of brittleness and internal despair and females—more likely now than in adolescence to show behavioral signs of interpersonal antagonism.[8] That is, as young adults, both males and females with depressive symptoms were beginning to be characterized by a mixture of both allocentric and autocentric modes of symptom expression. Despite this beginning convergence of depression-related personality attributes in young adulthood, the early antecedents of these attributes are substantially different, especially prior to puberty (Gjerde, 1995a; see the discussion of these findings and their interpretation later in this chapter).

Looking Backward: Earlier Gender Differences in the Allocentric-Autocentric Balance of Impulse Expression in Dysphoric Adolescents

Our longitudinal data were then evaluated to trace the childhood origins of the depressive symptoms observed in late adolescence and young adulthood. Are dysphoric adolescent and young adult males characterized in childhood by an allocentric pattern of impulse expression? Are dysphoric adolescent and young adult females characterized in childhood by an autocentric pattern of impulse expression?

The prospective correlates obtained that foretold adolescent dysphoria closely parallel the concurrent gender differences obtained in our initial concurrently

---

[8] Complicating this age-related interpretation is the fact that different measures of depressive symptoms were used at ages 18 and 23. At age 18, we employed the Center for Epidemiological Studies Depression Scale (CESD, Radloff, 1977); at age 23, we employed the General Behavior Inventory (GBI) Depression scale (Depue, 1987). The CESD was designed as a brief screening device for depressive state and asks how frequently an individual has experienced depressive symptoms during the previous week only. In contrast, the GBI, a trait-based inventory, was developed to identity chronic depressive conditions and to tap mood, cognitive, vegetative, psychomotor, and motivational domains of depressive conditions. Hence, elevated GBI scores do not simply reflect depressed mood but either frequent episodic or chronic-intermittent affective conditions or both (Depue, Krauss, Spoont, & Arbisi, 1989). The GBI, a more elaborate, trait-based measure, is more likely to identify depressive symptoms as an

oriented study (J. Block et al., 1991). In 14-year-old adolescent boys, an allocentric pattern of impulse expression (i.e., the open expression of aggression, lack of impulse control, and interpersonal antagonism) anticipated depressive symptoms at age 18. In 14-year-old girls, an autocentric pattern of impulse expression (rumination, low self-esteem, bodily symptoms of anxiety, and self-concern) foretold dysphoria four years later.

Going back even earlier in time, to childhood and preschool years, the personality correlates of depressive symptoms were similar to those observed subsequently in adolescence: from their nursery-school years, subsequently dysphoric males were viewed by observers as aggressive, self-aggrandizing, and undercontrolling of impulse early—an allocentric pattern. In contrast, dysphoric 18-year-old females during their nursery-school years were evaluated by observers as relatively intropunitive, oversocialized, and overcontrolling of impulse—an autocentric pattern. These prospective findings, spanning 15 years, provide evidence for the coherence over time in the personality attributes that prefigure depressive symptoms in late adolescence.

Our next analysis examined the personality context in childhood and adolescence that anticipate chronic depressive symptoms in 23-year-old young adults (Gjerde, 1995a). This study implemented analytic procedures to reduce the possibility of false positive findings.[9] Many prepubertal personality characteristics foretelling later depressive symptoms were obtained for males. Young men who acknowledged chronic depressive symptoms manifested early as in preschool (ages 3-4) allocentric behaviors: undersocialization of impulse and interpersonal antagonism. The nature of these prospective correlates remained highly stable over the following 15 years. Prospective correlates of depressive symptoms in young women, not reliably identified until adolescence (ages 14 and 18), were more likely than in young men to express such autocentric concerns as oversocialization of impulse, introspection, and self-concern. Based on both sets of findings (J. Block et al., 1991; Gjerde, 1995a), we conclude that gender moderates the prospective relations between early personality characteristics and subsequently acknowledged depressive symptoms in adolescents and young adults. The second prospective analysis (Gjerde, 1995a) further indicates that reliable personality antecedents of subsequently acknowledged depressive symptoms are stronger and emerge earlier (i.e., prior to puberty) in males than in females.

However, because autocentric patterns of impulse expression are unlikely to be associated with strong behavioral markers, our reliance on observer evaluations (Gjerde, (1995a) may have underestimated the strength of the early characterological roots of depressive symptoms in girls. Because dysphoric females are less

---

enduring characteristic with roots in early personality. As described below, the behavioral expressions of antagonism seen in the concurrent correlates of GBI depression scores at age 23 were not observable in the age-23 correlates of age-18 CESD scores (Gjerde, 1995b). Hence the GBI scale may provide a more sensitive assessment of this domain in females.

[9] See Gjerde (1995a) for further description of these procedures.

likely than dysphoric boys to emit clear behavioral cues about their inner states, self-report data may be required to identify the early antecedents of depressive disorders in girls.[10] This conjecture is further supported by the view that a passive, ruminative coping style (i.e., heightened self-focus) represents a risk factor for depression in women (Nolen-Hoeksema, 1990).

Of particular interest are the substantial gender differences obtained in relations between IQ and depressive symptoms. Intelligence, as measured at ages 4, 11, and 18 by standard intelligence tests, correlated positively with depressive symptoms in females and negatively with depressive symptoms in males (J. Block et al., 1991; Gjerde, 1995a). Dysphoric females—both as adolescents and young adults—were significantly more intelligent than dysphoric males as early as in preschool. This finding raises the intriguing and hitherto unexplored possibility that intelligence plays a different role in the origins of depressive symptoms in males and females. We provide a more extensive conceptual analysis of this finding below.

Gjerde (1995a) proposed several explanations for why depressive symptoms in young adulthood are associated, prospectively, with different behavioral tendencies in very young males and females. First, parents may react differently to early problem behaviors in boys and girls. Mothers, by backing away from oppositional behaviors in boys, may perpetuate their sons' negative behaviors (Maccoby & Jacklin, 1983). Evidence also indicates that shy, withdrawn girls receive affection from their parents while shy, withdrawn boys face parental discontent (Hinde & Stevenson-Hinde, 1987; Radke-Yarrow, Richters, & Wilson, 1988). Second, boys and girls differ in their perceptions of the repercussions of aggressive behaviors. Compared to girls, boys expect less guilt, less harmful consequences, less suffering by the victim, and less peer and parental displeasure for overt aggression (Perry, Perry, & Rasmussen, 1986; Perry, Perry, & Weiss, 1989; Smetana, 1989). Third, the different conduct rules distinguishing groups of boys and girls may also influence the manifestation of psychological distress. Maccoby (1986) has argued that "a girl risks ostracism and the breaking off of highly valued friendships if she shows hostility or even disagreement too openly. The control of aggressive behaviors appears to be stronger in girls' groups than boys" (p. 274). Finally, Zahn-Waxler (1993) notes that females are better then males in detecting others' affective cues, an ability that is likely to facilitate relationship formation and reduce the probability of overt transgressions of social rules.

---

[10] A related explanation of the lack of prepuberty precursors of depressive symptoms in girls would suggest that because childhood behaviors are associated with such symptoms are less disruptive in girls than in boys they are less likely to be identified. Zahn-Waxler et al. (1991) suggest that the antecedents of depression can be particularly difficult to identify because children at risk may appear "eager to please, competent in social relationships, mature, and sensitive to the needs of others" (p. 245). This conjecture may be more valid for females than for males and is consistent with our hypothesis about gender differences in the autocentric-allocentric balance of depression-related personality characteristics. For example, the greater autocentric tendency of females may obscure the early precursors of psychopathology in women; oversocialized behaviors are less socially disruptive and may therefore be harder to identify than undersocialized behaviors (see Gjerde, 1995a, for further discussion of this issue).

To summarize our concurrent and prospective findings: Depressive symptoms during adolescence and young adulthood reflect a personality structure vulnerable to depression. This vulnerability can be identified early in life. Its nature differs for boys and girls. These gender differences exhibit longitudinal consistency over time. The direction of anger expression (allocentric vs. autocentric) differentiates in a highly consistent manner between dysphoric adolescent males and females: Dysphoric males experience greater difficulties than do dysphoric females with impulse regulation and the ability to modulate expression of emotions according to social rules; however, as the subjects approach young adulthood, the strength of these gender differences is diminished. Equifinality (i.e., a pattern where different developmental trajectories lead to a similar outcome) therefore seems to characterize the development of depressive symptoms in males and females: despite noticcably different antecedents in childhood and (to a somewhat lesser extent) in adolescence, the concurrent age-23 correlates of depressive symptoms show considerable gender convergence.

Looking Forward:
Dysphoric Adolescents as Young Adults

Our next analysis (Gjerde, 1995b) examined the psychological characteristics of 18-year-old dysphoric adolescents five years later, as young adults. Psychopathology research has neglected the study of young adulthood (e.g., Bemporad, Ratey, & Hallowell, 1986; Fredrichs, Aneshansel, & Clark, 1981; Jessor et al., 1991; Looney, 1989). Nonetheless, existing prospective findings, however few, suggest that depressive symptoms in adolescence are related subsequently to a difficulty in establishing intimate relationships (Kandel & Davies, 1986), are associated subsequently with low self-esteem and emotional problems (Fleming et al., 1993), and are a risk factor for the subsequent diagnosis of clinical depression (Lewinsohn et al., 1988). These findings are congruent with the research of Jessor et al. (1991), showing substantial longitudinal consistency of individual differences in behavior problems from adolescence through young adulthood. However, previous studies of the psychological sequelae in young adults of adolescent dysphoria have tended to rely primarily on self-report data. In contrast, the present analyses employed both observer evaluations and subjects' self-perceptions of psychological status and are therefore less vulnerable to the possible biases that beset exclusive reliance on self-report methodologies in the study of depressive symptoms (see, e.g., Hirschfeld et al., 1983).

With respect to the female sample, we were particularly interested in learning whether the anger and interpersonal alienation acknowledged, but not expressed, by the dysphoric 18-year-old girls would continue to be withheld in young adulthood or whether these feelings would have "burst through" and manifested themselves in their interactions with others by the time these young women reached age

23. Adolescent girls' withholding of anger may represent an adolescence-specific intensification of gender-role differentiation (cf. J. H. Block, 1976; Hill & Lynch, 1983) and therefore might not persist into young adulthood when an easing of gender-role proscriptions might permit a more open expression in young women of negative feelings. More generally stated, does the allocentric-autocentric distinction, which so aptly distinguished the personalities of dysphoric male and female adolescents, continue to summarize the gender differences observed five years later when these individuals have become young adults?

As indicated by both observer data and subjects' own self-perceptions, depressive symptoms in adolescence, rather then merely reflecting an ephemeral state of despondency, continued to foretell psychological vulnerability and emotional distress in young adulthood. Dysphoric 18-year-olds of both sexes were more likely five years later to suffer from chronic depressive symptoms, to have engaged in suicidal ideation, and to be psychologically more brittle than nondysphoric 18-year-olds. These gender-neutral expressions in young adults of age-18 depressive symptoms may reflect an attenuation of the gender-divergent role pressures particularly characteristic of adolescence.

Despite these gender similarities, substantial gender differences nonetheless continued to characterize the age-23 correlates of age-18 depressive symptoms. These differences are consistent with our previously observed gender differences in the allocentric-autocentric mode of impulse expression. For males, depressive symptoms at age 18 predicted observer evaluations of undercontrol of impulse and inter- personal antagonism five years later and self-perceptions of hostility, inadequate anger control, stress reactivity, and absence of introspective tendencies. For females, depressive symptoms at age 18 predicted later observer evaluations of anxious rumi- nation about self-adequacy and self-perceptions of unfulfilled relationship aspirations (in particular, preoccupation with, or the experience of, a strong desire for intimacy mixed with worries that others are reluctant to get close) and low self-esteem.

The absence of interpersonal antagonism in age-23 females who had been dysphoric at age 18 may derive from the continuing influence of socialization practices and social norms that emphasize impulse control and acceptable conduct in women. This absence of outer-directed negative affect fits with the suggestion by Zahn-Waxler et al. (1991) that depression in women may be related to the withholding of aggression and self-assertiveness. Girls may be at risk for depressive mood when they are socialized early into roles that minimize self-expression and maximize role performance aimed at pleasing others. For girls, an inner-directed, anxiously introspective response to depressive feelings, although distressing and self-denying, is a socially untroublesome way of coping. In contrast, the lower threshold for impulse expression and antagonism characterizing dysphoric males may reflect the intensification, under distress, of socialization-based and socialization-rewarded gender differences in behavior patterns—the achievement of goals by aggressive means (e.g., Eron, 1980).

In young women, depressive feelings at age 18 were associated with unfulfilled relationship strivings five years later. These findings support the conjecture that emotional crises in adolescent girls are related to the issue of communion versus isolation, with depression emerging when interpersonal relatedness seems unattainable. Similarly, Leadbeater, Blatt, and Quinland (1995) argue that girls are particularly vulnerable to stressful events involving others (e.g., fear of abandonment, lack of nurturance). Emotional crises among adolescent boys, on the other hand, may be more closely associated with frustrated agency and unattainable achievement motives (Gjerde et al., 1988). This gender difference may explain why girls are more likely than boys to approach others when feeling depressed and therefore report less loneliness than do dysphoric boys (Koening, Isaacs, & Schwartz, 1994), whose interpersonal antagonisms are more likely to alienate others and produce a cycle of increasing isolation.

The finding that dysphoric 18-year-old girls described themselves at age 23 as harboring unfulfilled relational strivings also relates to theories emphasizing females' relationship orientation (e.g., Gilligan, 1982) and the early origins of depression as "rooted in patterns of responsibility and overinvolvement" (Zahn-Waxler et al., 1991, p. 259). Interpersonally sensitive individuals may be unduly preoccupied with the needs of others to the detriment of fulfillment of their own needs and desires. Gore, Aseltine, and Colten (1993), examining the implications for girls' mental health of an interpersonal caring orientation and overinvolvement in the lives of significant others, report that adolescent girls became more distressed by stressful family events if they were closely involved in their mother's problems and emotionally invested in relationships in general.

In view of the findings reported by Gore et al. (1993), it is particularly noteworthy that—in our longitudinal study—dysphoric 18-year-old females were more likely to have grown up in families characterized by maternal authoritarian control in conjunction with positive maternal engagement with the daughter (Gjerde et al., 1991). This maternal style resembles the parental "binding" strategy described by Stierlin (1973), in which the child is kept unduly psychologically captive by a dominant parent and deprived of opportunities to become autonomous. This "double-bind" may prevent the daughter from moving beyond the mother-daughter dyad and toward secure engagement with other individuals. Also emphasizing the role of the mother-daughter dyad in the origin of these relationship difficulties, Kandel and Davies (1982, 1986) note that depressed adolescents, especially girls, are likely to feel the same disaffection within a concurrent close relationship that they experienced in their early family situation. In sum, the findings from several independent studies converge in suggesting that dysphoric girls re-experience, within the context of intimate relationships, the feelings of insecurity they earlier felt in their relationship to their mother.

To summarize, depressive symptoms in adolescence predict further psychological distress during young adulthood. In addition, such symptoms anticipate both gender different and gender similar types of maladjustment. When gender differences are found, they continue to conform to our theoretical expectations regarding the directionality of symptom expression.

Issues for Further Consideration

Some overarching issues are raised by our studies of male and female depressive symptoms: (1) equifinality of pathways in the development of depressive symptoms in males and females; (2) the implications of the origin and continuance of depressive symptoms in males and females due to differential behavioral visibility associated with allocentric and autocentric symptom patterns; (3) the role of introspection in the regulation of impulse expression; and (4) the gender-related role of intelligence in the formation of depressive symptoms.

Equifinality in Development: Gender Differences in Developmental Pathways to Depression

Understanding psychopathology from a developmental perspective necessitates the examination of pathways toward (as well as away from) psychological maladaptation. Developmental pathways "are reconstructions of stages of development, normally not directly observable to researchers, but inferred from repeated assessments of the same subjects over time" (Loeber, p. 97). Loeber (1991b) correctly notes that, within the same diagnostic category, all individuals do not necessarily manifest identical developmental trajectories. Understanding such developmental pathways is especially complicated when different sets of individuals (in our studies, males and females) are moving toward a similar psychological outcome (depressive symptoms), but along different paths. The endpoint of this developmental pattern has been called *equifinality*, as opposed to *multipotentiality* (or multifinality) where a common starting point leads to different outcomes (e.g., Bertalanffy, 1968; Cicchetti, 1990; Loeber, 1991b). The ideas of equifinality and multipotentiality are necessary recognitions in attempts to understand the origins and course of depressive conditions.

The longitudinal research reviewed in this chapter identified depressive symptoms in young adults and traced backward in time their different personality and intellectual antecedents in males and females. These analyses provide evidence for movement toward equifinality in the development of depressive symptoms for the sexes: Despite substantial similarity in the ways that, as young adults, males and females experienced and reflected depressive symptoms, the prepubertal antecedents of these symptoms differed markedly. This gender difference in antecedents could still be discerned, albeit to a slightly lesser degree, in adolescence.

We further note that within each gender, further differences in pathways toward depressive symptoms may exist. Analyses of such within-gender trajectories, based on more homogeneous subgroupings of subjects, may provide an even more differentiated view of the development of depression. However, such analyses necessitate a change from the dominant emphasis on *variable-centered* methodologies (which examine intercorrelations among variables across individuals over time or the degree of consistent ordering of individuals over time) to a greater emphasis on *person-centered* methodologies (which examine patterns of characteristics within persons and identify subsets of individuals who share similar configurations of attributes). Person-centered studies identify the degree of consistent patterning of an individual's characteristics over time and set the stage for the recognition of different developmental types (or subgroups). The developmental course through time of these developmental types can then be studied (see e.g., Block, 1971; Ozer & Gjerde, 1988). Wohlwill (1980) referred to the distinction between variable-centered versus person-centered approaches as "the constancy of relative position" versus "the stability of ipsative relations," respectively. Preliminary results suggest the usefulness of establishing such "types." Using this analytical approach, Gjerde (1995c) identified a group of unresilient and overcontrolling preschool girls who described themselves as more vulnerable to dependency-related depression than their peers.[11]

To achieve a richer understanding of the factors underlying gender convergence in the manifestations of depressive symptoms after puberty, the following analytical possibilities are suggested: (1) the internal psychological dynamics of depressive symptoms differs between adolescence and young adulthood; (2) there are gender-related differences between these two ages in the kinds of life stressors contributing to depressive symptoms; (3) once adolescence is left behind, the expressions of depressive symptoms begin to converge for the sexes due to an easing of gender-role restrictions. The gender-role hypothesis holds special promise in our view. For example, Huselid and Farmer (1994) suggest that "differential socialization of young men and women [in adolescence] at least partly explain specific gender-linked vulnerabilities to the experience and / or expression of symptoms" (p. 602). With the relative freedom of adulthood, individuals may feel freer to express their internal despair in ways that are less limited by gender-role proscriptions.

Directionality of Impulse Expression, Behavioral Markers, and Gender Differences in the Dynamics of Depression.

As noted earlier, allocentric and autocentric patterns are likely to differ in the degree to which they are associated with clear behavioral markers. This difference in "behavioral penetration" (Tellegen, 1991) is relevant both to the measurement of depressive symptoms as well as to our understanding of the processes underlying the origins and perseverance of such symptoms. In general, the allocentric pattern,

---

[11] Blatt (1974) distinguished between two types of vulnerability to depression: dependency (or anaclitic depression) versus self-criticism (or introjective depression). (For a more recent and theoretically expanded description of this distinction, see Blatt, 1990, and Blatt & Shichman, 1983.)

with its emphasis on problems of external adaptation, is likely to have more potential behavioral markers and therefore be socially more visible and societally more disruptive than the autocentric pattern, with its unobtrusive emphasis on problems of internal adaptation.

As suggested by applications of Brunswik's (1956) lens model to person perception (e.g., Funder & Sneed, 1993; Gangestad, DiGeronimo, Simpson, & Biek, 1992; Harkness, Tellegen, & Waller, 1995), the inability of observers to discern clear behavioral markers of a trait may stem from two processes: (1) the failure of a target person to emit sufficient valid behavioral cues regarding internal psychological states (i.e., low trait visibility or reduced cue emission), or (2) the failure of others to recognize or correctly interpret accessible behavioral cues (i.e., reduced or incorrect cue utilization). The former process (i.e., reduced cue emission), we suggest, is most likely to explain why the psychological despondency of dysphoric girls is less noticed by others than the psychological despondency of dysphoric boys.

Differences in the behavioral visibility associated with allocentric and autocentric symptom patterns may influence the origins and continuance of depressive symptoms. Caspi and Bem (1990) describe three types of person-environment interactions that are especially significant in regulating life trajectories:

> Reactive interactions occur when different individuals exposed to the same environment experience it, interpret it, and react to it differently. Evocative interaction occurs when an individual's personality evokes distinctive responses in others. Proactive interactions occur when individuals select or create environments of their own (p. 565)

We suggest the concept of *evocative* interaction is particularly useful in explaining depressive symptoms in males—the aversive interactional style of dysphoric males (e.g., disruptive anger and uncontrolled aggression) may, through the alienation of others, reduce the availability of the very social supports that can help lessen depressive affect. By evoking reciprocal antagonistic or rejecting responses from others, aggressive dysphoric males may further increase their social isolation and thereby also their level of depression (e.g., Rook, Pietromonaco, & Lewis, 1994; see also Coyne, Downey, & Boergers, 1992, for a review of the literature on the aversive stimulus value of depressed persons). There already is evidence indicating that behavioral problems place males at increased risk for depressive symptoms (Capaldi, 1991; Hops, Lewinsohn, Andrews, & Roberts, 1990; Rohde, Lewinsohn, & Seeley, 1991) and that unsocialized behavior of boys with relatively low intelligence eventually leads to interpersonal rejection and depressive mood (Patterson & Capaldi, 1990; Patterson & Stoolmiller, 1991). For females, in contrast, persistent rumination regarding the origin and meaning of depressive episodes may increase depressive affect (Nolen-Hoeksema, 1990). This autocentric symptom pattern is less likely to evoke negative responses from others (except perhaps in very close relationships). Due to their autocentric focus, dysphoric females may withdraw from

social engagement and seek isolation and seclusion. This evasion of social participation—illustrative of a *proactive* interaction pattern—may contribute to the origin
and continuance of depressive symptoms in females.

Introspective Tendencies as Mediating Mode of Impulse Expression in Dysphoric Individuals.

Depression is usually related to heightened self-awareness, especially after failure
(cf. e.g., Greenberg & Pyszcynski, 1986). Our own findings indicate that introspection and self-awareness were more characteristic of dysphoric females than of
dysphoric males. This gender difference in tendency to evoke the self-percept under
conditions of distress may mediate the relation between the direction of impulse
expression and psychological dysfunction. Greenberg, Kusche, and Speltz (1991),
in their discussion of impulse regulation and self-awareness, note that impulsive
acts are characterized by a lack of self-awareness:

Although the one does not imply the other, when self-control is not evident, it is likely
that self-awareness (as a conscious process) is also minimal or absent. Thus, from a
developmental perspective, it is unlikely that treatments that attempt to alter self-
control or impulse regulation can be successful unless the individual also becomes
aware of affective signals. (p. 48)

Analogous views have been advanced for many years by psychodynamic theory
(e.g., Shapiro, 1965) and more recently by self psychology (e.g., Baumeister, 1990).
Shapiro (1965) observes that impulsive action is designated by "speediness, abrupt-
ness, and lack of planning," characteristics that seems "to reflect a deficiency in
certain mental processes that are normally involved in the translation of incipient
motives into actions" (p. 140). Baumeister (1990) notes that negative affects often
derive from awareness of the self as failing to achieve desired standards. The
immediate response to psychologically aversive states, he suggests, is a "subjective
shift to less meaningful, less integrative forms of thought and awareness" that he
labels as "cognitive deconstruction" (p. 92). This subjective state entails a failure to
recognize meaningful, long-term connections between present action and future
outcomes; the behavioral consequences of this lack of inner life are impulsive and
destructive acts that soon have adverse consequences for the individual.

While excessive introspection, especially in the form of rumination on depressive episodes, may be maladaptive (e.g., Nolen-Hoeksema, 1990), a set of connected characteristics (e.g., lack of introspective ability, low self-awareness, absence
of a subjective sense of intentionality, and constricted temporal focus) may set the
stage for undersocialized behavior. As already noted, such societally disruptive
behavior patterns are likely to promote interpersonal rejection, and its consequence, depressive symptoms (e.g., Patterson & Stoolmiller, 1991). The greater
autocentric tendency of dysphoric females may stem from socialization orientations

that emphasize the development of self-consciousness and inhibition of aggression in girls (J. H. Block, 1983).

Intelligence, Gender, and the Development of Depressive Symptoms.

We conclude this chapter by considering connections between depressive symptoms and intelligence. Because intelligence can be viewed as a basis for adaptive behavior, it is relevant to evaluate its relation to depressive symptoms.

For males, the relation between IQ and depressive symptoms is negative (J. Block et al., 1991; Gjerde, 1995a). This finding fits the expectation that lack of competence in early life contributes cumulatively to depression. Failure to master early developmental tasks reduces the boys' ability to cope with subsequent challenges and sets the stage for subsequent failures. In time, the accumulation of failures and the recognition of how failures are proving to be life limiting lead to vulnerability to depression (cf. e.g., Cicchetti & Schneider-Rosen, 1986, for a similar view). Research also suggests that high intelligence is likely to operate as a protective factor (e.g., Kandel et al., 1988; Masten, Garmezy, Tellegen, Pellegrini, Larkin, & Larsen, 1988). Repeatedly, in ongoing life, low intelligence evokes negative reinforcement from the environment, which diminishes the individual's overall sense of self-worth.

For the female sample, however, we find the relation between intelligence and depressive symptoms to be positive. This reliable positive relation between IQ and depressive symptoms in girls has few antecedents in research on psychopathology. However, several other studies also have observed that the relation between intelligence and psychological adaptation may not be invariably positive. Luthar (1991) reports that, under levels of stress, intelligence may increase vulnerability rather than operate as a protective mechanism. Intelligent children have also been found to be more responsive and therefore susceptible to the qualities—negative as well as positive—of their environment (Zigler & Farber, 1985), to interpersonal interactions (Pellegrini, Masten, Garmezy, & Ferrarese, 1987), and to intrapsychic experiences (Zigler & Glick, 1986). Of particular importance is the prospective finding that intelligent youth who report symptoms of depression and anxiety are more likely than less intelligent youth to show decreases in social competence over time. Under adversity, intelligent youth, perhaps because of their greater sensitivities and articulation of experience, may have psychological vulnerabilities that do not occur in less intelligent, "mindless" youth (Luthar & Ripple, 1994). For reasons involving societal pressures regarding sex-role development, we suggest this possibility is more likely for girls than for boys.

In our society, intelligence has had greater implication for psychological adaptation in males than females. Throughout childhood and adolescence, our findings indicate, the negative correlations of IQ with depressive symptoms in males were in absolute terms stronger than the positive correlations of IQ with depressive symptoms in females. Finding that men with high IQ seem to have developed a sense of self as competent and agentic, J. Block and Kremen (1995) note that "intelligence represents a means through which they can implement the agentic, masculine role prescribed by sex-linked socialization patterns and therefore, when intelligence is present, it serves to shore up self-esteem" (p. 358). In contrast, J. Block and Kremen suggest that IQ is less crucial to female life options as compared to male life options. For women,

> [I]ntelligence as measured by IQ tests may be less of a prerequisite for competence in the interpersonal world, the area of competence which is especially valued in feminine, communal models of self. Thus, high intelligence may have less afformative significance for a sense of self in young women than does, for a prime example, physical and social attractiveness (p. 358).

In contrast, intelligence may be more centrally involved in the kinds of striving for success and respect that characterize male life paths.

Our own interpretation of these different implications of IQ per se is that adherence to their societally designated sex-roles creates greater stress for intelligent girls than it does for intelligent boys. Both females and males, during their transition to and through adolescence, appear to intensify identification with their own gender (Hill & Lynch, 1983). This transitional intensification has been recognized as a period of conflict and therefore special vulnerability for girls (Archer, 1984; Newcomb, Huba, & Bentler, 1986; Petersen, 1979, Petersen et al., 1991; Siddique & D'Arcy, 1984). Traditionally, the female gender role has been one that has discouraged competence and assertiveness on the grounds that such characteristics are "unfeminine." Relatively intelligent adolescent girls can be expected to be less traditional than their peers; they may retain, even with intensified gender-role expectations, their interests in traditionally "unfeminine" activities. Pressured to accept conventional mores, they may find such pushes and pulls to be conflict inducing. On the one hand, they wish indeed to be feminine; on the other hand, society suggests and may even dictate constraining the particular interests and activities that their native intelligence has evoked. Therefore, female sex-role expectations may be experienced by intelligent girls as ego dystonic; for intelligent males, however, male sex-role expectations may be experienced as ego syntonic. As a possible consequence of pressures to accept narrow cultural assumptions, relatively bright girls may experience particular sadness over the way their lives are being shaped and the criteria by which they are being judged. Over time, the cumulative effect of these recognitions may result in introspectiveness, problems of self-esteem, and a persistent melancholia. Support for this interpretation comes from a variety of studies indicating that females who do not conceal their competence and assertiveness are at risk for depression (e.g., Gove & Tudor, 1973; McGrath, Keita, Strickland, & Russo, 1990; Radloff, 1975, 1978; Teti, 1974).

Conclusion

Building on principles of developmental psychopathology, psychodynamic theories regarding the direction of impulse expression / regulation, and examination of gender differences in socialization, this chapter has illustrated how longitudinal research can increase understanding of gender differences in the development of depressive symptoms in adolescence and young adulthood. Differences in the direction of impulse expression—allocentric versus autocentric—were introduced as a framework to summarize prospective as well as concurrent depression-related personality characteristics in males and females. In particular, depressive symptoms in females were suggested as reflecting problems of internal adaptation—an autocentic pattern associated with few easily visible behavioral markers—and depressive symptoms in males were suggested as reflecting problems of external adaptation—an allocentric pattern associated with many and obvious behavioral markers. The following issues were discussed: equifinality of pathways in the development of depressive symptoms, the implications for the development of depressive symptoms of the differential behavioral visibility associated with allocentric and autocentric behavior patterns, the role of introspection in the regulation of impulse expression in dysphoric individuals, and the gender related role of intelligence in the formation of depressive symptoms. The ramifications of allocentric and autocentric modes of impulse expression for measurement strategies in the study of depression were also briefly evaluated. Overall, our longitudinal findings indicate that the pathways to depressive symptoms in young males and females differ in a theoretically meaningful and predictable manner. As such, they provide substantial support for the view that one must not prematurely assume gender equivalence in the processes underlying the development of psychopathology. Whether these gender differences are specific to adolescence and young adults or will also persist as the subjects pass through their third decade of life is an issue for further research.

REFERENCES

Abraham, K. (1927). Notes on the psychoanalytical investigation and treatment of manic-depressive insanity and allied conditions. In K. Abraham (Ed.), *Selected papers on psychoanalysis* (pp. 137–156). London: Hogarth Press. (Original work published in 1911.)

Abramson, L. Y., Seligman, M. E. P., & Teasdale, J. D. (1978). Learned helplessness in humans: Critique and reformulation. *Journal of Abnormal Psychology, 87,* 49–74.

Achenbach, T. M. (1991). *Manual for the Child Behavior CBLC / 4–18, YSR, and TRF Profiles.* Burlington: Department of Psychiatry, University of Vermont.

Achenbach, T. M., & Edelbrock, C. S. (1978). The classification of child psychopathology: A review and analysis of empirical efforts. *Psychological Bulletin, 85,* 1275–1301.

Akiskal, H. S., Hirschfeld, R. M., & Yerevanian, B. I. (1983). The relationship of personality to depressive disorders. *Archives of General Psychiatry, 40,* 801–810.

Allgood-Merten, B., Lewinsohn, P. M., & Hops, H. (1990). Sex differences and adolescent depression. *Journal of Abnormal Psychology, 99*, 55–63.

Anderson, J. C., Williams, S., McGee, R., & Silva, P. A. (1987). *DSM-III* disorders in pre-adolescent children. *Archives of General Psychiatry, 44*, 69–76.

Arieti, S., & Bemporad, J. (1980). The psychological organization of depression. *American Journal of Psychiatry, 137*, 1360–1365.

Barnett, P. A., & Gotlib, I. H. (1988). Psychosocial functioning and depression: Distinguishing among antecedents, concomitants, and consequences. *Psychological Bulletin, 104*, 97–126.

Baumeister, R. F. (1990). Suicide as escape from self. *Psychological Review, 97*, 90–113.

Beck, A. T. (1967). *Depression: Clinical, experimental, and theoretical aspects*. New York: Hoeber.

Bemporad, J. (1980). The psychodynamics of mild depression. In S. Arieti & J. Bemporad (Eds.), *Severe and mild depression: The psychotherapeutic approach* (pp. 156–184). London: Tavistock.

Bemporad, J., Ratey, J. J., & Hallowell, E. M (1986). Loss and depression in young adults. *Journal of the American Academy of Psychoanalysis, 14*, 167–179.

Bemporad, J., & Wilson, A. (1978). A developmental approach to depression in childhood and adolescence. *Journal of the American Academy of Psychoanalysis, 6*, 325–352.

Blatt, S. J. (1974). Levels of object representation in anaclitic and introjective depression. *Psychoanalytic study of the child: Vol. 29* (pp. 107–159). New York: International Universities Press.

Blatt, S. J. (1990). Interpersonal relatedness and self-definition: Two personality configurations and their implications for psychopathology and psychotherapy. In J. Singer (Ed.), *Repression and dissociation: Implications for personality theory, psychopathology, and health* (pp. 299–335). Chicago: University of Chicago Press.

Blatt, S. J., & Shichman, S. (1983). Two primary configurations of psychopathology. *Psychoanalysis and Contemporary Thought, 6*, 187–254.

Block, J. (1950). *An experimental investigation of the construct of ego-control*. Unpublished doctoral dissertation, Department of Psychology, Stanford University, Stanford, CA.

Block, J. (1971). *Lives through time*. Berkeley, CA: Bancroft Books.

Block, J. (1978). *The Q-sort method in personality assessment and psychiatric research*. Palo Alto, CA: Consulting Psychologists Press. (Original work published 1961.)

Block, J. (1993). Studying personality the long way. In D. Funder, R. Parke, C. Tomlinson-Keasey, & K. Widaman (Eds.), *Studying lives through time: Personality and development* (pp. 9–44). Washington, DC: American Psychological Association.

Block, J., & Gjerde, P. F. (1990). Depressive symptoms in late adolescence: A longitudinal perspective on personality antecedents. In J. E. Rolf, A. Masten, D. Cicchetti, K. Neuchterlein, & S. Weintraub (Eds.), *Risk and protective factors in the development of psychopathology* (pp. 334–360). New York: Cambridge University Press.

Block, J., Gjerde, P. F., & J. H. Block. (1991). Personality antecedents of depressive tendencies in 18-year-olds: A prospective study. *Journal of Personality and Social Psychology, 60*, 726–738.

Block, J., & Kremen, A. (1995). *Ego-resilience and intelligence: Differential personality implications*. Manuscript in preparation.

Block, J. H. (1951). *An experimental study of a topological representation of ego structure*. Unpublished doctoral dissertation, Department of Psychology, Stanford University, Stanford, CA.

Block, J. H. (1973). Conceptions of sex role: Some cross-cultural and longitudinal perspectives. *American Psychologist, 28*, 512–526.

Block, J. H. (1983). Differential premises arising from differential socialization of the sexes: Some conjectures. *Child Development, 54*, 1335–1354.

Bertalanffy, L. (1968). *General systems theory: Foundations, development, applications*. New York: Braziller.

Block, J. H., & Block, J. (1980). The role of ego-control and ego-resiliency in the organization of behavior. In W. A. Collins (Ed.), *Minnesota Symposia on Child Psychology:* Vol. 13 (pp. 39–101). Hillsdale, NJ: Erlbaum.

Bowlby, J. (1988). Developmental psychiatry comes of age. *American Journal of Psychiatry,* 145, 1–10.

Brunswik, E. (1956). *Perception and the representative design of psychological experiments.* Berkeley: University of California Press.

Buck, R. W., Savin, V. J., Miller, R. E., Caul, W. F. (1972). Communication of affect through facial expression in humans. *Journal of Personality and Social Psychology,* 23, 362–371.

Capaldi, D. M. (1991). Co-occurrence of conduct problems and depressive symptoms in early adolescence boys: I. Familial and general adjustment at grade 6. *Development and Psychopathology,* 3, 277–300.

Capaldi, D. M. (1992). Co-occurrence of conduct problems and depressive symptoms in early adolescence boys: II. A 2-year follow-up at grade 8. *Development and Psychopathology,* 4, 125–144.

Caspi, A., & Bem, D. J. (1990). Personality continuity and change across the life course. In L. A. Pervin (Ed.), *Handbook of personality: Theory and research* (pp. 549–575). New York: Guilford Press.

Chodoff, P. (1972). The depressive personality. *Archives of General Psychiatry,* 27, 666–673.

Cicchetti, D. (1989). A historical perspective on the discipline of developmental psychopathology. In J. Rolf, A. Masten, D. Cicchetti, S. Weintraub & K. H. Nuechterlein, (Eds.), *Risk and protective factors in the development of psychopathology* (pp. 2–28). New York: Cambridge University Press.

Cicchetti, D., & Schneider-Rosen, K. (1986). An organizational approach to childhood depression. In M. Rutter, C. E. Izard, & P. B. Read (Eds.), *Depression in young people* (pp. 71–134). New York: Guilford Press.

Colvin, C. R. (1993). Judgable persons: Personality, behavior, and competing explanations. *Journal of Personality and Social Psychology,* 64, 861–873.

Coyne, J. C., Downey, G., & Boergers, J. (1992). Depression in families: A systems perspective. In D. Cicchetti and S. L. Toth (Eds.), *Rochester Symposium on Developmental Psychopathology: Vol. 4. Developmental perspectives on depression* (pp. 211–250). Rochester, NY: University of Rochester Press.

Csikszentmihalyi, M., & Larson, R. (1984). *Being adolescent: Growth and conflict in the teenage years.* New York: Basic Books.

Dawes, R. M. (1993). Prediction of the future versus understanding of the past: A basic asymmetry. *American Journal of Psychology,* 106, 1–24.

Depue. R. A. (1987). *The General Behavior Inventory.* Department of Psychology, University of Minnesota, Minneapolis.

Depue, R. A., Krauss, S., Spoont, M. R., & Arbisi, P. (1989). General behavior inventory identification of unipolar and bipolar affective conditions in a nonclinical university population. *Journal of Abnormal Psychology,* 98, 117–126.

Dweck, C. S., Davidson, W., Nelson, S., & Enna, B. (1978). Sex differences in learned helplessness: II. The contingencies of evaluative feedback in the classroom; III. An experimental analysis. *Developmental Psychology,* 14, 268–276.

Edelbrock, C., & Achenbach, T. M. (1980). A typology of child behavior profile patterns: Distribution patterns and correlates for disturbed children aged 6–16 years. *Journal of Abnormal Child Psychology,* 8, 441–470.

Eron, L. D. (1980). Prescription for reduction of depression. *American Psychologist,* 35, 244–252.

Farmer, R., & Nelson-Gray, R. O. (1990). Personality disorders and depression: Hypothetical relations, empirical findings, and methodological considerations. *Clinical Psychology Review, 10*, 453–476.

Fenichel, O. (1945). *The psychoanalytic theory of neurosis*. New York: Norton.

Fischoff, B., & Beyth, R. (1975). "I knew it would happen"—Remembered probabilities of once-future things. *Organizational Behavior and Human Performance, 13*, 1–16.

Fleming, J.E., Boyle, M.H., and Offord, D. R. (1993). The outcome of adolescent depression in the Ontario child health study follow-up. *Journal of the American Academy of Child and Adolescent Psychiatry, 32*, 2833.

Fredrichs, R. R., Aneshansel, C. S., & Clark, V. A. (1981). Prevalence of depression in Los Angeles County. *American Journal of Epidemiology, 113*, 691–699.

Freud, S. (1968). The psychogenesis of a case of homosexuality in woman. In J. Strachey (Ed.), *The standard edition of the complete works of Sigmund Freud: Vol. 18* (pp. 146–174). London: Hogarth Press.

Friedman, A. S. (1970). Hostility factors and clinical improvements in depressed patients. *Archives of General Psychology, 23*, 524–537.

Funder, D. C., & Sneed, C. D. (1993). Behavioral manifestations of personality: An ecological approach to judgmental accuracy. *Journal of Personality and Social Psychology, 64*, 479–490.

Gangestad, S. W., DiGeronimo, K., Simpson, J. A., & Biek, M. (1992). Differential accuracy in person perception across traits: An ecological approach to judgmental accuracy. *Journal of Personality and Social Psychology, 62*, 688–698.

Gershon, E. S., Cromer, M., & Klerman, G. L. (1968). Hostility and depression. *Psychiatry, 31*, 224–235.

Gilligan, C. (1982). *In a different voice*. Cambridge, MA: Harvard University Press.

Gjerde, P. F. (1993). Depressive symptoms in young adults: A developmental perspective on gender differences. In D. Funder, R. Parke, C. Tomlinson-Keasey, & K. Widaman (Eds.), *Studying lives through time: Personality and development* (pp. 255–288). Washington, DC: American Psychological Association.

Gjerde, P. F. (1995a). Alternative Pathways to Chronic Depressive Symptoms in Young Adults: Gender Differences in Developmental Trajectories. *Child Development*, in press.

Gjerde, P. F. (1995b). *Gender Differences in Young Adulthood Sequelae of Depressive Symptoms During Adolescence: A Five-Year Prospective Study*. Manuscript in review.

Gjerde, P. F. (1995c, March). *A typological analysis of girls' personality: A longitudinal study of developmental pathways*. Paper presented at the Symposium "Resilience beyond adolescence: Longitudinal studies of transition for adolescence to young adult years" (S. Hauser and J. Brooks-Gunn, Chairs) at the meeting of the Society for Research on Child Development, Indianapolis.

Gjerde, P. F., & Block, J. (1991). Preadolescent antecedents of depressive symptomatology in late adolescence: A prospective study. *Journal of Youth and Adolescence, 20*, 215–230.

Gjerde, P. F., Block, J., & Block, J. H. (1988). Depressive symptoms and personality during late adolescence: Gender differences in the externalization-internalization of symptom expression. *Journal of Abnormal Psychology, 97*, 475–486.

Gjerde, P. F., Block, J., & Block, J. H. (1991). The Preschool Family Context of 18-Year-Olds with Depressive Symptoms: A Prospective Study. *Journal of Research on Adolescence, 1*, 63–91.

Gore, S., Aseltine, R. H., & Colten, M. E. (1993). Gender, social-relational involvement, and depression. *Journal of Research on Adolescence, 3*, 101–125.

Gove, W. R., & Tudor, J. F. (1973). Adult sex roles and mental illness. *American Journal of Sociology*, 78, 812–835.

Greenberg, J., & Pyszcynski, T. (1986). Persistent high self-focus after failure and low self-esteem after success: The depressive self-focus style. *Journal of Personality and Social Psychology*, 50, 1039–1044.

Greenberg, M., Kusche, C. A., & Speltz, M. (1988). Emotional regulation, self-control, and psychopathology: The role of relationships in early childhood. In D. Cicchetti and S. L. Toth (Eds.), *Rochester Symposium on Developmental Psychopathology: Vol. 2. Internalization and externalization expression of dysfunction* (pp. 21–56). Rochester, NY: University of Rochester Press.

Harkness, A. R., Tellegen, A., & Waller, N. (1995). Differential convergence of self-report and informant data for Multidimensional Personality Questionnaire traits: Implications for the construct of negative emotionality. *Journal of Personality Assessment*, 64, 185–204).

Harter, S., & Marold, D. B. (1994). The directionality of the link between self-esteem and affect: Beyond causal modeling. In D. Cicchetti and S. L. Toth (Eds.), *Rochester Symposium on Developmental Psychopathology: Vol. 5. Disorders and dysfunctions of the self* (pp. 333–370). Rochester, NY: University of Rochester Press.

Henry, B., Moffitt, T. E., Caspi, A., Langley, J., & Silva, P. A. (1994). On the *Remembrance of Things Past*: A longitudinal evaluation of the retrospective method. *Psychological Assessment*, 6, 92–101.

Hill, J. P., & Lynch, M. E. (1983). The intensification of gender-related role expectations during early adolescence. In J. Brooks-Gunn & A. Petersen (Eds.), *Girls at puberty: Biological and psychosocial perspectives* (pp. 175–201). New York: Plenum Press.

Hinde, R. A., & Stevenson-Hinde, J. (1987). Implications of a relationship approach for the study of gender differences. *Infant Mental Health Journal*, 8, 221–235.

Hirschfeld, R. M. A., Klerman, G. L., Chodoff, P., Korchin, S., & Barrett, J. (1976). Dependency—self–esteem—clinical depression. *Journal of the American Academy of Psychoanalysis*, 4, 373–388.

Hirschfeld, R. M. A., Klerman, G. L., Clayton, P. J., Keller, M. B., McDonald-Scott, P., & Larkin, B. H., (1983). Assessing personality: The effects of the depressive state on trait measurement. *American Journal of Psychiatry*, 140, 695–699.

Hops, H., Lewinsohn, P. M., Andrews, J. A., & Roberts, R. E. (1990). Psychosocial correlates of depressive symptomatology among high school students. *Journal of Clinical Child Psychology*, 19, 211–220.

Huselid, R. F., & Cooper, M. L. (1994). Gender roles as mediators of sex differences in expressions of pathology. *Journal of Abnormal Psychology*, 103, 595–603.

Ingram, R. E., Cruet, D., Johnson, B. R., & Wisnicki, K. S. (1988). Self-focused attention, gender, gender role, and vulnerability to negative affect. *Journal of Personality and Social Psychology*, 55, 967–978.

Jessor, R., Donovan, J. E., & Costa, F. M. (1991). *Beyond adolescence: Problem behavior and young adult development*. Cambridge, MA. Cambridge University Press.

Jones, H. E. (1935). The galvanic skin response as related to overt emotional expression. *American Journal of Psychology*, 47, 241–251.

Kahn, J., Coyne, J. C., & Margolin, G. (1985). Depression and marital disagreement: The social construction of despair. *Journal of Social and Personal Relationships*, 2, 447–461.

Kandel, D. B., & Davies, M. (1986). Sequelae of adolescent depressive symptoms. *Archives of General Psychiatry*, 43, 255–262.

Kaplan, A. (1986). The "self-in-relation": Implications for depression in women. *Psychotherapy, 23*, 234–242.

Kavanagh, K., & Hops, H. (1994). Good girls? Bad boys?: Gender and development as contexts for diagnosis and treatment. In T. H. Ollendick & R. J. Prinz (Eds.), *Advances in clinical child psychology* (Vol. 16, pp. 45–79). New York: Plenum Press.

Kobak, R. R., Sudler, N., & Gamble, W. (1992). Attachment and depressive symptoms during adolescence: A developmental pathways analysis. *Development and Psychopathology, 3*, 461–474.

Koening, L. J., Isaacs, A. I., & Schwartz, J. A. J. (1994). Sex differences in depression and loneliness: Why are boys lonelier if girls are more depressed? *Journal of Research in Personality, 28*, 27-43.

Leadbeater, B. J., Blatt, S. J., & Quinland, D. M. (1995). Gender-linked vulnerabilities to depressive symptoms, stress, and problem behaviors in adolescents. *Journal of Research on Adolescence, 5*, 1–30.

Lewin, K. (1935). *A dynamic theory of personality*. New York: McGraw-Hill.

Lewin, K. (1951). Behavior and development as a function of the total situation. In *Field theory in the social sciences* (pp. 238–303). New York: Harper.

Lewinsohn, P. M., Hoberman, H. M., & Rosenbaum, M. (1988). A prospective study of risk factors for unipolar depression. *Journal of Abnormal Psychology, 97*, 251–264.

Lewinsohn, P. M., Hops, H., Roberts, R. E., Seeley, J. R., & Andrews, J. A. (1993). Adolescent psychopathology: I. Prevalence and incidence of depression and other *DSM-III-R* disorders in high-school students, *Journal of Abnormal Psychology, 102*, 133–144.

Lewinsohn, P. M., Roberts, R. E., Seeley, J. R., & Rohde, P., Gotlib, I. H., & Hops, H. (1994). Adolescent psychopathology: II. Psychosocial risk factors for depression. *Journal of Abnormal Psychology, 103*, 302–315.

Lewinsohn, P. M., Steinmetz, J. L, Larson, D. W., & Franklin, J. (1981). Depression-related cognitions: Antecedents or consequence? *Journal of Abnormal Psychology, 90*, 213–219.

Loeber, R. (1991a). Antisocial behavior: More enduring than changeable. *Journal of American Academy for Child and Adolescent Development, 30*, 393–397.

Loeber, R. (1991b). Questions and advances in the study if developmental pathways. In D. Cicchetti and S. L. Toth (Eds.), *Rochester Symposium on Developmental Psychology: Vol. 3. Models and intergrations* (pp. 97–116). Rochester, NY: University of Rochester Press.

Loeber, R., & Farrington, D. P. (1994). Problems and solutions in longitudinal and experimental treatment studies of child psychopathology and delinquency. *Journal of Consulting and Clinical Psychology, 62*, 887–900.

Looney, J. G. (1989). Editor's introduction. *Adolescent Psychiatry: Developmental and Clinical Studies, 16*, 121–126.

Luthar, S. S. (1991). Vulnerability and resilience: A study of high-risk adolescents. *Child Development, 62*, 600–616.

Luthar, S. S., & Ripple, C. H. (1994). Sensitivity to emotional stress among intelligent adolescents: A short-term prospective study. *Development and Psychopathology, 6*, 343–357.

Maccoby, E. E. (1986). Social groupings in childhood: Their relationship to prosocial and antisocial behavior in boys and girls. In D. Olweus, J. Block, & M. Radke-Yarrow (Eds.), *Development of prosocial and antisocial behavior* (pp. 263–284). London: Academic Press.

Maccoby, E. E. (1990). Gender and relationships. *American Psychologist, 45*, 513–520.

Maccoby, E. E., & Jacklin, C. N. (1983). The "person" characteristics of children and the family environment. In D. Magnusson & V. Allen (Eds.), *Human development: An interactional perspective*. New York: Academic Press.

Masten, A., & Braswell, L. (1991). Developmental psychopathology: An integrative frame-work for understanding behavior problems in children and adolescents. In P. R. Martin (Ed.), *The handbook of behavior therapy and psychological science: An integrative approach* (pp. 35–56). New York: Pergamon.

Masten, A., Garmezy, N., Tellegen, A., Pellegrini, D. S., Larkin, K., & Larsen, A. (1988). Competence and stress in school children: The moderating effects of individual and family qualities. *Journal of Child Psychology and Psychiatry, 29,* 745–764.

McCranie, E. J. (1971). Depression, anxiety, and hostility. *Psychiatric Quarterly, 45,* 117–133.

McGrath, E., Keita, G. P., Strickland, B. R., & Russo, N. F. (1990). *Women and depression.* Washington, DC: American Psychological Association.

Moffitt, T. E. (1993). Adolescence-limited and life-course-persistent antisocial behavior: A developmental taxonomy. *Psychological Review, 100,* 674–701.

Newcomb, M. D., Huba, G. J., & Bentler, P. M. (1986). Desirability of various life change events among adolescents: Effects of exposure, sex, age, and ethnicity. *Journal of Research on Personality, 20,* 207–227.

Nolen-Hoeksema, S. (1990). *Sex differences in depression.* Stanford, CA: Stanford University Press.

Nolen-Hoeksema, S., Girgus, J. S., & Seligman, M. E. (1992). Predictors and consequences of childhood depressive symptoms: A 5-year longitudinal study. *Journal of Abnormal Psychology, 101,* 405–422.

Ollendick, T. H., & King, N. J. (1994). Diagnosis, assessment, and treatment of internalizing problems in children: The role of longitudinal data. *Journal of Consulting and Clinical Psychology, 62,* 918–927.

Ostrov, E., Offer, D., & Howard, K. I. (1989). Gender differences in adolescent symptomatology: A normative study. *Journal of the American Academy of Child and Adolescent Psychiatry, 28,* 394–398.

Ozer, D. (1989). Construct validity in personality measurement. In D. M. Buss & N. Cantor (Eds.), *Personality psychology: Recent trends and emerging directions* (pp. 224–234). New York: Springer-Verlag.

Ozer, D., & Gjerde, P. F. (1989). Patterns of personality consistency and change from childhood through adolescence. *Journal of Personality, 57,* 483–507.

Patterson, G. R., & Capaldi, D. M. (1990). A mediational model for boys' depressed mood. In J. Rolf, A. S. Masten, D. Cicchetti, K. H. Nuechterlein, & S. Weintraub (Eds.), *Risk and protective factors in the development of psychopathology* (pp. 141–163). New York: Cambridge University Press.

Patterson, G. R., & Stoolmiller, M. (1991). Replications of a dual failure model for boys' depressed mood. *Journal of Consulting and Clinical Psychology, 59,* 491–498.

Pellegrini, D. S., Masten, A., Garmezy, N., & Ferrarese, M. J. (1987). Correlates of social and academic competence in middle childhood. *Journal of Child Psychology and Psychiatry, 28,* 699–714.

Perry, D. G., Perry, L. C., & Rasmussen, P. (1986). Cognitive social learning mediators of aggression. *Child Development, 57,* 700–711.

Perry, D. G., Perry, L.C., & Weiss, R. J. (1989). Sex differences in the consequences that children anticipate for aggression. *Developmental Psychology, 25,* 312–319.

Petersen, A. C., Compas, B. E., Brooks-Gunn, J., Stemmler, M., Ey, S., Sydney, S., & Grant, K. E. (1993). Depression in adolescence. *American Psychologist, 48,* 155–168.

Petersen, A. C., Sarigiani, P. A., & Kennedy, R. E. (1991). Adolescent depression: Why more girls? *Journal of Youth and Adolescence, 20,* 247–271.

Puig-Antich, J. (1982). Major depression and conduct disorder in prepuberty. *Journal of the American Academy of Child Psychiatry, 21*, 118–128.

Radke-Yarrow, M., Richters, J., & Wilson, W. E. (1988). Child development in the network of relationships. In R. A. Hinde & J. Stevenson-Hinde (Eds.), *Relationships within families: Mutual influences* (pp. 48–67). New York: Oxford University Press.

Radloff, L. S. (1977). The CESD scale: A self-report depression scale for research in the general population. *Applied Psychological Measurement, 3*, 385–401.

Radloff, L. S., & Rae, D. S. (1979). Susceptibility and precipitating factors in depression: Sex differences and similarities. *Journal of Abnormal Psychology, 88*, 174–181.

Rohde, P., Lewinsohn, P. M., & Seeley, J. R. (1991). Comorbidity of unipolar depression: II. Comorbidity with other mental disorders in adolescents and adults. *Journal of Abnormal Psychology, 100*, 214–222.

Rook, K. S., Pietromonaco, P. R., & Lewis, M. A. (1994). When are dysphoric individuals distressing to others and vice versa: Effects of friendship, similarity, and interaction task. *Journal of Personality and Social Psychology, 67*, 548–559.

Rutter, M. (1986). The developmental psychopathology of depression: Issues and perspectives. In M. Rutter, C. E. Izard, & P. B. Read (Eds.), *Depression in young people* (pp. 3–32). New York: Guilford Press.

Rutter, M. (1988). Longitudinal data in the study of causal processes: Some uses and some pitfalls. In M. Rutter (Ed.), *Studies of psychosocial risk* (pp. 1–28). New York: Cambridge University Press.

Rutter, M., & Garmezy, N. (1983). Developmental psychopathology. In P. Mussen (Ed.), *Handbook of child psychology* (4th ed., Vol. 4, pp. 775–911). New York: Wiley.

Rutter, M., Tizard, J., & Whitmore, K. (1970). *Education, health, and behavior*. Harlow, England: Longman.

Shapiro, D. (1965). *Neurotic styles*. New York: Basic Books.

Siddique, C. M., & D'Arcy, C. (1984). Adolescence, stress, and psychological well-being. *Journal of Youth and Adolescence, 13*, 459–473.

Smetana, J. G. (1989). Toddlers' social interactions in the context of moral and conventional transgressions in the home. *Developmental Psychology, 25*, 499–509.

Sroufe, L. A., & Rutter, M. (1984). The domain of developmental psychopathology, *Child Development, 55*, 17–29.

Stierlin, H. (1973). Interpersonal aspects of internalizations. *International Journal of Psychoanalysis, 54*, 203–213.

Tellegen, A. (1991). Personality traits: Issues of definition, evidence, and assessment. In D. Cicchetti & W. Grove (Eds.), *Thinking clearly about psychology: Essays in honor of Paul Everett Meehl* (Vol. 2, pp. 10–35). Minneapolis: University of Minnesota Press.

Teti, L. (1982). Depression in adolescence: Its relationship to assertion and various aspects of self-image. *Journal of Clinical Child Psychology, 11*, 101–106.

Weissman, M. M., & Klerman, G. L. (1979). Sex differences and the etiology of depression. In E. S. Gomberg & V. Franks (eds.), *Gender and disordered behavior* (pp. 381–425). New York: Brunner / Mazel.

Weissman, M. M., Klerman, G. L., & Paykel, E. S. (1971). Clinical evaluation of hostility in depression. *American Journal of Psychiatry, 128*, 261–266.

Wierson, M., & Forehand, R. (1994). Introduction to special section: The role of longitudinal data with child psychopathology and treatment: Preliminary comments and issues. *Journal of Consulting and Clinical Psychology, 62*, 883–886.

Wohlwill, J. F. (1980). Cognitive development in childhood. In O. G. Brim & J. Kagan (Eds.), *Constancy and change in human development* (pp. 359–444). Cambridge, MA: Harvard University Press.

Zahn-Waxler, C. (1993). Warriors and worriers: Gender and psychopathology. *Development and Psychopathology, 5,* 79–89.

Zahn-Waxler, C., Cole, P. M., & Barrett, K. C. (1991). Guilt and empathy: Sex differences and implications for the development of depression. In J. Garber & K. A. Dodge (Eds.), *The development of emotion regulation and dysregulation* (pp. 243–272). Cambridge: Cambridge University Press.

Zigler, E., & Farber, E. A. (1985). Commonalties between the intellectual extremes: Giftedness and mental retardation. In F. Horowitz & M. O'Brien (Eds.), *The gifted and the talented: Developmental perspectives* (pp. 387–408). Washington, DC: American Psychological Association.

Zigler, E., & Glick, M. (1986). *A developmental approach to adult psychopathology.* New York: Wiley.

VII   The Psychobiology of Adolescent Depression

Ronald E. Dahl & Neal D. Ryan

This chapter focuses on psychobiologic research on adolescent depression with emphasis on a developmental perspective. One might question the reason for this particular intersection of approaches to study a disorder like depression: Why biologic research? Why adolescence? Why emphasize development? This approach reflects several premises that have driven a series of investigations by our research group: (1) Biologic vulnerability encompasses an important component to understanding early-onset affective disorders. (2) The expression of this biologic vulnerability occurs in the context of normal maturational changes. (3) Adolescence appears to represent a critical period in at least the expression of dysregulation in affective systems. (4) Psychosocial events and circumstances, and their cognitive appraisal, are not independent from these biologic regulatory systems (e.g., a frightening mental image results in immediate changes in limbic and autonomic function, particularly in an environment and context appraised as threatening; conversely, emotion and arousal influence cognitive processing). (5) This interface between cognitive processes and emotional regulation also shows a developmental trajectory (e.g., the types of cognitive images that cause limbic arousal change significantly from early childhood to puberty while the ability to consciously modulate or suppress emotional responses undergoes profound changes over this interval).

Our research group, initially under the direction of the late Joaquim Puig-Antich, M.D., has investigated these biologic systems, their development, and their role in affective illness over the past decade. This work has included psychobiologic studies with hundreds of children and adolescents with major depressive disorder and normal controls. Over the past five years, the controls have included *supernormals* (free of medical problems, psychiatric disorders, *and* negative family histories for affective disorders) as well as *high-risk controls* (free of medical or affective illness but very high family loading for depression). On the basis of these studies, the initial framing questions have been reformulated more specifically, as follows:

1. Are there specific, measurable biologic changes associated with a vulnerability toward developing early onset major depressive disorders? That is, do *trait markers* exist that may help to identify individuals at high risk for early onset affective illness who may then be targeted for prevention / early intervention efforts?

2. How do psychobiologic changes interact with normal maturational changes in these regulatory systems, particularly with respect to sleep and stress systems?
3. What is the role of pubertal maturation per se on these regulatory systems and interactions?
4. What are the implications of these findings regarding opportunities for intervention during adolescence?

Preliminary answers to these questions will be considered, as well as presentation of relevant data supporting the basic premises underlying this work. An initial picture is beginning to emerge from ongoing studies (including longitudinal studies and follow-up of earlier samples currently in progress). However, the full answers to these questions, including direct clinical correlates relevant to treatment and prevention, may require a new generation of studies. This next generation of studies, in our opinion, would benefit from even more developmental focus and greater interdisciplinary breadth. The interactions of biologic vulnerability with development, cognitive processes, family interactions, social support, concepts of self, the role of adverse life events, and the success of coping strategies represent a very complex unfolding process; however, many of these domains are inextricably overlapping. Thus, one of the major goals of this chapter is not so much to provide answers but to help bring attention to the need for further developmental and interdisciplinary efforts in future studies.

Background

Before moving to the psychobiologic studies, it is valuable to frame these laboratory investigations in the larger context of epidemiologic, family history, longitudinal, and treatment experience with early-onset affective illness. For a more complete review of these topics, see Brent, Ryan, Dahl, and Birmaher (1995).

The epidemiologic evidence suggests that affective illness increases from a prevalence of approximately 2% in prepubertal children to about 5% in adolescents. Lifetime rates for major depression in adolescents are estimated at 8% (Fleming & Offord, 1990). However, in addition to the general increase near puberty, there is also the onset of gender differences. That is, before puberty, males and females appear to be affected approximately equally (or slightly greater rates for males), however after puberty and through adulthood, there is approximately a 2:1 female to male preponderance of major depressive disorder (Fleming & Offord, 1990).

There is also evidence that depression and suicide have been occurring more frequently among youth over the past few decades. Evidence for either a birth cohort effect or a secular trend for depression have included studies by Hagnell, Lanke, Rorsman, and Öjesjö (1982); Klerman, Lavori, Rice, Reich, Endicott,

Andreasen, Keller, and Hirschfeld (1985); Lavori, Klerman, Keller, Reich, Rice, and Endicott (1987); and Ryan, Birmaher, Perel, Dahl, Meyer, Al-Shabbout, Iyengar, and Puig-Antich (1992). The increased rate of suicide in adolescents and young adults has also been well documented (Brent, Crumrine, Varma, Allan, & Allman, 1987; Shaffer, 1988).

Risk factors for the development of early-onset affective illness have been examined through a number of investigations. The strongest risk factor is having a parent with depression. It is unclear, however, whether this has more to do with familial loading or the experience of growing up with a depressed parent (Hammen, Adrian, Gordon, Burge, & Jaenicke, 1990; Rutter & Quinton, 1984). Early experiences of adverse life events have also been associated with risk for depression including loss of a parent (Weller, Weller, Fristad, & Bowes, 1991), loss of a sibling or friend to suicide (Brent, Poling, McKain, & Baugher, 1993), and physical or sexual abuse (Kaufman, 1991; Toth, Manly, & Cicchetti, 1992). Following an adverse experience, depression is more likely to occur when there is a personal or family history of depression (Brent et al., 1993; Weller et al., 1991). Chronic medical conditions have also been associated with increased risk of affective disorders, including juvenile onset diabetes (Kovacs, Feinberg, Crouse-Novak, Paulauskas, & Finkelstein, 1984a), inflammatory bowel disease (Burke, Kocoshis, Chandra, Whiteway, & Sauer, 1990; Burke, Meyer, Kocoshis, Grenstein, Chandra, Nord, Sauer, & Cohen, 1989), and epilepsy (Brent et al., 1987).

The natural history of early-onset affective illness has been well delineated. Major depression in childhood and adolescence is a chronic and recurrent condition with significant morbidity and mortality. Naturalistic follow-up of untreated depressed children ages 8 to 13 years showed mean episode duration of seven months (Kovacs, Feinberg, Crouse-Novak, Paulauskas, & Finkelstein, 1984a), and the probability of recurrence of depression was 40% in two years and 72% within five years (Kovacs, Feinberg, Crouse-Novak, Paulauskas, Pollock, & Finkelstein, 1984b). There is substantial evidence that childhood depression persists into adulthood (Fleming & Offord, 1993; Harrington, Fudge, Rutter, Pickles, & Hill, 1991). Prepubertal and adolescent depression are also associated with significant social impairment at home, in school, and with peers (Puig-Antich, Kaufman, Ryan, Williamson, Dahl, Lukens, Todak, Ambrosini, Rabinovich, & Nelson, 1993; Puig-Antich, Lukens, Davies, Goetz, & Todak, 1985a) with some functional impairment persisting after treatment and recovery (Puig-Antich, Lukens, Davies, Goetz, & Todak, 1985b). Nearly a decade of follow-up of adolescents with depressive symptoms have indicated that depression is a persistent illness with high rates of school dropout, substance use, delinquent behavior, and interpersonal difficulties (Rao, Dahl, Ryan, Birmaher, Williamson, & Rao, in press-a). However, the most significant outcome of early-onset depression is completed suicide. Depression is the single most important risk factor for completed suicide in adolescents across a number of studies (Brent, Kolko, Wartella, Boylan, Moritz, Baugher, & Zelenak, 1993; Brent, Perper, Golstein, Kolko, Allan, Allman, & Zelenak, 1988; Shaffer, 1988; Shafii, Carrigan, Whittinghill, & Derrick, 1985; Shaffi, Steltz-Lenarsky,

Derrick, Beckner, & Whittinghill, 1988). Longitudinal studies of early-onset depression have also shown a high rate for attempted and completed suicide (Harrington, Fudge, Rutter, Pickles, & Hill, 1990; Kovacs, Goldston, & Gatsonis, 1993; Rao, Weissman, Martin, & Hammond, 1993).

Treatment studies of early-onset affective illness have been quite disappointing in general. Several control studies have failed to find significant differences between antidepressants and placebo in depressed children and adolescents (reviewed in Ambrosini, Bianchi, Rabinovich, & Elia, 1993; Ryan, 1990). With respect to psychosocial and cognitive treatments, a number of early studies are under way; however, here too, there are as yet few convincing data for effective interventions.

In summary then, early-onset depression affects large numbers of children and adolescents, appears to be increasing in our society, is associated with significant morbidity and mortality, and there are no proven effective treatments. In this context, psychobiologic research represents a promising approach to advance the understanding of etiologic mechanisms and *predictors of differential course*. Such advances are likely to inform future strategies for affective early intervention of these serious disorders.

Psychobiologic Studies in Early-Onset Depression

Historically, psychobiologic studies first helped to establish the *validity* of childhood depression within a context of skepticism. The general approach that dominated most psychobiologic research in early onset depression was an attempt to replicate in children and adolescents the findings from biologic studies of adult affective disorders. As will be presented in this chapter, these investigations have found areas of continuity across child, adolescent, and adult depression, as well as areas where maturational factors appear to exert large influences. The field, however, has been shifting in conceptual framework away from the downward extension of adult studies and a greater focus on a developmental approach. The value of a framework emphasizing developmental psychopathology has been well argued by others (Cichetti, 1984; Rutter, 1986; Zigler & Glick, 1986). A strong case can be made for the need to better integrate psychobiologic research in early-onset depression with recent advances in developmental neurobiology and neurodevelopmental psychology to better understand the relevant psychobiologic changes in early-onset affective disorders.

An overview of psychobiologic research in early-onset depression will be presented in three sections: (1) some discussion of conceptual and methodologic issues of performing these types of investigations in young subjects; (2) an overview of growth hormone and serotonergic studies emphasizing maturational *continuities* across adolescence; (3) a focus on psychobiologic studies of EEG sleep and stress hormones with emphasis on maturational variation across child, adolescent, and adult findings.

Conceptual and Methodologic Issues

Psychobiologic studies of affective disorders in young populations are promising as a method to understand *similarities and differences* between child and adult forms of these disorders. Clearly, the early work of Puig-Antich, Davies, Novacenko, Davies, Tabrizi, Ambrosini, Goetz, Bianca, Goetz, and Sachar, 1984c; Puig-Antich, Goetz, Davies, Fein, Hanlon, Chambers, Tabrizi, Sachar, and Weitzman, 1984b; Puig-Antich, Goetz, Davies, Tabrizi, Novacenko, Hanlon, Sachar, and Weitzman, 1984d; and Puig-Antich, Novacenko, Davies, Fein, Hanlon, Chambers, Tabrizi, Sachar, and Weitzman, 1984a, showing some adult-pattern biologic abnormalities in prepubertal depression helped to validate the diagnoses of depression in children and adolescents. In addition, these biologic markers may help to distinguish various subtypes of depression which might permit more accurate predictions of clinical course as has been shown in adult depression (Franz, Kufper, Miewald, Jarrett, & Grachacinski, in press; Giles, Biggs, Rush, & Roffwarg, 1988). For example, Rao et al. (1995) has performed a seven-year follow-up on an initial adolescent sample studied in our laboratory and has found promising results regarding EEG sleep and cortisol changes as predictors of recurrent unipolar depression (Rao, Ryan, Birmaher, Dahl, Williamson, Kaufman, Rao, & Nelson, in press). Further, these psychobiologic tests may in the future, if refined, prove to have value as *diagnostic or confirmatory* tests. Others have discussed the potential of identifying particularly *informative subgroups* of depressed children who are at high risk for violence or suicide (emphasizing the ties to serotonergic systems). However, the primary long-term goal of such studies is to help to understand the developmental *pathophysiology* of early-onset depressive disorders and provide insights into the mechanisms involved in the normal and abnormal regulation of affect and arousal relevant to the development of these disorders.

Biologic studies in young subjects provide unique opportunities and challenges compared to similar studies in adult populations. One advantage is that young depressed patients are relatively "clean" with respect to frequent confounding variables in adult depressives, including alcohol use, drug use, and chronic medical illnesses. Finally, young populations offer a unique opportunity with respect to a high-risk paradigm. That is, studies of unaffected children with very high family loading for affective disorders will have a relatively high rate of developing the disorder over the following 5 to 10 years. In contrast, if one identifies an individual with high family loading in his 40s who has never had a lifetime episode of depression, it is much less likely that this individual would experience a first episode of depression over the same 5–10-year time span.

Perhaps the most intriguing opportunities presented by the studies of young populations, however, is the opportunity of examining *developmental progression*. That is, there are a variety of parallels between normal maturational changes and changes associated with psychobiology of depression. The *sequence* of the development of psychobiologic changes within and across individuals is likely to be informative with respect to potential mechanisms of dysregulation. Specifically, the earliest markers may reflect more proximal elements of dysregulation with affective disorder, while later appearing markers may be more closely related to dysregulatory mechanisms secondary to chronic stress as a result of the disorder.

On the other hand, psychobiologic studies in children and adolescence are also much more challenging than adult studies for the following reasons: (1) Depression occurs at lower rates in young ages and the relative age bands available for developmental studies are narrow (the pool of 40–50-year-old depressed subjects is much larger than 8–10-year-old depressed subjects). Thus, extensive efforts and expenses are required for subject recruitment within this relative scarcity. (2) Consent and assent for these complicated studies must be obtained from both the family and the child. If either refuses, obviously the study cannot be performed. (3) The process of performing these studies in young children creates considerable complexities, both *psychologically* and *physiologically* for systems related to stress and arousal (since procedures such as starting IVs, adapting to a new environment, and sleeping away from home can cause stress responses with some children, particularly in the presence of comorbid anxiety). Addressing these complexities requires extensive efforts to create a low stress and enjoyable environment and the need for longer term studies to address adaptation. Combine the behavioral management and entertainment of these subjects and the need for skilled and well-trained staff, and the result is that these studies are much more costly and complex than comparable studies of adults. A more thorough discussion of methodologic issues is presented elsewhere (Dahl & Ryan, in press).

In summary, then, there are compelling reasons to perform these studies; however, the methodologic challenges are immense. Because the process of obtaining the measures can significantly disrupt the systems of interest, extensive efforts must be employed to address issues of psychological stress, adaptation, behavior, and cognitive appraisal—and the solution to these challenges must be developmentally appropriate.

Growth Hormone and Serotonergic Regulation

In this section, we will discuss a body of research focused on growth hormone regulation and serotonergic function. The themes that will be emphasized in this section are *continuity* across child, adolescent, and adult major depressive disorder (MDD) and evidence for *trait marker status* for these abnormalities. Although there are some developmental variations, in general, findings in these measures with affective disorders are similar across the lifespan. Further, the evidence suggests that altered growth hormone regulation and serotonergic changes occur not only *during the depressive episode and following recovery* from the episode, but also appear to be present in young subjects *at high familial risk* for depressive disorders who currently have no affective symptoms.

Growth Hormone Regulation

Growth hormone (GH) is a single chain peptide with 191 amino acids secreted by the pituitary and involved in metabolic regulation on target cells throughout the body. Its role in promoting growth and stimulating somatamedins is central to its name. In young children, the majority of growth hormone secretion occurs during sleep. By the period of peak growth velocity during puberty, GH pulses occur throughout the 24-hour period. However, at all ages, the largest growth hormone secretion is most likely to occur during the first few hours of nocturnal sleep. Early studies associated this GH secretion to stage 4 or delta sleep (Takahaski, Kipnis, & Daughaday, 1968); however, other studies suggest that it is *sleep onset* itself that stimulates GH secretion (Born, Muth, & Fehm, 1988). Growth hormone is regulated primarily by two hypothalamic hormones: Growth Hormone Releasing Hormone (GHRH), which stimulates GH release, and somatastatin, which inhibits GH release.

A number of studies in adult major depressive disorder have found blunted growth hormone response to a variety of provocative stimuli including clonidine, dextroamphetamine, meta-amphetamine, L-dopa, insulin-induced hypoglycemia, and desimipramine. Early work emphasized a probable link to adrenergic dysregulation since many of these agents stimulate postsynaptic alpha-2 receptors in the hypothalamus, causing a release of GHRH (Amsterdam & Maislin, 1990; Ansseau, Von-Frenckell, Cerfontaine, Papart, Franck, Timsit-Berthier, Geenen, & Legros, 1988; Calil, Lesieur, Gold, Brown, Zavadil, & Potter, 1984; Charney, Heninger, Sternberg, Hafstad, Giddings, & Landis, 1982; Checkley, Slade, & Shur, 1981; Gruen, Sachar, Altman, & Sassin, 1975; Koslow, Stokes, Mendels, Ramsey, & Casper, 1982; Langer, Heinze, Reim, & Matussek, 1976; Matussek, Ackenheil, Hippius, Muller, Schroder, Schultes, & Wasilewski, 1980; Mueller, Heninger, & McDonald, 1969; Sachar, Finkelstein, & Hellman, 1971; Sachar et al., 1973; Siever, Uhde, Silberman, Jimerson, Aloi, Post, & Murphy, 1982). Clonidine is the best studied of these agents and blunted GH response to clonidine has been well replicated across adult studies of affective disorder (Amsterdam et al., 1990; Ansseau et al., 1988; Charney et al., 1982; Checkley, Slade, & Schur, 1981; Matussek et al., 1980; Siever et al., 1982).

In landmark papers, Puig-Antich and his colleagues present parallel results in prepubertal children in very carefully performed studies (1984a,b,c,d). These results showed blunted growth hormone response to insulin-induced hypoglycemia in prepubertal endogenously depressed children compared to a nondepressed psychiatric control group (there were no normal controls studied at that time). This finding persisted after sustained medication-free recovery. Parallel results were later reported by Jensen and Garfinkel (1990) in prepubertal boys with major depressive disorder in response to oral clonidine and oral L-dopa and by Meyer, Richards, Cavallo, Holt, Hejazi, Wigg, and Rose (1991) in response to insulin-induced hypoglycemia, arginine, and oral clonidine.

In more recent work by our group (Ryan, Dahl, Birmaher, Williamson, Iyengar, Nelson, Puig-Antich, & Perel, 1994), Puig-Antich's basic findings have been replicated and extended, showing blunted growth hormone response to insulin-induced hypoglycemia in 38 depressed prepubertal children compared to 19 normal prepubertal controls. In addition, these controls were free of family history of

Figure 1.

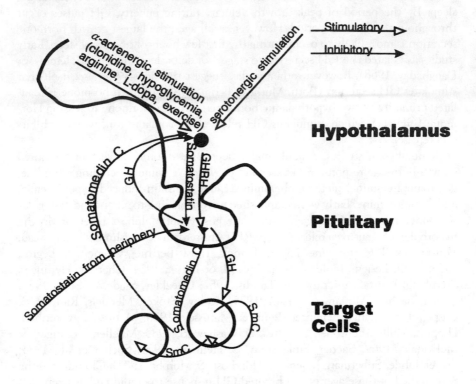

affective disorders. Closely parallel results were found for intravenous clonidine, although the results failed to reach statistical significance. Surprisingly, however, significantly blunted growth hormone response was also found for GHRH infusion that bypasses the alpha adrenergic systems, causing direct secretion at the level of the pituitary. These results were not hypothesized and suggest that the dysregulation is elsewhere in the system from the initial adrenergic hypotheses.

Results in the high-risk children (those with very high familial loading for affective disorders but not depressed) show GH responses intermediate between the depressed and control subjects. Specifically, approximately half of the high-risk subjects reveal findings similar to the depressed group while many of the high-risk subjects show normal growth hormone secretion. Longitudinal follow-up of these subjects is currently underway with the prediction that abnormal growth hormone secretion may predict onset of affective disorders among the individuals with high familial loading for depression.

The evidence of growth hormone dysregulation, however, does not appear to be specific for MDD. Particularly in the adult studies, anxiety disorders also appear to have similar evidence of blunted growth hormone response to provocative stimuli. These results are also convergent with a larger body of evidence indicating frequent

Figure 2. GH Response to ITT.

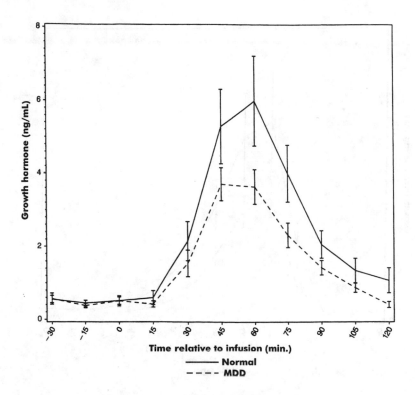

comorbidity between anxiety disorders and depression (Weissman, Leckman, Merikangas, Gammon, & Prusoff, 1984). There are also data in young subjects (Kovacs,
Gatsonis, Paulauskas, & Richards, 1989) to suggest that anxiety disorders often
predate depressive disorders in young subjects. Thus, *one hypothesis is that the blunted
growth hormone response may reflect a more general vulnerability in affect regulation
relevant to both anxiety disorders and depression.* The particular factors that may result
in the development of specific disorders would then be relevant questions emerging
secondary to this vulnerability.

An obvious clinical question that emerges is whether these children with
apparent changes in GH regulation show changes in physical growth. There are
very few controlled data to address these issues. There are preliminary data in
anxiety disorders to suggest that decreased growth velocity and decreased adult
height could be sequelae of these changes. However, there are also reasons to
believe that compensatory changes in the system could result in little or no change
in actual growth velocity and normal height despite significant central changes in
these regulatory systems. These hypotheses will require well-controlled studies that
include careful longitudinal measures of height in conjunction with GH measures.

Figure 3. GH Response to Clonidine

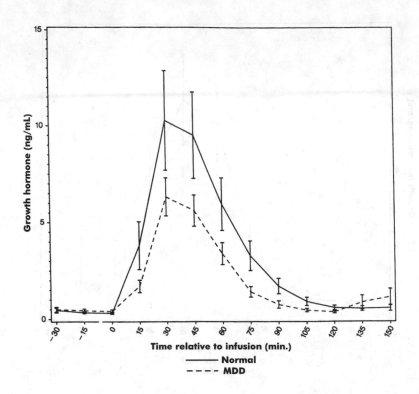

Serotonergic Regulation

Numerous investigations have pointed toward a central rule for serotonergic neural systems in relation to suicide and depression in adults (Asberg, Edman, & Rydin, 1984; Banki, Arato, & Papp, 1984; Coppen & Wood, 1982; Korpi, Klein-man, & Goodman, 1986; Lopez-Ibor, Saiz Ruiz, & Perez De Los Cobos, 1985; Mann, Arango, Underwood, Baird, & McBride, 1990; Murphy, Campbell, & Costa, 1978; Van Praag, 1982). From a clinical perspective, it is clear that serotonergic agents have a role in the treatment of depression (as has been evident not only in the scientific literature, but also in the popular press). Only one published study to our knowledge has examined serotonergic regulation in depressed children (Ryan et al., 1992). That investigation examined 37 depressed children compared to 23 normal control children (healthy medically, psychiatrically, and with negative family histories for affective disorders). The research approach was the use of a challenge study; that is, challenging the serotonergic neural systems with an agent that has a predictable and measurable physiological response. In this case, a sero-tonergic agonist was used (an intravenous challenge of L-5-hydroxytryptophan

Figure 4. GH Response to GHRH

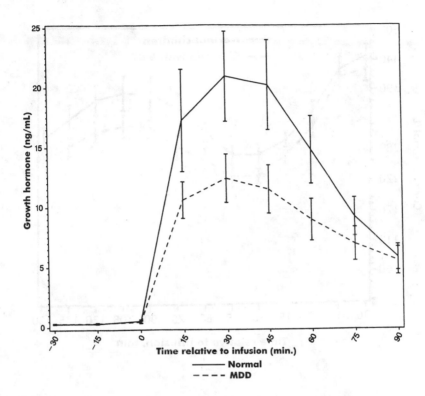

Normal
MDD

[L-5HTP] which is a precursor to serotonergic production). In addition, an oral preloading with carbidopa was utilized to inhibit peripheral metabolism of L-5HTP, thus maximizing the central serotonergic increase by the infusion. The measured physiologic effect of serotonin increase from this infusion was secretion of prolactin and cortisol into the blood. This challenge resulted in group differences and gender differences in this study. The prolactin response was significantly greater in the depressed children; however, virtually all of this group difference was accounted for by the depressed females. Cortisol response, however, was blunted in the depressed children (with males *and* females) compared to control children. In this sample, suicidal ideation or attempts within the depressed sample was not significantly correlated with the endocrine responses to the L-5 HTP.

Comparison with adult studies is complicated because of the variability of serotonergic agents that have been used across different studies (these issues are beyond the scope of this review) (Ryan et al., 1992). A review of these studies (Lopez-Iber et al., 1989) has shown abnormalities in the release of prolactin or cortisol or both after stimulation of central serotonergic systems in adult depression. The results in children are consistent with those in adults (Ryan & Dahl,

Figure 5.

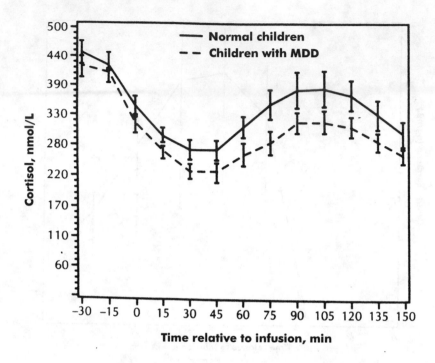

1993). Preliminary analyses from these serotonergic studies in the high-risk sample appear to parallel the findings described in growth hormone, namely, nonaffectively ill children with high familial loading show results intermediate to the depressed and normals, consistent with a trait marker status for serotonergic changes. The gender differences in the prolactin findings also raise a variety of questions regarding *maturation / gender interactions* with depressive disorders described earlier in the chapter. That is, the significant increase in the rates of major depressive disorder observed in adolescence shows disproportionately higher rates in females during this interval. In the L-5HTP study, it was only the depressed females who demonstrated exaggerated prolactin responses to the serotonergic probe. Although these children were technically prepubertal or at the very early stages of puberty (Tanner I or II), brain changes of pubertal maturation began years before gonadal changes. Thus, it is possible that the early pubertal changes in serotonergic regulation in these depressed girls may contribute to abnormal responses to the L-5HTP. A variety of additional studies will be necessary to further understand these data with regard to trait status and the developmental/maturational component.

Figure 6.

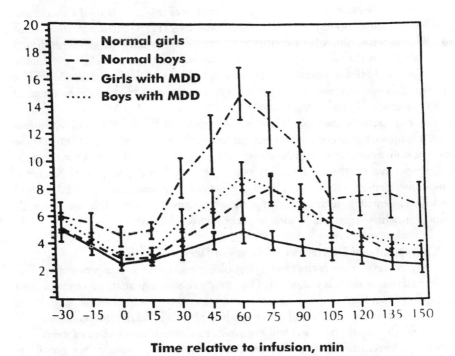

Time relative to infusion, min

In summary, however, growth hormone challenge studies and serotonergic challenge studies in child and adolescent depression are consistent with a model of trait abnormalities.

By "trait" status, we are referring to a biologic change that is independent of the state of the illness. In the strictest sense, a trait marker would be evident before a depressive episode, during the episode, and following the episode, as a stable indicator of the illness across clinical states. Conceptually, a trait marker is thought of as a marker of vulnerability or predisposition toward the disorder. In contrast, a "scar" marker is conceptualized as the pathophysiologic *sequelae* of the illness. Scar markers would be observed in the recovered state following an episode of depression, but not apparent before the first episode. Finally, a "state" marker refers to a biologic change that is only evident *during* a depressive episode and disappears with other signs of the illness between episodes.

Although preliminary at this point, changes in growth hormone regulation and serotonergic function appear to be promising candidates as trait markers since they are apparent in many children at high risk for depression (but without any sign of illness), occur during an episode and following clinical recovery, and show continuity across adult studies. Currently, longitudinal follow-up of these samples are underway to test the hypothesis that these markers will help to predict clinical course.

Sleep and Stress Measures

In this next section, we will present an overview of research focused on measures of sleep and the hypothalamic-pituitary-adrenal-axis (HPA) measures. In contrast to GH and serotonin, where continuities were emphasized with adult studies, a major theme in this section is *maturational variation*. In adult biologic psychiatry, EEG sleep and HPA abnormalities are among the best-studied markers. As will be presented, however, these psychobiologic changes appear to have a developmental progression in early-onset depression.

Before presenting specific clinical investigations of EEG sleep and HPA measures, it is important to consider, in broader terms, the physiologic regulation of these systems from a conceptual and developmental perspective. This somewhat theoretical perspective on the evolutionary psychology of sleep, arousal, and perceived threat may seem a bit tangential. It leads, however, to an emphasis on social context and development as a framework for these regulatory systems that may provide critical insights—particularly with respect to dysregulation.

Conceptual and Developmental Issues in Sleep and Stress

Sleep is a mysterious entity from a physiologic and evolutionary perspective. The basic function of sleep is unknown. The commonsense explanations of simple rest or replenishment fail to provide meaningful answers to this mystery because *sleep is much more than just rest*. Rest alone (including physical and mental relaxation and the avoidance of all activities) will not result in the restorative state of sleep. The specific requirement of sleep is, essentially, a *cessation of vigilance*, or giving up awareness and the ability to respond to the immediate environment. Now the story becomes more intriguing: Why would evolution have selected for individuals who spend a third of their life unresponsive to what is going on around them? More specifically, consider the conditions surrounding our prehuman ancestors: (1) with loss of the ability to sleep in trees, physically safe sleep sites were minimal or nonexistent; (2) major predators at that time (including large cats and hyenas) were primarily nocturnal hunters. Thus, sleeping at night posed a dangerous activity for primitive humans. One adaptation, however, involves the inherent cycling nature of sleep, that is, although some stages of sleep cause a virtual absence of any external awareness (particularly stage 4 sleep), the depth of sleep cycles back to wakefulness every 60–90 minutes. This cycling creates the opportunity to briefly "check out" the environment before returning to sleep for the next cycle. This system may be particularly effective for *group safety*; that is, given the irregularity of cycling across individuals, a group of 20–30 would tend to have one or more individuals awake (or in light sleep between deep stages of sleep) at any given moment. The evolutionary psychology of this system may have considerable relevance to our understanding of sleep dysregulation in modern humans along the following lines: (1) *A global sense of feeling safe should be a requirement for going to sleep (and for going back to sleep after normal night arousals)*. (2) The subjective sense of

Figure 7.

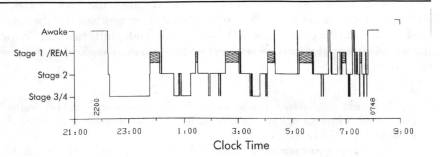

safety relies as much on social constructs (a *sense of belonging to a stable social group*) as to physical safety.

Not only were these concepts consistent with anthropologic evidence from the past, but they are also supported by more recent observations of primitive humans. Consider the following description from a recent anthropologic look at sleep (Ellis, 1991):

> The most primitive people studied to date are the Bushmen of Southwest Africa. The classic study of the !Kung San is the *Harmless People,* written by Elizabeth Marshall Thomas in [1958] . . . . The Bushmen are hunter-gatherers, people who do not grow crops or herd animals but who move about constantly in search of food . . . . A band generally is no larger than 20 people and usually consists of: an old man and his wife, their daughters, and their daughter's children . . . . At night the group clears an area of ground, pulling up the grass and dead tree limbs. This is called their werf. Each family unit may make a shelter, called a scherm, by staking four limbs in the ground and stretching some grass over the top, which the wind eventually blows away. Or, they may simply put four stakes upright in the ground to mark off their territory. The women then dig a shallow depression in the ground and cover it with grass. They build a fire in front of each scherm . . . . Though the desert is quite hot by day, temperatures drop well below freezing at night. Thomas described how the bushmen sleep.
>
> . . . they lay down, naked as they were on the bare ground, close to the fire, with their knees bent, letting as much skin as possible be exposed to the heat. The warm smoke and ashes blew over them and they went to sleep on their sides as Bushmen always must, with one ear on the ground but with the other up and listening, to hear what comes along.
>
> . . . one evening a grass fire moves across the veld, burning all grass and the few baobab trees in its path. As the band finishes eating, they are relieved to see the wall of fire move away from their area. They lie awake through the night listening to the cries of birds . . .
>
> . . . she hears something large run past the group; then, later on, she hears the sharp roar of a lion who must have singed his paws crossing a patch of burned ground. The lion tests the earth carefully and walks in a huge half-circle around the werf murmuring a steady growl.

In another section, Thomas describes how the Bushmen find the trail of a 14-foot-long snake, a mamba, and track it to its hole. The mamba is the deadliest snake in Africa and its poison can kill an adult in 10 minutes. They conclude that the mamba has been circling the camp, attracted by a colony of bushy little rats that live near the werf. Unable to locate the snake, and finding still more tracks near the camp, they are fearful.

> "That night we all stayed close to our fire, in our circles of light, and kept out of the dark veld where the snake might be mousing, lying quiet in the grass where one of us might step on him . . . ."
>
> . . . In all the higher primates and among the !Kung Bushmen, mother and child sleep in close physical proximity, if not in direct physical contact. "This pattern was probably selected early in higher primate evolution. An infant sleeping alone, even among most human hunter-gatherers would be subject to almost certain death from predators . . . ." Thomas's account of her nights spent among the Bushmen of Southwest Africa certainly supports [the view] that sleep is a time of increased vulnerability and danger for primitive people. Perhaps there continues to be a universal need to bond with others at night.

Thus the evolutionary psychology of human sleep and arousal would appear to be closely intertwined with issues of both physical safety and social context.

These same principles apply to the physiology of stress responses. Most definitions of stress are framed in terms of the *appraisal of threat*. That is, a stress response is a physiological adaptation to a perceived threat for the individual. Here again, however, the equation changes as one moves from usual mammalial physiology (simple "fight or flight" in the face of physical threat) to a more complex system in higher primates and humans. Specifically, social context and social cues appear to have the most powerful influence on stress physiology in primates (Sapolsky, 1990). This may make sense from a survival perspective since social acceptance and rejection probably represented the most serious threat to survival. Field studies among higher primates have shown that individuals who are part of a stable troop in the wild are rarely taken by predators; however, if an individual becomes old, injured, or sick and cannot keep up with the group or is rejected from the group (as happens to all rhesus males when they reach adolescence), the risk of death increases dramatically (Suomi, 1991).

Here is the point relevant to psychobiology: The basic neurobehavioral systems involved in the regulation of sleep and stress are interrelated, opponent processes on an axis of perceived threat. This axis is strongly framed by social context (e.g., social status, a sense of "belonging" to a relevant social group, and threats of aggression from other members) as this was crucial to survival and reproductive success throughout the period of evolution when these brain mechanisms were undergoing the most recent refinements.

These principles appear to be quite relevant to modern humans. For example, why is it so hard to quell our internal physiologic signs of threat (and arousal overshoot) when giving a public speech among peers, but so easy to fall asleep driving an automobile on the interstate? Clearly, tens of thousands of people are killed each year on the highway and rarely is anyone physically injured at public meetings; yet our "stress wiring" retains old relationships to the historic buttons: social acceptance and social status were the keys to individual and genetic survival for hundreds of thousands of years during our brain's evolution. Threat (as a stimulus for stress response) and safety (as a cue to permit sleep) are built around social constructs as well as physical safety. Thus, for example, not only is it more difficult to fall asleep in a strange environment but also most people show stress reactions and sleep disturbances following experiences of rejection, loss, grief, or personal failures. This may also be related to the reason that loss of a loved one is among the most powerful stresses experienced in a lifetime.

It is also valuable to consider the explicit *developmental aspects* of these systems. Whatever the physiologic function of sleep, it appears to be *particularly important during early brain development*—across species, young individuals require more sleep than older individuals. By the age of two years, a typical child has spent 10,000 hours asleep and 8,000 hours in all waking activities combined. Thus, in many ways, the main activity of the young brain is sleeping. From an evolutionary biology perspective, such huge amounts of time with vigilance turned off was not a problem for survival since young children were continuously carried by a parent and the appraisal of threat and safety was performed by this adult. Thus, there is a *developmental progression* of threat appraisal influences on sleep and stress. In the first few months of life, infants are relatively oblivious to external cues influencing sleep cycles. By about six to nine months the presence of a parent or attachment figure begins to have significant influences on sleep / wake cycles, and separation anxiety can markedly disturb sleep. As children begin to become independently mobile, they begin to need to make their own judgments about safety and threat. However, they initially continue to rely on parent or adult cues (i.e., social referencing) in situations of ambiguity (Sorce & Emde, 1985). As toddlers and young children develop further, there is a greater degree of independent cognitive appraisal of safety. In modern society, where children are often sleeping alone without the immediate presence of a parent or other family member, the "good sleepers" have learned techniques for "self-comforting," which essentially indicate safety to the individual. This process can involve a behavior (such as thumb sucking, body rocking, twirling a piece of hair) or may involve a "security object" (such as a favorite blanket, stuffed animal, etc.) that is necessary for the child to go to sleep. Interestingly, children will usually need the same behavior or object following normal arousals between sleep cycles in order to return to sleep. The "bad sleepers" often simply use "maladaptive" self-comforting techniques (such as requiring a *parent* to hold or rock them or needing to be fed) in order to experience the sense of comfort and safety to go to sleep. In these children, the only way they can return to sleep following normal nocturnal arousals is to have a repetition of the same "safety button" as they used to go to sleep the first time.

By school age and adolescence, however, concepts of safety and threat incorporate much more complex cognitive appraisals. The psychological issues involved in judgments of feeling safe, social and family context, sense of self, and the experience of stressors begins to approach adult levels of complexity. Early experiences, attachment, traumatic events, and cognitive style may all interact in the development of sleep and stress regulation—and dysregulation.

A final conceptual component to frame these studies is the issue of *individual differences* regarding threat appraisal. There is considerable evidence that the threshold for regarding situations as threatening appears to have an important genetic / temperamental component. This has been well-established in animal studies (Rasmussen & Suomi, 1989; Suomi, 1987) as well as human studies (Kagan, Reznick, & Snidman, 1987). Some individuals are very prone to regard novel or neutral stimuli as threatening from very early ages. These individuals are also more prone to sleep problems (Bernal, 1973), separation anxiety, and the development of anxiety disorders (Biederman, Rosenbaum, Bolduc-Murphy, Faraone, Chaloff, Hirshfeld, & Kagan, 1993). These patterns also occur more frequently in the children of parents with anxiety disorders (Rosenbaum, Biederman, Gersten, Hirshfeld, Meminger, Herman, Kagan, Reznick, & Snidman, 1988). As presented earlier, there is also reason to hypothesize that this group will be at higher risk for affective disorders.

Now this general conceptual framework can be extended to the psychobiology of pubertal maturation. From an evolutionary biology perspective, it makes sense that this sleep / stress system regarding appraisal of threat and safety *should* undergo fundamental changes with sexual maturation. Across human history, adolescence and pubertal maturation signaled a time of fundamental change in social role for both males and females. In many species of higher primates and in many cultures from anthropologic studies, males at puberty must take on new roles that require major changes in vigilance, aggression, and actual danger. For example, as mentioned earlier, all rhesus males must leave the primary troop in which they grew up at puberty and find their way into a new troop. The process entails a period of forming adolescent male "gangs" and then subsequently either fighting their way into a new social group or gradually affiliating their way from the periphery into the new social group (Suomi, 1991). Either method entails significant risks for the individual with approximately a 50% mortality (accounting for the 2:1 female to male ratio in rhesus). During the time that the adolescent leaves the safety of the immediate family and major social troop, the demands on the vigilance system are categorically different, and one would predict that sleep regulation would undergo fundamental changes with a greater cognitive component of appraisal (compared to earlier developmental stage where family and social troop provided the essential framework of safety). Similarly, in most higher primates and throughout early human history, puberty for females was soon followed by the role of parenting. The transition to parenthood for a nursing mother may also require fundamental changes in the vigilance system since the responsibility of caring for and responding to a young infant requires altered demands on the regulation of sleep (as any new parent can strongly attest). It is not surprising, then, that sleep studies show dramatic changes in sleep regulation associated with adolescent development in human studies. During delta sleep, young children are completely unresponsive to external stimuli (one study recorded no arousal during stage 4 sleep in preadolescent boys receiving 123 decibel noise bilaterally through earphones! (Busby & Pivik, 1983). At adolescence, there is a significant decrease in the depth of sleep, the amount of deep sleep, and a lower threshold of arousal within sleep. There are also changes in REM, including decreased REM latency. Further, there are fundamental changes in daytime sleepiness (Carskadon, Keenan, & Dement, 1987) and evidence for a circadian shift at puberty (Carskadon, Vieira, & Acebo, 1993). Taken together, then, it should not be surprising that the psychobiologic changes associated with depression regarding stress hormones and sleep changes also show fundamental changes at or near puberty.

EEG Sleep Measures in Depression

Among adult studies of depression, EEG sleep changes are among the most robust psychobiologic markers, demonstrating abnormalities in 90% of subjects with major depressive disorder (Reynolds & Kupfer, 1987). These changes include: (1) difficulty going to sleep and problems staying asleep across the night; (2) less deep delta (stage 3 and 4) sleep, especially in the first 100 minutes of the night; (3) reduced latency to the first REM period; and (4) an altered pattern of REM sleep with more REM and greater density of eye movements early in the night. EEG studies of child and adolescent MDD, however, have revealed a contrasting picture. Despite similar subjective complaints about disturbed sleep among depressed children and adolescents (Ryan, Puig-Antich, Rabinovich, Robinson, Ambrosini, Nelson, & Iyengar, 1987), objective EEG studies have not found consistent sleep changes paralleling adult MDD studies. Among the four studies of prepubertal MDD subjects that included normal controls, three found no significant group differences in any sleep variables (Dahl, Ryan, Birmaher, Al-Shabbout, Williamson, Neidig, Nelson, & Puig-Antich, 1991; Puig-Antich, Goetz, Hanlon, Tabrizi, Davies, & Weitzman, 1982; Young, Knowles, MacLean, Boag, & McConville, 1982), while one study found decreased REM latency and increased sleep latency in a sample of inpatient MDD children (Emslie, Rush, Weinberg, Rintelmann, & Roffwarg, 1990). Puig-Antich, Goetz, Hanlon, Tabrizi, Davies, and Weitzman (1983) reported evidence of reduced REM latency and improved sleep efficiency in depressed subjects restudied after recovery from depression compared to themselves in earlier studies. Dahl, Ryan, Perel, Birmaher, Al-Shabbout, Nelson, and Puig- Antich (1994) found reduced REM latency in MDD children following infusion of arecoline (a cholinergic agonist) indicating sleep dysregulation at an apparently subthreshold level among some prepubertal depressed subjects that was uncovered with the cholinergic challenge. Among adolescent studies using normal control subjects (Appleboom-Fondu, Kerkhofs, & Mendlewicz, 1988; Dahl, Puig-Antich, Ryan, Nelson, Dachille, Cunningham, Trubnick, & Klepper, 1990; Goetz, Puig-Antich, Ryan, Rabinovich, Ambrosini, Nelson, Krawiec, 1987; Kahn & Todd, 1990;

Kutcher, Williamson, Szalai, & Marton, 1992; Lahmeyer, Poznanski, & Bellur, 1983), two of six studies found reduced REM latency, three reported increased sleep latency, and two of six found decreased sleep efficiency in the MDD subjects.

Taken together, these results suggest that the adult pattern of sleep abnormalities in depression occurs only infrequently in early-onset depression. In further support of this concept are the result of two meta-analyses. Knowles and MacLain (1990), found that age and depression appeared to interact with respect to sleep dysregulation with the smallest differences (or no differences) at early ages. Benca, Obermeyer, Thisted, and Gillin (1992) conducted a meta-analysis showing that in young subjects only sleep latency differentiated depressed from control subjects.

Most interpretations of these data have focused on the role of maturational changes contributing to the infrequency of sleep abnormalities in early-onset depression. A variety of clinical factors have also been considered as potentially contributing to the discrepancies among studies of young depressed subjects. In our adolescent study, we found that suicidality or inpatient status or both appeared to be a critical factor in the early emergence of sleep dysregulation (Dahl et al., 1990). Naylor, Greden, and Alessi, (1990) found that REM latency was reduced in depressed adolescents showing psychotic features compared to nonpsychotic depressed adolescents. Severity of depression has been suggested as an important factor and has been examined in a variety of studies; however, most have not supported the concept that more severe depression is associated with EEG sleep abnormalities in young subjects. In the prepubertal study by Dahl et al. (1991), there was a small subgroup (8 of 36 depressed children) with adult-pattern sleep findings and these individuals did show more severe depression; however, they did not in any way seem to be clinically distinct from the rest of the MDD sample. However, long-term clinical follow-up of this sample will be necessary to see if this early expression of sleep abnormality portends altered clinical course.

Another factor that has appeared as potentially relevant to these findings is the influence of hospitalization per se on adolescent sleep findings. That is, EEG sleep differences have been found more frequently in samples studying inpatient depressed subjects (Dahl et al., 1990; Emslie et al., 1990) and in one adolescent study using rigid control over sleep / wake schedules (Lahmeyer et al., 1983). This pattern of findings was also seen in a study of young adults with MDD (Goetz, Puig-Antich, Dahl, Ryan, Asnis, Rabinovich, & Nelson, 1991). The association with inpatient status and EEG sleep abnormalities raised a set of questions: (1) Since inpatients follow an imposed rigid sleep / wake schedule in the hospital (while outpatients and normals often have erratic schedules), what is the role of sleep/wake schedules on EEG sleep measures in these studies? (2) What is the role of adaptational effects, which may be different in outpatients who are coming to the lab from their home environment compared to inpatients who are already acclimatized to sleeping in an unfamiliar environment? (3) What are the possible effects of access to caffeine, nicotine, alcohol, and other substances among outpatients?

In the most recent study of adolescent depression by our group, a great deal of effort was extended to address these methodologic questions by (1) controlling sleep / wake schedules rigidly with careful validation including actigraphy measures across inpatients, outpatients, and normal controls; (2) verifying the avoidance of caffeine, nicotine, alcohol, medications, and other substances with urine tests in the home environment; and (3) examining a multiple-night study in the sleep lab, including a sleep restriction challenge to the sleep regulation system with early morning awakening.

The results of that study (Dahl, Matty, Birmaher, Al-Shabbout, & Ryan, 1996) indicate that, with precise control over these other sources of variance on sleep measures, there were no signs of delta sleep differences and no evidence of altered sleep continuity after sleep onset; however, REM latency was significantly reduced in the depressed sample compared to controls, and there were robust differences in sleep latency between groups. The sleep latency differences stood out dramatically since they occurred across conditions. That is, the depressed adolescents took longer to fall asleep at home, in the laboratory (despite uniform control over sleep/wake schedules and total amounts of sleep), and even more so after being partially sleep deprived. Normal adolescents responded to mild sleep deprivation with very short sleep latencies and very small standard deviations, while the depressed adolescents continued to show long sleep latencies with large standard deviations. Further, subjective ratings of the adolescents' subjective feelings of tiredness and sleepiness, and staff ratings of the appearance of tiredness and sleepiness, showed that the depressed adolescents were significantly more tired but required twice as long to fall asleep as the control adolescents.

Integrating findings across studies, these results (including the meta-analyses) suggest that the *first sign of dysregulation of sleep in early-onset depression is difficulty initiating sleep*. Developmentally, reduced REM latency appears to occur as a second level of dysregulation, particularly in a subgroup of adolescents. Delta sleep differences, which are robust in adult depression, are not evident at younger ages. This pattern of developmental changes in sleep regulation with depression may also offer some insights into the sequence and etiology of sleep dysregulation in all depression. Along these lines, the study by Ford and Kammerow (1989) from an epidemiologic sample suggested that difficulty falling asleep may predict the new onset of depression in nondepressed adults.

One theoretical framework for these findings is that sleep is relatively protected at young ages. That is, children are very deep sleepers and it is quite difficult to disrupt sleep at young ages (Busby & Pivik, 1983). Thus, even when there exists a level of dysregulation of the sleep / wake system associated with depression, the overall architecture of sleep may remain unchanged at young ages. With increasing age, particularly at or near adolescence, this protective aspect of sleep may decrease, uncovering underlying abnormalities. *Dysregulation in sleep and arousal may also be a pathway to dysregulation in other related systems*. That is, the interval near sleep onset appears to represent a vulnerable transition moving from a state of self-control (where prefrontal cortex and executive control dominate) to a sleep state where subcortical neural systems predominate. The transition into sleep requires a balancing and coordination across multiple regulatory systems and may be more easily disrupted than the actual sleep states themselves. However, if the entry into sleep is chronically disturbed, it appears likely that this could result in further disruption in the regulation of sleep and arousal could result. Thus, this could create a pathway to more serious neurobiologic dysregulation. Although this possibility will require further investigation, it may create opportunities for early intervention. This may be particularly true in adolescents with a large cognitive component to these physiological changes. That is, negative ruminations or recurrent negative thoughts can clearly cause limbic arousal. For some individuals who have experienced traumatic events or suffer from negative self-images and poor self-esteem, negative thoughts and images at bedtime may significantly disrupt the ability to go to sleep. For individuals who use distraction strategies (i.e., the children and adolescents who are successful at avoiding negative or distressing thoughts during the day by high levels of activity, playing, and other active distractions), the process of going to sleep creates particular difficulties. That is, behavioral distraction strategies to displace stressful experiences and memories, is ineffective at bedtime (since distraction activities are usually incompatible with going to sleep). Individual differences in tendencies toward ruminative thinking and negative cognitive styles may thus interact with these physiological systems and lead to chronic dysregulation. Investigations to further explore these issues are needed.

HPA Measures

Among the physiological measures of stress and stress hormones, the HPA axis has occupied a central role in biological studies of depression (Gold, Goodwin, & Chrousos, 1988). Among children and adolescents, these have included baseline measures of cortisol secretion and response of cortisol to provocative or stressful stimuli, the ability to suppress the HPA axis (e.g., the dexamethasone-suppression test), and physiological challenges to the axis usually using a CRH stimulation test. Each of these will be discussed briefly.

Baseline Cortisol Measures. Although a preliminary report by Puig-Antich and his colleagues with a very small sample indicated cortisol hypersecretion in association with prepubertal depression, the results from the final report of that study with a large sample found no significant group differences (Puig-Antich, Dahl, Ryan, Novacenko, Goetz, Goetz, Twomey, & Klepper, 1989). A parallel study in outpatient depressed adolescents (Dahl, Puig-Antich, Ryan, Nelson, Novacenko, Twomey, Williamson, Goetz, & Ambrosini, 1989) also failed to find any significant group differences in 24 hour cortisol measures using samples every 20 minutes for 24-hours and subjects who were well adapted to a sleep laboratory setting. A second sample of adolescent depressed subjects (with a higher rate of inpatients included in the sample) revealed elevated cortisol levels associated with depression in the interval near sleep onset (Dahl et al., 1991). Sleep onset is a time that is normally associated with suppression of cortisol and a virtual cessation of activity in the HPA axis. A number of the depressed subjects continued to show HPA activity, both before sleep onset and after sleep onset. This is a very nonphysiologic observation as normally cortisol secretion turns off in the interval before and after sleep onset, suggesting dysregulation at this time in some depressed adolescents. Further analysis found that the same individuals with cortisol dysregulation near sleep onset (the subgroup of inpatient and suicidal depressed adolescents) also showed the greatest disruption in sleep early in the night, including difficulty falling asleep and early REM latency (Dahl et al., 1990, 1991). These results were also considered in the context of growth hormone dysregulation that appears to be most evident near sleep onset (Dahl, Ryan, Williamson, Ambrosini, Rabinovich, Novacenko, Nelson, & Puig-Antich, 1992). Further as presented in the sleep section, the most recent studies with the greatest degree of control over sleep / wake sleep schedules and other methodologic areas in adolescence suggest that differences in sleep onset are the most robust finding in adolescent depression (Dahl et al., 1996). Taken together, these results suggest that abnormalities in baseline cortisol secretion are infrequent in depressed children and may become more evident during adolescence. In particular, it appears that the earliest dysregulation in the HPA axis occurs at or near sleep onset and may reflect a more fundamental dysregulation in the gating of sleep onset. In addition, the possibility of a cognitive component resulting in limbic arousal contributing to chronic as well as acute changes in regulation of arousal and stress hormones must be considered. That is, chronic ruminations or intrusive thoughts that cause emotional arousal at a time when the sleep / wake system is attempting to turn down arousal for sleep onset, over repeated episodes, result in chronic changes in arousal regulation influencing sleep *and* stress hormones.

Dexamethasone-Suppression Test. There have been numerous studies of dexamethasone-suppression tests (DST) in depressed children and adolescents. These have included 10 studies of children in inpatient settings (Doherty, Madansky, Kraft, Carter-Ake, Rosenthal, & Coughlin, 1986; Freeman, Poznanski, Grossman, Buchsbaum, & Banegas, 1985; Fristad, Weller, Teare, & Preskorn, 1989; Livingston & Martin-Cannici, 1987; Livingston, Reis, & Ringkahl, 1984; Naylor et al., 1990; Petty, Asarnow, Carlson, & Lesser, 1985; Pfeffer, Stokes, Weiner, Shindledecker, Faughnan, Mintz, Stoll, & Heiligenstein, 1989; Weller, Weller, Fristad, Preskorn, & Teare, 1984; Weller, Weller, Fristad, Preskorn, & Teare, 1985), as well as 4 child studies in outpatient settings (Birmaher, Ryan, Dahl, Rabinovich, Ambrosini, Williamson, Novacenko, Nelson, Sing Lo, & Puig-Antich, 1992; Geller, Rogol, & Knitter, 1983; Poznanski, Carroll, Banegas, Cook, & Grossman, 1982; Steingard, Biederman, Keenan, & Moore, 1990). Among adolescent samples, there are 8 inpatient studies (Appleboom-Fondu et al., 1988; Evans, Nemeroff, Haggerty, & Perdersen, 1987; Ha, Kaplan, & Foley, 1984; Kahn, 1987; Robbins, Alessi, Yanchyshyn & Colfer, 1982; Robbins, Alessi, Yanchyshyn, & Colfer, 1983; Targum & Capodanno, 1983; Woodside, Brownstone, & Fisman, 1987) and 2 outpatient studies (Birmaher, Puig-Antich, Dahl, Ryan, Williamson, Rabinovich, Ambrosini, & Novacenko, 1990; Bramibilla, Musetti, Tacchini, Fontanilla, & Guareshi- Cazzulo, 1989), as well as one outpatient study with an even mixture of inpatient and outpatient adolescents (Dahl et al., 1990). Diagnostic and clinical procedures were quite consistent across studies; however, the methods of performing the DST showed some variance. Most studies utilize repeated venipunctures to obtain serum specimens; however, 4 studies

Figure 8. 24-Hour Cortisol Suicidal vs. Nonsuicidal vs. Normal

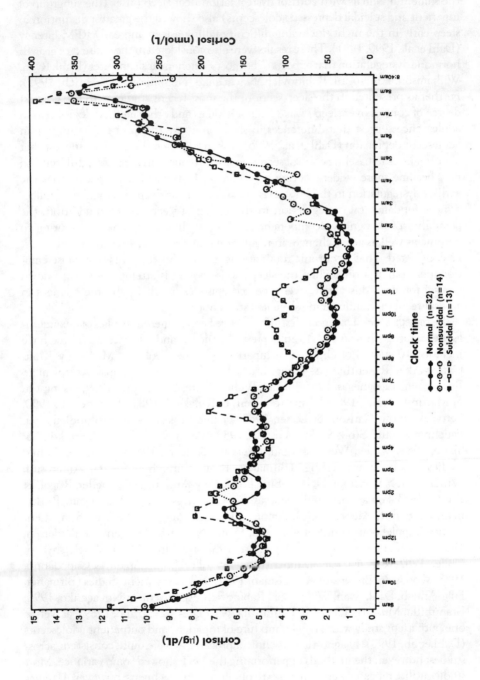

used indwelling venous catheters. Most child studies used 0.5 mg. of dexamethasone, whereas most adolescent studies utilized a 1 mg dose. All studies defined post-dex-amethasone nonsuppression as cortisol values greater than 5 mcg/dl.

Summarized across studies, the results indicate that (1) the sensitivity of the DST appears to be higher among children than adolescents (58% vs. 44% sensitivity; (2) sensitivity was higher in inpatient settings compared to outpatient (61% vs. 29%); (3) specificity compared to normal controls was higher in adolescent samples (84% vs. 74%); and (4) specificity compared to other psychiatric controls for child inpatients was approximately 60%, whereas and for adolescent inpatients it was approximately 85%.

There has been considerable discussion about factors in addition to age and inpatient status that appear to be important contributing to the variance among DST studies. There is some evidence that suicidality may be associated with higher rates of DST nonsuppression (Pfeffer, Stokes, & Shindledecker, 1991; Robbins & Alessi, 1985), and as mentioned earlier, nocturnal cortisol changes also are associated with suicidality. There are also procedural factors that may contribute to variance across studies. Three of the four studies with sensitivity rates below 20% used an indwelling catheter rather than repeated venipunctures to draw blood samples. Although initially it may appear that indwelling catheters would be more stressful, they eliminate the anticipatory anxiety of needle sticks and the actual process of drawing blood is completely painless once the individual has adapted to the catheter. Further, in these studies, all the individuals, patients and controls, had adapted to the same relatively low stress environment for multiple-day biological studies (Birmaher, Puig-Antich, Dahl, Ryan, Williamson, Rabinovich, Ambrosini, Novacenko, 1990; Dahl et al., 1990). It is possible that the quantity of exogenous stressors present at or near the time of the study are a critical factor in DST nonsuppression. That is, the dose of stress may need to be sufficient to create escape from dexamethasone for the vulnerable individuals and hospitalization (inpatient status) and repeated venipunctures may contribute to this finding in some of the positive studies.

Corticotropin-Releasing Hormone Challenge. Corticotropin releasing hormone (CRH) is a synthetic peptide analog to the endogenously produced hypothalamic peptide that physiologically stimulates pituitary release of ACTH (corticotropin). ACTH subsequently causes release of cortisol from the adrenal gland. Among adult studies (Gold et al., 1988), CRH infusion tests have provided further evidence for HPA axis abnormalities associated with depression, revealing blunted ACTH response to CRH despite normal to high cortisol responses (reviewed in Birmaher, Dahl, Perel, Williamson, Nelson, Stull, Kaufman, Waterman, Rao, Nguyen, Puig-Antich, Ryan, 1996). There is only one CRH challenge study in prepubertal depressed children. This study found CRH infusion tests revealed no overall group differences in ACTH or cortisol response compared to age-matched normal control children. There were, however, subgroups of depressed subjects: Blunted ACTH response was seen in a subset of depressed subjects with melancholia (Birmaher et al., 1996) and depressed subjects with a history of sexual abuse (Kaufman, Brent, Dahl, Perel, Birmaher,

Williamson, & Ryan, 1993). These findings are also consistent with the recent study by DeBellis, Chrousos, Dorn, Burke, Helmers, Kling, Trickett, and Putnam (1994), showing blunted ACTH in depressed adolescent females with a history of abuse. These results suggest that HPA dysregulation may only occur in a subset of depressed children, perhaps in association with extreme psychosocial stress. However, it will be necessary to investigate these issues in future studies.

Integrative Summary

The underlying premise to this chapter is that psychobiologic investigations into adolescent depression are a promising approach to furthering our understanding of the development of these disorders and to guide strategies for effective early intervention. The first point to be developed was the evidence that changes in GH secretion and serotonergic regulation represent trait markers of early-onset depression. The findings suggest that changes in these neural systems are biologic flags, or markers, associated with a vulnerability toward affective disorders. Although this work is at a very early point, ultimately such markers may help to target early interventions for individuals at very high risk for developing depression.

A second point addressed the evidence that alternative biologic changes associated with depression demonstrate maturational variance (from child to adolescent through adult depression). Sleep and stress hormone changes fit this general model, suggesting that maturational changes in these neural systems are closely linked to the dysregulation associated with depression. There are, however, some depressed children and adolescents who demonstrate sleep and cortisol abnormalities, and some data to indicate that early expression of sleep and cortisol abnormalities may predict a recurrent unipolar course in adulthood (Rao, Ryan, Birmaher, Dahl, Williamson, Kaufman, Rao, & Nelson, in press). It is important to re-emphasize the preliminary nature of any conclusions regarding the specificity of these findings. There are relatively sparse amounts of data and virtual absence of biologic studies of early-onset bipolar disorders. Similarly, early-onset anxiety disorders may also share some features with affective dysregulation and have not received sufficient psychobiologic study.

A third point was the emphasis on the transition from wakefulness into sleep as a critical window for early dysregulation associated with early-onset depression. A conceptual framework was put forth focusing on the transition from a cognitively dominated waking state to a subcortically driven sleep state. The developmental implications of this framework also appear to be critical to our understanding. Increased cognitive influences over other brain regions, which occurs during normal child and adolescent development, may create a greater vulnerability for dysregulation in sleep and stress hormones by cognitive influences associated with depression. This discussion has also emphasized the cognitive-emotional interface. It is our belief that interactions between prefrontal cortex and limbic systems are a critical focus for understanding dysregulation in early-onset depression. The balance or hierarchical relationship between these systems undergoes changes with normal development, shifts dramatically at sleep onset, and appears to be linked to early dysregulation associated with depression.

Implicit in this discussion is the acknowledgement of the nascent status of our understanding of these neural-behavioral systems and their normal and pathologic development. An explicit assumption, however, is the clinical potential of advancing our knowledge in these areas regarding pathophysiologic mechanisms. As has been presented in other conceptual models of the developmental psychopathology of early-onset affective disorders (see Cicchetti & Schneider-Rosen, 1986; Collins & Depue 1992; Dawson, Hessel, & Frey, 1994; and Post, Weiss, & Leverich, 1994), the biologic roots of early depression are likely to involve some interaction between vulnerability and stress, occurring in a sensitive periods in the development of neural systems underlying affect regulation. Key steps in advancing our knowledge include localizing the specific neural systems, identifying the most sensitive periods of development for those systems, and delineating the specific types of stressors (or nonoptimal environment) contributing to the early development of these disorders. It seems very likely that some of these biologic markers may be linked to vulnerability factors in those at risk for depression, while other markers are linked to the expression of dysregulation in the face of stressors during early development.

Clearly, maturational changes in these systems may influence the picture at multiple levels: affecting the *expression* of dysregulation (that is, youth may *mask* underlying changes in biologic systems); affecting the *occurrence* of dysregulation (youth may be protective against disruption of some systems); or affecting the *measurement* of dysregulation (e.g., the stress created by the process of obtaining biologic measures may change with development obscuring maturational changes of scientific interest). Clearly, these are complex questions, whose answers may be elusive to our current methods of investigation. Further method development in localizing brain functions involved in affect regulation in development may be a keystone of progress. In addition, multidisciplinary efforts to bring together and integrate the rapid advances in developmental neurobiology, developmental psychology, and behavioral neuroscience appear to be crucial to address these complex issues. Based on our research, we speculate that shining the light of some of these efforts on the interface between cognitive-emotional regulation, at the point of early pubertal maturation, with an emphasis on the sleep onset transition, will be a particularly fruitful approach.

REFERENCES

Ambrosini, P. J., Bianchi, M. D., Rabinovich, H., & Elia, J. (1993). Antidepressant treatments in children and adolescents I. Affective disorders. *Journal of the American Academy of Child and Adolescent Psychiatry, 32*, 1–6.

Amsterdam, J. D., & Maislin, G. (1990). Comparison of growth hormone response after clonidine and insulin hypoglycemia in affective illness. *Biological Psychiatry, 28*, 308–314.

Ansseau, M., Von-Frenckell, R., Cerfontaine, J. L., Papart, P., Franck, G., Timsit-Berthier, M., Geenen, V., & Legros, J. J. (1988). Blunted response of growth hormone to clonidine and apomorphine in endogenous depression. *British Journal of Psychiatry, 153*, 65–71.

Appleboom-Fondu, J., Kerkhofs, M., & Mendlewicz, J. (1988). Depression in adolescents and young adults-polysomnographic and neuroendocrine aspects. *Journal of Affective Disorders, 14*, 35–40.

Asberg, M., Edman, G., & Rydin, E. (1984). Biological correlates of suicidal behavior. *Clinical Neuropharmacology, 7*, 758–759.

Banki, C. M., Arato, M., & Papp, Z. (1984). Biological markers in suicide patients. *Journal of Affective Disorders, 6*, 345–350.

Benca, R. M., Obermeyer, W. H., Thisted, R. A., & Gillin, J. C. (1992). Sleep and psychiatric disorders: a meta-analysis. *Archives of General Psychiatry, 49*, 651–668.

Bernal, J. F. (1973). Night waking in infants during the first 14 months. *Developmental and Medical Child Neurology, 15*, 760–769.

Biederman, J., Rosenbaum, J. F., Bolduc-Murphy, E. A., Faraone, S. V., Chaloff, J., Hirshfeld, D. R., & Kagan, J. (1993). Behavioral inhibition as a temperamental risk factor for anxiety disorders. *Child and Adolescent Psychiatric Clinics of North America, 2*, 667–683.

Birmaher, B., Dahl, R. E., Perel, J., Williamson, D. E., Nelson, B., Stull, S., Kaufman, J., Waterman, G. S., Rao, U., Nguyen, N., Puig-Antich, J., Ryan, N. D. Corticotropin releasing hormone challenge in prepubertal major depression. *Biological Psychiatry*. (in press)

Birmaher, B., Puig-Antich, J., Dahl, R. E., Ryan, N. D., Williamson, D. E., Rabinovich, H., Ambrosini, P., & Novacenko, H. (1990). Dexamethasone suppression test in children and adolescents. It is clinically relevant? Paper presented at American Academy of Child and Adolescent Psychiatry, Chicago.

Birmaher, B., Ryan, N. D., Dahl, R. E., Rabinovich, H., Ambrosini, P., Williamson, D. E., Novacenko, H., Nelson, B., Sing Lo, E., & Puig-Antich, J. (1992). Dexamethasone suppression test in children with major depressive disorder. *Journal of the American Academy of Child and Adolescent Psychiatry, 31*, 291–297.

Born, J., Muth, S., & Fehm, H. L. (1988). The significance of sleep onset and slow wave sleep for nocturnal release of growth hormone (GH) and cortisol. *Psychoneuroendocrinology, 13*, 233–243.

Brambilla, F., Musetti, C., Tacchini, C., Fontanilla, S., & Guareshi-Cazzulo, A. (1989). Neuroendocrine investigation in children and adolescents with dysthymic disorder: The DST, TRH, and clonadine test. *Journal of Affective Disorders, 17*, 279–284.

Brent, D. A., Crumrine, P. K., Varma, R. R., Allan, M., & Allman, C. (1987). Phenobarbital treatment and major depressive disorder in children with epilepsy. *Pediatrics, 80*, 909–917.

Brent, D. A., Kolko, D. J., Wartella, M. E., Boylan, M. A., Moritz, G., Baugher, M., Zelenak, J. (1993). Adolescent psychiatric inpatients' risk of suicide attempt upon six-month follow-up. *Journal of the Academy of Child and Adolescent Psychiatry, 32*, 95–105.

Brent, D. A., Perper, J., Goldstein, C. E., Kolko, D. J., Allan, M. J., Allman, C. J., & Zelenak, J. P. (1988). Risk factors for adolescent suicide: A comparison of adolescent suicide victims with suicidal inpatients. *Archives of General Psychiatry, 45*, 581–588.

Brent, D. A., Poling, K., McKain, B., & Baugher, M. (1993). A psychoeducational program for families of affectively ill children and adolescents. *Journal of the American Academy of Child and Adolescent Psychiatry, 32*, 770–774.

Brent, D. A., Ryan, N. D., Dahl, R. E., & Birmaher, B. (1995). Early-onset affective illness. In F. E. Bloom & D. J. Kupfer (Eds.), *Psychopharmacology: The fourth generation of progress* (pp. 1631–1642). New York: Raven Press.

Burke, P., Kocoshis, S., Chandra, R., Whiteway, M., & Sauer, J. (1990). Determinants of depression in recent onset pediatric inflammatory bowel disease. *Journal of the American Academy of Child and Adolescent Psychiatry, 29*, 608–610.

Burke, P., Meyer, V., Kocoshis, S., Grenstein, D., Chandra, R., Nord, D., Sauer, J., & Cohen, E. (1989). Depression and anxiety in pediatric inflammatory bowel disease and cystic fibrosis. *Journal of the American Academy of Child and Adolescent Psychiatry, 28*, 948–951.

Busby, K. & Pivik, R. T. (1983). Failure of high intensity auditory stimuli to affect behavioral arousal in children during the first sleep cycle. *Pediatric Research, 17*, 802–805.

Calil, H. M., Lesieur, P., Gold, P. W., Brown, G. M., Zavadil, A., & Potter, W. Z. (1984). Hormonal responses to zimelidine and desipramine in depressed patients. *Psychiatry Research, 13*, 231–242.

Carskadon, M. A., Keenan, S., & Dement, W. C. (1987). Nighttime sleep and daytime sleep tendency in preadolescents. In Christian Guilleminault (Ed.), *Sleep and its disorders* (pp. 43–52). New York: Raven Press.

Carskadon, M. A., Vieira, C., & Acebo, C. (1993). Association between puberty and delayed phase preference. *Sleep, 16*, 258–262.

Charney, D. S., Heninger, G. R., Sternberg, D. E., Hafstad, K. M., Giddings, S., & Landis, D. H. (1982). Adrenergic receptor sensitivity in depression. Effects of clonidine in depressed patients and healthy subjects. *Archives of General Psychiatry, 39*, 290–294.

Checkley, S. A. , Slade, A. P., & Schur, E. (1981). Growth hormone and other responses to clonidine in patients with endogenous depression. *British Journal of Psychiatry, 138*, 51–55.

Cicchetti, D. (1984). The emergence of developmental psychopathology. *Child Development, 55*, 1–7.

Cicchetti, D., & Schneider-Rosen, K. (1986). An organizational approach to childhood depression. In M. Rutter, C. Izard, & P. Read (Eds.), *Depression in young people: Clinical and developmental perspective* (pp. 239–252). New York: Guilford Press.

Collins, J. F. & Depue, R. A. (1992). A neurobehavioral systems approach to developmental psychopathology: Implications for disorders of affect. In D. Cicchetti & S. L. Toth (Eds.), *Rochester Symposium on Developmental Psychopathology: Vol. 4. Developmental perspectives on depression* (pp. 29–101). Rochester, NY: University of Rochester Press.

Coppen, A., & Wood, K. (1982). 5-Hydroxytryptamine in the pathogenesis of affective disorders. In B. T. Ho, J. C. Schoolar, E. Usdin (Eds.), *Serotonin in Biological Psychiatry* (pp. 249–258). New York: Raven Press.

Dahl, R. E., Matty, M. K., Birmaher, B., Al-Shabbout, M., & Ryan, N. D. (1996). Sleep onset abnormalities in depressed adolescents. *Biological Psychiatry, 39*: 400–410.

Dahl, R. E., Puig-Antich, J., Ryan, N., Nelson, B., Dachille, S., Cunningham, S., Trubnick, L., &Klepper, T. (1990). EEG sleep in adolescents with major depression: The role of suicidality and inpatient status. *Journal of Affective Disorders, 19*, 63–75.

Dahl, R. E., Puig-Antich, J., Ryan, N. D., Nelson, B., Novacenko, H., Twomey, J., Williamson, D., Goetz, R., & Ambrosini, P. J. (1989). Cortisol secretion in adolescents with major depressive disorder. *Acta Psychiatrica Scandinavica, 80*, 18–26.

Dahl, R. E. & Ryan, N. D. (in press). Sleep and neuroendocrine measures. In D. Shaffer and J. Richters (Eds.), *Assessment in child psychopathology.* New York: Guilford Press.

Dahl, R. E., Ryan, N. D., Birmaher, B., Al-Shabbout, M., Williamson, D. E., Neidig, M., Nelson, B., & Puig-Antich, J. (1991). EEG sleep measures in prepubertal depression. *Psychiatry Research, 38,* 201–214.

Dahl, R. E., Ryan, N. D., Perel, J., Birmaher, B., Al-Shabbout, M., Nelson, B., & Puig-Antich, J. (1994). Cholinergic REM induction test with arecoline in depressed children. *Psychiatry Research, 51,* 269–282.

Dahl, R. E., Ryan, N. D., Williamson, D. E., Ambrosini, P. J., Rabinovich, H., Novacenko, H., Nelson, B., & Puig-Antich, J. (1992). The regulation of sleep and growth hormone in adolescent depression. *Journal of the American Academy of Child and Adolescent Psychiatry, 31,* 615–621.

Dawson, G., Hessel, D., & Frey, K. (1994). Social influences on early developing biological and behavioral systems related to risk for affective disorder. In D. Cicchetti & D. Tucker (Eds.), *Development and psychopathology: Vol. 6:4* (pp. 759–779). New York: Cambridge University Press.

DeBellis, M. D., Chrousos, G. P., Dorn, L. D., Burke, L., Helmers, K., Kling, M., Trickett, P. K. & Putnam, F. (1994). Hypothalamic-pituitary-adrenal axis dysregulation in sexually abused girls. *Journal of Clinical Endocrinology and Metabolism, 78,* 249–255.

Doherty, M., Madansky, D., Kraft, J., Carter-Ake, L., Rosenthal, P., & Coughlin, B. (1986). Cortisol dynamics and test performance of the dexamethasone suppression test in 97 psychiatrically hospitalized children aged 3–16 years. *Journal of the American Academy of Child Psychiatry 25,* 400–408.

Ellis, E. M. (1991). Watchers in the night: An anthropological look at sleep disorders. *American Journal of Psychotherapy, 45*(2), 211–220.

Emslie, G. J., Rush, A. J., Weinberg, W. A., Rintelmann, J. W., & Roffwarg, H.P. (1990). Children with major depression show reduced rapid eye movement latencies. *Archives of General Psychiatry, 47,* 119–124.

Evans, D., Nemeroff, C., Haggerty, J., & Perdersen, C. (1987). Use of the dexamethasone suppression test with *DSM-III* criteria in psychiatrically hospitalized adolescents. *Psychoneuroendocrinology, 12,* 203–209.

Fleming, J. E., Boyle, M. H., & Offord, D. R. (1993). The outcome of adolescent depression in the Ontario child health study follow-up. *Journal of the American Academy of Child and Adolescent Psychiatry, 32,* 28–33.

Fleming, J. E. & Offord, D. R. (1990). Epidemiology of childhood depressive disorders: A critical review. *Journal of the American Academy of Child and Adolescent Psychiatry, 29,* 571–580.

Ford, D. E. & Kamerow, D. B. (1989). Epidemiological studies of sleep disturbances and psychiatric disorders: An opportunity for prevention? *Journal of the American Medical Association, 262,* 479–1484.

Franz, B., Kupfer, D. J., Miewald, J. M., Jarrett, D. B., & Grochocinski, V. J. (in press). Growth hormone secretion timing in depression: Clinical outcome comparisons. *Biological Psychiatry.*

Freeman, L., Poznanski, E., Grossman, J., Buchsbaum, Y., & Banegas, M. (1985). Psychotic and depressed children: A new entity. *Journal of the American Academy of Child Psychiatry, 24,* 195–202.

Fristad, M., Weller, R., Teare, M., & Preskorn, S. (1989). Self report vs. biological markers in assessment of childhood depression. *Journal of Affective Disorders, 15,* 339–345.

Geller, B., Rogol, A., & Knitter, E. (1983). Preliminary data on the dexamethasone suppression test in children with major depressive disorder. *American Journal of Psychiatry, 140,* 620–622.

Giles, D. E., Biggs, M. M., Rush, A. J., & Roffwarg, H. P. (1988). Risk factors in families of unipolar depression I. Incidence of illness and reduced REM latency. *Archives of General Psychiatry, 14*, 51–59.

Goetz, R., Puig-Antich, J., Dahl, R. E., Ryan, N. D., Asnis, G., Rabinovich, H., & Nelson, B. (1991). EEG sleep of young adults with major depression: A controlled study *Journal of Affec Disorders, 22*, 91–100.

Goetz, R., Puig-Antich, J., Ryan, N., Rabinovich, H., Ambrosini, P. J., Nelson, B., & Krawiec, V. (1987). Electroencephalographic sleep of adolescents with major depression and normal controls. *Archives of General Psychiatry, 44*, 61–68.

Gold, P., Goodwin, F., & Chrousos, G. (1988). Clinical and biochemical manifestations of depression. Relation to the neurobiology of stress. *New England Journal of Medicine, 319*, 348–353.

Gruen, P. H., Sachar, I. J., Altman, N., & Sassin, J. (1975). Growth hormone response to hypoglycemia in postmenopausal depressed woman. *Archives of General Psychiatry, 32*, 31–33.

Ha, H., Kaplan, S., & Foley, C. (1984). The dexamethasone suppression test in adolescent psychiatric patients. *American Journal of Psychiatry, 141*, 421–423.

Hagnell, O., Lanke, J., Rorsman, R., & Öjesjö, L. (1982). Are we entering an age of melancholy? Depressive illness in a prospective epidemiologic study over 25 years; the Lundby study, Sweden. *Psychological Medicine, 12*, 279–289.

Hammen, C., Adrian, C., Gordon, D., Burge, D., & Jaenicke, C. (1990). Longitudinal study of diagnoses in children of women with unipolar and bipolar affective disorder. *Archives of General Psychiatry, 96*, 190–198.

Harrington, R., Fudge, H., Rutter, M., Pickles, A., & Hill, J. (1990). Adult outcomes of childhood and adolscent depression: I. Psychiatric status. *Archives of General Psychiatry, 47*, 465–473.

Harrington, R., Fudge, H., Rutter, M., Pickles, A., & Hill, J. (1991). Adult outcomes of childhood and adolescent depression: II. Links with antisocial disorders. *Journal of the American Academy of Child and Adolescent Psychiatry, 30*, 434–439.

Jensen, J. B. & Garfinkel, B. D. (1990). Growth hormone dysregulation in children with major depressive disorder. *Journal of the American Academy of Child and Adolescent Psychiatry, 29*, 295–301.

Kagan, J., Reznick, J. S., & Snidman, N. (1987). The physiology and psychology of behavioral inhibition in young children. *Child Development, 58*, 1459–1473.

Kahn, A. (1987). A biochemical profile of depressed adolescents. *Journal of the American Academy of Child and Adolescent Psychiatry, 26*, 873–878.

Kahn, A. U., & Todd, S. (1990). Polysomnographic findings in adolescents with major depression. *Psychiatry Research, 33*, 313–320.

Kaufman, J. (1991). Depressive disorders in maltreated children. *Journal of the American Academy of Child and Adolescent Psychiatry , 30*, 257–265.

Kaufman, J., Brent, D., Dahl, R. E., Perel, J., Birmaher, B., Williamson, D., Ryan, N. D. (1993). *Measures of family adversity, clinical symptomatology, and cortisol secretion in a sample of preadolescent depressed children.* Paper presented at the annual meeting of the Society for Research In Child and Adolescent Psychopathology, Santa Fe, NM.

Klerman, G. L., Lavori, P. W., Rice, J., Reich, T., Endicott, J., Andreasen, N. C., Keller, M. B., & Hirschfeld, R. M. A. (1985). Birth cohort trends in rates of major depressive disorder among relatives of patients with affective disorder. *Archives of General Psychiatry, 42*, 689–695.

Knowles, J. B. & MacLean, A. W. (1990). Age-related changes in sleep in depressed and healthy subjects. *Neuropsychopharmacology*, 3, 251–259.

Korpi, E. S., Kleinman, J. E., & Goodman, S. I. (1986). Serotonin and 5-Hydroxyindoleacetic acid in brains of suicide victims. *Archives of General Psychiatry*, 43, 594–600.

Koslow, S. H., Stokes, P. E., Mendels, J., Ramsey, A., & Casper, R. (1982). Insulin tolerance test: human growth hormone response and insulin resistance in primary unipolar depressed, bipolar depressed and control subjects. *Psychol Medicine*, 12, 45–55.

Kovacs, M., Feinberg, T. L., Crouse-Novak, M. A., Paulauskas, S., & Finkelstein, R. (1984a). Depressive disorders in childhood I. A longitudinal prospective study of characteristics and recovery. *Archives of General Psychiatry*, 41, 229–237.

Kovacs, M., Feinberg, T. L., Crouse-Novak, M. A., Paulauskas, S., Pollock, M., & Finkelstein, R. (1984b). Depressive disorders in childhood. II. A longitudinal study of the risk for a subsequent major depression. *Archives of General Psychiatry*, 41, 643–649.

Kovacs, M., Gatsonis, C., Paulauskas, S. L., & Richards, C. (1989). Depressive disorders in childhood: IV. A longitudinal study of comorbidity with and risk for anxiety disorders. *Archives of General Psychiatry*, 46, 776–782.

Kovacs, M., Goldston, D., & Gatsonis, C. (1993). Suicidal behaviors and childhood-onset depressive disorders: A longitudinal investigation. *Journal of the American Academy of Child and Adolescent Psychiatry*, 32, 8–20.

Kutcher, S., Williamson, P., Szalai, J., & Marton, P. (1992). REM latency in endogenously depressed adolescents. *British Journal of Psychiatry*, 161, 399–402.

Lahmeyer, H. W., Poznanski, E. O., & Bellur, S. N. (1983). EEG sleep in depressed adolescents. *American Journal of Psychiatry*, 140, 1150–1153.

Langer, G., Heinze, G., Reim, B., & Matussek, N. (1976). Reduced growth hormone responses to amphetamine in "endogenous" depressive patients: Studies in normal "reactive" and "endogenous" depressive, schizophrenic, and chronic alcoholic subjects. *Archives of General Psychiatry*, 33, 1471–1475.

Lavori, P. W., Klerman, G. L., Keller, M. B., Reich, T., Rice, J., & Endicott, J. (1987). Age-period cohort analysis of secular trends in onset of major depression: Findings in siblings of patients with major affective disorder. *Journal of Psychiatric Research*, 21, 23–35.

Livingston, R., & Martin-Cannici, C. (1987). Depression, anxiety and the dexamethasone suppresion test in hospitalized prepubertal children. *Hillside Journal of Clinical Psychiatry*, 9, 55–63.

Livingston, R., Reis, C., & Ringkahl, I. (1984). Abnormal dexamethasone suppression test results in depressed and nondepressed children. *American Journal of Psychiatry*, 141, 106–108.

Lopez-Ibor, J. J., Saiz Ruiz, J., & Perez De Los Cobos, J. C. (1985). Biological correlations of suicide and aggressivity in major depressions (with melancholia): 5-Hydroxi-indoleacetic acid and cortisol in cerebral spinal fluid. Dexamethasone suppression test and therapeutic response to 5-hydroxytryptophan. *Neuropsychobiology*, 14, 67–74.

Lopez-Ibor, J. J., Saiz-Ruiz, J., & Inglecino, L. M. (1989). Neuroendocrine challenges in the diagnosis of depressive disorders. *British Journal of Psychiatry*, 154 (Suppl. 4), 73–76.

Mann, J. J., Arango, V., Underwood, M. D., Baird, F., & McBride, P. A. (1990). Neurochemical correlates of suicidal behavior: Involvement of serotonergic and non-serotonergic systems. *Pharmacological Toxicology*, 66, 37–60.

Matussek, N., Ackenheil, M., Hippius, H., Muller, F., Schroder, H. T., Schultes, H., & Wasilewski, B. (1980). Effect of clonidine on growth hormone release in psychiatric patients and controls. *Psychiatry Research*, 2, 25–36.

Meyer, W., Richards, G. E., Cavallo, A., Holt, K. G., Hejazi, M. S., Wigg, C., & Rose, D. (1991). Depression and growth hormone [letter comment]. *Journal of the American Academy of Child and Adolescent Psychiatry, 30,* 2.

Mueller, P. S., Heninger, G. R. & McDonald, R. K. (1969). Insulin tolerance test in depression. *Archives of General Psychiatry, 21,* 587–594.

Murphy, D. L., Campbell, I., & Costa, J. L. (1978). Current status of the indoleamine hypothesis of affective disorders. In M. A. Lipton, A. DiMascio, & K. F. Killam (Eds.), *Psychopharmacology: A generation of progess.*

Naylor, M., Greden, J., & Alessi, N. (1990). Plasma dexamethasone levels in children given the dexamethasone suppression test. *Biological Psychiatry, 27,* 592–600.

Petty, L., Asarnow, J., Carlson, G., & Lesser, L. (1985). The deamethasone suppression test in depressed, dysthymic, and nondepressed children. *American Journal of Psychiatry, 142,* 631–633.

Pfeffer, C., Stokes, P., Weiner, A., Shindledecker, R., Faughnan, L., Mintz, M., Stoll, P., & Heiligenstein, E. (1989). Psychopathology and plasma cortisol responses to dexamethasone in prepubertal inpatients. *Biological Psychiatry, 26,* 677–689.

Pfeffer, C. R., Stokes, P., & Shindledecker, R. (1991). Suicidal behavior and hypothalamic-pituitary-adrenocrotical axis indices in child psychiatric patients. *Biological Psychiatry, 29,* 909–917.

Post, R. M., Weiss, S. R., & Leverich, G. S. (1994). Recurrent affective disorder: Roots in developmental neurobiology and illness progression based on changes in gene expression. In D. Cicchetti & D. Tucker (Eds.), *Development and psychopathology: Vol. 6:4* (pp. 781–813). New York: Cambridge University Press.

Poznanski, E., Carroll, B., Banegas, M., Cook, S., & Grossman, J. (1982). The dexamethasone suppression test in prepubertal depressed children. *American Journal of Psychiatry, 139,* 321–324.

Puig-Antich, J., Dahl, R., Ryan, N., Novacenko, H., Goetz, D., Goetz, R., Twomey, J., & Klepper, T. (1989). Cortisol secretion in prepubertal children with major depressive disorder. Episode and recovery. *Archives of General Psychiatry, 46,* 801–809.

Puig-Antich, J., Davies, M., Novacenko, H., Davies, M., Tabrizi, M. A., Ambrosini, P., Goetz, R., Bianca, J., Goetz, D., & Sachar, E. J. (1984). Growth hormone secretion in prepubertal major depressive children: III. Response to insulin induced hypoglycemia in a drug-free fully recovered clinical state. *Archives of General Psychiatry, 41,* 471–475.

Puig-Antich, J., Goetz, R., Davies, M., Fein, M., Hanlon, C., Chambers, W. J., Tabrizi, M. A., Sachar, E. J., & Weitzman, E. D. (1984). Growth hormone secretion in prepubertal major depressive children: II. Sleep related plasma concentrations during a depressive episode. *Archives of General Psychiatry, 41,* 463–466.

Puig-Antich, J., Goetz, R., Davies, M., Tabrizi, M. A., Novacenko, H., Hanlon, C., Sachar, E. J., & Weitzman, E. D. (1984). Growth hormone secretion in prepubertal major depressive children: IV. Sleep related plasma concentrations in a drug-free fully recovered clinical state. *Archives of General Psychiatry, 41,* 479–483.

Puig-Antich, J., Goetz, R., Hanlon, C., Tabrizi, M. A., Davies, M., & Weitzman, E. (1982). Sleep architecture and REM sleep measures in prepubertal major depressives during an episode. *Archives of General Psychiatry, 39,* 932–939.

Puig-Antich, J., Goetz, R., Hanlon, C., Tabrizi, M. A., Davies, M. A., & Weitzman, E. D. (1983). Sleep architecture and REM sleep measures in prepubertal major depressives. Studies during recovery from the depressive episode in a drug-free state. *Archives of General Psychiatry, 40,* 187–192.

Puig-Antich, J., Kaufman, J., Ryan, N. D., Williamson, D. E., Dahl, R. E., Lukens, E., Todak, G., Ambrosini, P., Rabinovich, H., & Nelson, B. (1993). The psychosocial functioning and family environment of depressed adolescents. *Journal of the American Academy of Child and Adolescent Psychiatry, 32,* 244–253.

Puig-Antich, J., Lukens, E., Davies, M., Goetz, D., & Todak, G. (1985a). Psychosocial functioning in prepubertal major depressive disorders: II. Interpersonal relationships after sustained recovery from the affective episode. *Archives of General Psychiatry, 42,* 511–517.

Puig-Antich, J., Lukens, E., Davies, M., Goetz, D., & Todak, G. (1985b). Psychosocial functioning in prepubertal major depressive disorders: II. Interpersonal relationships after sustained recovery from the affective episode. *Archives of General Psychiatry, 42,* 511–517.

Puig-Antich, J., Novacenko, H., Davies, M., Fein, M., Hanlon, C., Chambers, W. J., Tabrizi, M. A., Sachar, E. J., & Weitzman, E. D. (1984). Growth hormone secretion in prepubertal major depressive children: I. Response to insulin induced hypoglycemia. Final report. *Archives of General Psychiatry, 41,* 455–460.

Rao, U., Dahl, R. E., Ryan, N. D., Birmaher, B., Williamson, D. E., Giles D.E., Rao, R., Kaufman J., Nelson B., *The Relationship between longitudinal clinical course and sleep and cortisol changes in adolescent depression, Biological Psychiatry* (in press).

Rao, U., Ryan, N. D., Birmaher, B., Dahl, R. E., Williamson, D. E., Kaufman, J., Rao, R., & Nelson, B. (1995). Unipolar depression in adolescents: Clinical outcome in adulthood. *Journal of the American Academy of Child and Adolescent Psychiatry, 34: 5;* 566-578.

Rao, U., Weissman, M. M., Martin, J. A., & Hammond, R. W. (1993). Childhood depression and risk of suicide: A preliminary report of a longitudinal study. *Journal of the American Academy of Child and Adolescent Psychiatry, 32,* 21-27.

Rasmussen, K. R., & Suomi, S. J. (1989). Heart rate and endocrine responses to stress in adolescent male rhesus monkeys on Cayo Santiago. *Puerto Rico Health Sciences Journal, 8,* 65–71.

Reynolds, C. F., & Kupfer, D. J. (1987). State-of-the-art review: Sleep research in affective illness: State of the art circa. *Sleep, 10,* 199–215.

Robbins, D., Alessi, N., Yanchyshyn, G., & Colfer, M. (1982). Preliminary report on the dexamethasone suppression test in adolescents. *American Journal of Psychiatry, 22,* 467–469.

Robbins, D., Alessi, N., Yanchyshyn, G., & Colfer, M. (1983). The dexamethasone suppression test in psychiatrically hospitalized adolescents. *American Journal of Psychiatry, 22,* 467–469.

Robbins, D. R., & Alessi, N. E. (1985). Suicide and the dexamethasone suppression test in adolescence. *Biological Psychiatry, 20,* 107–110.

Rosenbaum, J. F., Biederman, J., Gersten, M., Hirshfeld, D. R., Meminger, S. R., Herman, J. B., Kagan, J., Reznick, J. S., & Snidman, N. (1988). Behavioral inhibition in children of parents with panic disorder and agoraphobia. *Archives of General Psychiatry, 45,* 463–470.

Rutter, M., & Quinton, D. (1984). Parental psychiatric disorder: Effects on children. *Psychological Medicine, 14,* 853–880.

Rutter, M. (1986). Child psychiatry: the interface between clinical and developmental research. *Psychological Medicine, 16,* 151–160.

Ryan, N. D. (1990). New research: Pharmacological treatment of adolescent depression. *Psychopharmacology Bulletin, 26,* 75–79.

Ryan, N. D., Birmaher, B., Perel, J. M., Dahl, R. E., Meyer, V., Al-Shabbout, M., Iyengar, S., & Puig-Antich, J. (1992). Neuroendocrine response to L-5-hydroxytryptophan challenge in prepubertal major depression: Depressed versus normal children. *Archives of General Psychiatry, 49,* 843–851.

Ryan, N. D., & Dahl, R. E. (1993). The biology of depression in children and adolescents. In J. J. Mann and D. J. Kupfer (Eds.), *The biology of depressive disorders* (pp. 37–58). New York: Plenum Press.

Ryan, N. D., Dahl, R. E., Birmaher, B., Williamson, D. E., Iyengar, S., Nelson, B., Puig-Antich, J., & Perel, J. M. (1994). Stimulatory tests of growth hormone secretion in prepubertal major depression: Depressed versus normal children. *Journal of the American Academy of Child and Adolescent Psychiatry, 33,* 824–833.

Ryan, N. D., Puig-Antich, J., Rabinovich, H., Robinson, D., Ambrosini, P. J., Nelson, B., & Iyengar, S. (1987). The clinical picture of major depression in children and adolescents. *Archives of Geneneral Psychiatry, 44,* 854–861.

Sachar, E. J., Finkelstein, J., & Hellman, L. (1971). Growth hormone responses in depressive illness. I. Response to insulin tolerance test. *Archives of General Psychiatry, 25,* 263–269.

Sachar, E. J., Frantz, A. G., Altman, N., & Sassin, J. (1973). Growth hormone and prolactin in unipolar and bipolar depressed patients: Responses to hypoglycemia and L-doppa. *American Journal of Psychiatry, 130,* 1362–1367.

Sapolsky, R. M. (1990). Adrenocortical function, social rank, and personality among wild baboons. *Biological Psychiatry, 28,* 862–878.

Shaffer, D. (1988). The epidemiology of teen suicide: an examination of risk factors. *Journal of Clinical Psychiatry, 49 (Suppl.),* 36–41.

Shaffi, M., Carrigan, S., Whittinghill, J. R., & Derrick, A. M. (1985). Psychological autopsy of completed suicide in children and adolescents. *American Journal of Psychiatry, 142,* 1061–1064.

Shafii, M., Steltz-Lenarsky, J., Derrick, A. M., Beckner, C., & Whittinghill, J. R. (1988). Comorbidity of mental disorders in the post-mortem diagnosis of completed suicide in children and adolescents. *Journal of Affective Disorders, 15,* 227–233.

Siever, L. J., Uhde, T. W., Silberman, E. K., Jimerson, D. C., Aloi, J. A., Post, R. M., & Murphy, D. L. (1982). Growth hormone response to clonidine as a probe of noradrenergic receptor responsiveness in affective disorder patients and controls. *Psychiatry Research, 6,* 171–183.

Sorce, J. F. & Emde, R. N. (1985). Maternal emotional signaling: Its effect on the visual cliff behavior of 1-year-olds. *Developmental Psychology, 21,* 195–200.

Steingard, R., Biederman, J., Keenan, K., & Moore, C. (1990). Comorbidity in the interpretation of the dexamethasone suppression test results in children: A review and report. *Biological Psychiatry, 28,* 193–202.

Suomi, S. J. (1987). Genetic and maternal contributions to individual differences in rhesus monkey biobehavioral development. In N.A. Krasnegor (Ed.), *Perinatal development: A psychobiological Perspective* (pp. 397–419). Orlando, FL: Academic Press.

Suomi, S. J. (1991). Primate separation models of affective disorders. In J. Madden IV (Ed.), *Neurobiology of Learning, Emotion, and Affect* (pp. 195–213). New York: Raven Press.

Takahashi, Y., Kipnis, D. M., & Daughaday, W. H. (1968). Growth hormone secretion during sleep. *Journal of Clinical Investigation, 47,* 2079–2084.

Targum, S., & Capodanno, A. (1983). The dexamethasone suppression test in adolescent psychiatric inpatients. *American Journal of Psychiatry, 140,* 589–591.

Toth, S. L., Manly, J. T., & Cicchetti, D. (1992). Child maltreatment and vulnerability to depression. *Development and Psychopathology, 4,* 97–112.

Van Praag, H. M. (1982). Serotonin precursors in the treatment of depression. In B. T. Ho, J. C. Schoolar, & E. Usdin (Eds.), *Serotonin in biological psychiatry* (pp. 259–286). New York: Raven Press.

Weissman, M. M., Leckman, J. F., Merikangas, K. R., Gammon, G. D., & Prusoff, B. A. (1984). Depression and anxiety disorders in parents and children. Results from the Yale family study. *Archives of General Psychiatry, 41*, 845–852.

Weller, R. A., Weller, E. B., Fristad, M. A., & Bowes, J. M. (1991). Depression in recently bereaved prepubertal children. *American Journal of Psychiatry, 148*, 1536–1540.

Weller, E., Weller, R., Fristad, M., Preskorn, S., & Teare, M. (1984). The dexamethasone suppresiion test in hospitalized prepubertal children. *American Journal of Psychiatry, 141*, 290–291.

Weller, E., Weller, R., Fristad, M., Preskorn, S., & Teare, M. (1985). The dexamethasone suppression test in hospitalized prepubertal children. *Journal of Clinical Psychiatry, 46*, 511–513.

Woodside, B., Brownstone, D., & Fisman, S. (1987). The dexamethasone suppression test and the children's depression inventory in psychiatric disorders in children. *Canadian Journal of Psychiatry, 32*, 2–4.

Young, W., Knowles, J. B., MacLean, A. W., Boag, L., & McConville B. J. (1982). The sleep of childhood depressives: Comparison with age-matched controls. *Biological Psychiatry, 17*, 1163–1169.

Zigler, E., & Glick, M. (1986). *A developmental approach to adult psychopathology.* New York: Wiley.

VIII Developmental Pathways to Adolescent Suicide

David A. Brent & Grace Moritz

The study of adolescent suicide is almost by nature a look backward. This is predicated by the relative rarity of this condition, rendering longitudinal developmental studies focussing on suicide as an outcome almost an impossibility. Therefore, most of the data on adolescent suicide are derived from case-control studies using psychological autopsy approaches, in which the life of the suicide victim is reconstructed via interviews with multiple key informants (Brent, Perper, Kolko, and Zelenak 1988a; Brent, Perper, Kolko, and Zelenak 1988b). Nevertheless, this information can be gathered and understood using a developmental framework. In this chapter, risk factors for adolescent suicide will be delineated. In keeping with the overall context of this volume, these risk factors will be discussed with respect to their contribution to the adolescent's developmental trajectory that ended in suicide. Two developmental concepts will be invoked: the development of the individual and the development of psychopathologic conditions associated with suicide.

Descriptive Epidemiology

Adolescent suicide is the third-leading cause of death in this age group (CDC, 1985). The rate has increased dramatically from 2.7 per 100,000 in 1950 to 11.1 per 100,000 in 1990 (CDC, 1990). The greatest increase has been among white males, although African American males have shown a dramatic increase in recent years (CDC, 1990). Males consistently show a higher rate than females. The rate is considerably higher among older than younger adolescents, and is extremely rare in childhood.

This work was supported by an award from the William T. Grant Foundation, #86-1063, and by NIMH grants MH44711, "Youth Exposed to Suicide," MH43366 "Adolescent Family Study," and MH46500 "Depressed Adolescent Suicide Attempters: A Clinical Trial," as well as by the Commonwealth of Pennsylvania state appropriation to Services for Teens at Risk (STAR) of Western Psychiatric Institute and Clinic. The authors gratefully acknowledge Jeanie Knox Houtsinger and Karen Rhinaman, who assisted in preparing this manuscript. In addition, a special note of thanks to the contributors to this work: Joshua Perper, M.D., Mary Beth Boylan, M.A., Kim Poling, M.S.W., Mary Wartella, M.S.W., Joy Schweers, M.Ed., Claudia Roth, M.S.W., Charles Goldstein, A.C.S.W., and Carl Bonner, M.A.

Age

Suicide is extremely rare in childhood and early adolescence, and becomes increasingly frequent with increasing age. There are at least four factors that may account for this developmental change: (1) the influence of puberty on the risk for psychopathology; (2) the increase in the frequency and severity of stressors associated with suicidal behavior; (3) developmental changes in the cognitive capacity of children versus adolescents; and (4) the role of children versus adolescents in the family.

It is well-known that the risk of various psychiatric conditions increases dramatically after puberty, namely, affective disorders (both unipolar and bipolar), and substance abuse (Costello, 1989; Fleming & Offord, 1990). While the rate of conduct disorder does not increase with puberty (at least in boys), the prevalence of severe, aggressive conduct disorder (e.g., delinquency) with societal, interpersonal, and legal consequences first emerges at around the time of puberty, often co-occurring with serious substance abuse (Farrington, 1986). Suicide attempts also become much more common after puberty (Velez & Cohen, 1988), presumably in part because of the above-noted developmental trends in psychiatric conditions.

A related factor associated with the onset of puberty is that there is an increased prevalence in the stressors associated with suicide, namely, interpersonal conflict with peers and parents and with romantic disappointments. There is evidence that certain factors such as interpersonal loss interact with specific disorders such as substance abuse to heighten risk for completed suicide in adolescence (Brent, Perper, Moritz, Baugher, Roth, Balach, and Schweers 1993a). There are other, more chronic stressors associated with puberty and adolescence that may also contribute to risk for suicide. One is the increased size and anonymity of middle and high schools compared to elementary schools, which has been documented to be associated with increased stress and depressive symptomatology in adolescence (Eccles, Midgley, Wigfield, Buchanan, Reuman, Flanagan, and Mac Iver 1993). One direct consequence of this social-ecological transition is a decrease in adult monitoring of the social and emotional development of adolescents, which may be particularly critical in those at-risk youth whose families are not supportive or involved. A related developmental factor is that later adolescence is associated with the formulation of life plans, in terms of both romantic attachments and school and career plans. The failure to qualify for entry into a special program of study, the receipt of poor school grades, or the occurrence of a disciplinary blot on one's record can lead to the perception on the part of the adolescent of derailment in his or her developmental life plan, subsequently leading to a spiral of hopelessness and despair. Not surprisingly, the above-noted stressors are often associated with both suicide attempts and completions (Brent, Perper, Goldstein, Kolko, Allan, Allman, and Zelenak 1988b, 1993a).

Another developmental aspect of risk for completed suicide is the role of cognitive development in the planning and execution of a lethal suicide attempt. Some posit that the relative cognitive immaturity of prepubertal children may be protective against suicide (Shaffer, 1974; Shaffer & Fisher, 1981). Among the cognitive prerequisites for completed suicide may be having a view of the future (hence, being able to feel hopeless), viewing death as irreversible, and finally, possessing the cognitive ability to plan and execute a lethal suicide attempt.

In addition to the epidemiologic data, what evidence is there that cognitive maturity is salient with respect to suicidal risk? First, there is the report of a study of 12–15-year-old suicide victims in the United Kingdom, in which there was an overrepresentation both of physically and intellectually precocious individuals (i.e., IQ≥130) (Shaffer, 1974). Shaffer (1974) interpreted this finding to mean that intellectual precocity in early adolescence could override the cognitive immaturity ordinarily protective against suicide normally associated with late childhood and early adolescence.

Second is the report of Carlson, Asarnow, and Orbach (1987), in which three groups of children and adolescents were compared as to their concept of death and knowledge of suicide: 8–13-year-old hospitalized inpatients; developmentally delayed, hospitalized adolescents; and 8–13-year-old normal controls. In this report, it was found that the majority of subjects, regardless of age, understood the irreversibility of death, but that the proportion of subjects who endorsed the irreversibility of death increased linearly as a function of age. However, the developmentally delayed youngsters did not differ from their normal IQ age-mates in this regard, except in the younger age group (e.g., 8–10 years of age). Insofar as the developmentally delayed adolescents had at least a mental age of 7, this is consistent with the view that the irreversibility of death is fairly well established by age 7 (Koocher, 1974).

In the study of Carlson et al. (1987), there was a trend for greater awareness of suicide with increasing mental age. More of the normal IQ 11–13 year olds and developmentally delayed adolescents endorsed overdose as a method of suicide, reflective of the actual methods used by adolescents in suicide attempts (Hawton & Catalan, 1987). When asked about motivations for suicide, more of the younger, developmentally delayed children were not able to endorse any motivation, but the majority of the normal IQ children aged 8–13 and developmentally delayed adolescents were able to endorse typical motivations for engaging in suicidal behavior. There was evidence that having a psychiatric condition was associated with a foreshortened view of one's own mortality, insofar as some of the hospitalized subjects thought their lives would end before old age, compared to none of the controls. The adolescents, despite their developmental disabilities, gave a more accurate picture of the most common methods of suicide, compared to the preadolescent children.

From these results, we can conclude that the concepts of irreversibility of death and hopelessness probably do not play a major role in explaining the developmental trends in the suicide statistics. Hopelessness is common in clinical populations regardless of age and is manifested in prepubertal populations well before the suicide rate begins to go up. It may be that hopelessness is correlated with suicidal ideation and intent, but not necessarily with actual suicidal activity, at least in cross-sectional studies of prepubertal children (Kazdin, French, Unis, Esveldt-Dawson, and Sherick 1983; Pfeffer, Conte, Plutchik, and Jerrett 1979).

Certain cognitive capabilities, such as knowledge of methods of suicide, awareness of suicide as an option, and understanding why someone would attempt suicide did appear to change with increasing age, with older children and developmentally delayed adolescents having the most knowledge about suicide. This would seem to confirm Shaffer's (1974) view that it is only in later childhood and early adolescence that the ability to plan and execute a lethal suicide attempt first emerges. The developmentally delayed adolescents appear to be quite capable of completing suicide on the basis of the study of Carlson et al. (1987). This could be a function of their exposure to psychiatric hospitalization and to other impaired youngsters, or instead might be attributable to their psychiatric conditions. This also raises the question of whether the risk for suicidal behavior increases with puberty *independent* of cognitive mediating factors.

Gender

Males complete suicide about five times more frequently than females. This is particularly puzzling, because adolescent females appear to have higher rates of affective illness and of suicide attempts (Lewinsohn, Rohde, and Seeley 1993), both major risk factors for completed suicide in adolescents (Brent, Perper, Moritz, Allman, Friend, Roth, Schweers, Balach, and Baugher 1993b; Garland, Gould, Fisher, and Trautman, 1988; Shafii, Steltz-Lenarsky, Derrick, Beckner, and Whittinghill 1988). Some explanations for the male preponderance among completed suicide victims include (1) males' choice of more lethal methods; (2) males' greater propensity to impulsive violence; (3) males' greater use and abuse of substances; (4) more frequent comorbidity of affective disorder with nonaffective comorbidity among males; (5) males' greater resistance to seeking help; and (6) males' greater vulnerability to stress. The evidence to support each of these hypotheses will briefly be reviewed.

There is consistent data that females are more likely to resort to overdose, and that suicidal males are more likely to resort to other more violent methods, such as hanging, both in attempts and completions (Brent & Moritz, 1993; Cairns, Peterson, and Neckerman 1988; Otto, 1972; Rich, Ricketts, Fowler, and Young 1988). Moreover, males are more likely to be intoxicated at the time of the suicidal episode, and unlike females, there is a relationship between intoxication and the use of firearms in male suicide victims (Brent, Perper, Moritz, Baugher, Schweers, and Roth 1993c). A handgun in the home was a risk factor for both males and females, but a long gun in the home was risk factor only for males (Brent, Perper, Moritz, Baugher, Schweers, and Roth 1993).

There is some evidence that substance abuse is more common in male than in female adolescent completers (Rich et al., 1988; Shaffer et al., 1988) and that substance abuse is a risk factor for suicide in males, but not in females (Shaffer et al.,

1988). However, other recent studies found that substance abuse was equally prevalent in both male and female completers and was a risk factor for suicide for both genders (Brent, Marttunen, Aro, Henriksson, and Longgquist 1991). However, there was trend in our data for drug abuse to be more prevalent and to be a risk factor for completed suicide in males, but not females. It may be that the issue is not only the differential *prevalence* of substance abuse between males and females, but the differential *impact* of substance abuse on males versus females. As noted above, when males are intoxicated during the suicidal episode, they are much more likely to commit suicide by firearms, whereas no such relationship exists for females. There are differences in central serotonin metabolism between males and females (Ryan, Birmaher, Perel, Dahl, Meyer, Al-Shabbout, Inyengar, and Puig-Antich 1992), and it is possible that these differences are further exaggerated by the effects of alcohol (Ballenger, Goodwin, Major, and Brown 1979), which in turn may lead to violent behavior in males, but not females.

Affective disorders are more common among female than male completers (Brent & Moritz, 1993; Marttunen et al., 1991; Rich et al., 1988; Shaffer et al., 1988), and there is some evidence that affective disorder is a greater risk factor for completed suicide in females than in males (Brent & Moritz, 1993; Shaffer et al., 1988). It is possible that affective disorder is qualitatively different in males and females. Rich et al. (1988) observed that affective disorder was more often comorbid with substance abuse in males than females, but this was not replicated in our sample (Brent & Moritz, 1993). It has been reported that depressive mood in males is more often accompanied by externalizing symptoms, whereas females tend to have more predominantly ruminative, internalizing symptoms (Gjerde, Block, and Block 1988; Kobak et al., 1991). However, this has not been specifically examined as a correlate of suicidal risk, nor would this gender difference in the expression of depressive symptomatology explain the stronger association of depression and suicide in females compared to males. One frequent correlate of affective illness in suicide victims is homicidality (Brent et al., 1993b), but there are no differences in the prevalence of significant homicidal ideation between male and female suicide victims (Brent & Moritz, 1993).

The greater propensity of males to aggression and disruptive disorders may also explain differences in suicidal risk (Offord, Boyle, Szatmari, Rae-Grant, Linhs, Cadman, Byles, Crawford, Blum, Byrne, Thomas, and Woodward 1987). The etiology of these differences are exceedingly complex and may relate to differences in socialization or impact of sex steroids on brain and personality development (Eccles et al., 1993; Maccoby & Jacklin, 1980). The rate of conduct disorder is higher among male than among female suicide completers (Brent & Moritz, 1993; Marttunen et al., 1991; Shaffer et al., 1988). However, the evidence as to whether conduct disorder is a greater risk factor for completed suicide in males than in females is contradictory, with one study finding conduct disorder to be a risk factor for suicide only in males (Brent & Moritz, 1993), whereas another found antisocial symptoms to be an equally potent risk factor for suicide in both genders (Shaffer et al., 1988). There is evidence that males are more vulnerable than females to develop symptoms of conduct disorder and, perhaps, depression in the face of family disruption (Brent & Moritz, 1993; Earls & Jung, 1987; Hetherington, Clingempell, Anderson, Deal, Hagan, Hollier, and Linder 1992; Jaffe, Wolfe, Wilson, and Zak 1986; Reinherz, Glaconia, Pakiz, Silverman, Frost, and Lefkowitz 1993). This in turn may explain why a male not living with both biological parents is at risk for completed suicide, whereas family constellation does not appear to affect suicidal risk in females. One possible explanation is that with family break-up, the children usually stay with the mother. The presence of a father in the home may be more critical to boys' development, particularly temperamentally difficult boys. A better understanding of how family stress differently affects males' and females' development may be critical to the prevention of conduct disorder and suicide.

There is evidence that males are more resistant to seek help for mental health problems than females (Overholser, Hemstreet, Spirito, and ?????? 1989; Shaffer, Vieland, Garland, Rojas, Underwood, and Busner 1990). Consequently, in the face of suicidality, females may be more willing to engage in behavior that would attract help and support. Rich et al. (1988) has observed that male suicide completions seem to be of higher intent, although we have not replicated this finding (Brent & Moritz, 1993).

With regard to vulnerability to stress, Marttunen et al. (1991) has hypothesized that males tended to complete suicide at a lower psychopathologic threshold than females. In this Finnish study of adolescent suicide, it was observed that one third of male suicide completers met criteria for an adjustment disorder versus none of the females. While this has not been our experience (Brent & Moritz, 1993), it is true that the vast majority of suicide victims who kill themselves without evidence of serious impairment or psychopathology are male (Brent, Perper, Moritz, Baugher, and Allman 1993d). The suicidal crises of those without apparent psychopathology are often precipitated by legal problems and facilitated by the presence of a loaded gun in the home (Brent et al., 1993d). As noted before, in our sample, coming from a "nonintact" home was a risk factor for males, but not females, lending further support to the notion of Marttunen et al. (1991).

Sexual Orientation

A report published by the U.S. Department of Health and Human Services (1989), often quoted, states that 30% of all adolescent suicide victims are lesbian or gay. However, the primary source of these data are unclear since there were no published psychological autopsy studies that actually examine this. Dr. David Shaffer (1993), in an article published in the *New Yorker*, cites his own work indicating that the rate of homosexual orientation is no greater in his extensive series of adolescent suicide victims than in his control sample. In further support of this view, Rich, Fowler, Young, and Blenkush (1988b) reported on the rate of homosexuality in a large, uncontrolled series of male suicides across the life span and found a prevalence of 10%, comparable to other population estimates of homosexuality. A report of an uncontrolled series of adolescent homosexual and bisexual youth sampled from a youth drop-in center, gay bars, and advertisements in gay publications showed a much higher suicide attempt rate than one might expect (Remafedi, Farrow, and Deisher 1991). However, this sample appeared to be highly skewed, as the rates of previous mental health treatment, prostitution, illicit drug use, and previous physical / sexual abuse in this sample were extraordinarily high. This in turn raises questions about whether this sample is truly representative of homosexual youth and whether their suicide attempt rate was a function of homosexual orientation or the sampling frame, the latter of which might have enriched the sample with a host of psychosocial risk factors. Nevertheless, in our homophobic society, the process of "coming out" can often be very frightening, and it is plausible that the threat of rejection by parents and peers might lead to a suicidal crisis. Further research is required in this challenging and important area, especially longitudinal studies of the development, psychosocial correlates, and consequences of sexual orientation.

Ethnicity

Suicide rates are highest in Micronesians and Native Americans, followed by whites, and finally, African Americans. The suicide rate in Micronesia varies greatly by location within the archipelago. Specifically, the rate is lowest in the areas that are either already completely Westernized (e.g., Guam) or have maintained a traditional orientation (at the periphery of the archipelago) (Rubenstein, 1983). A combined epidemiologic and ethnographic analysis of the situation in Micronesia indicates that the suicide rate is highest among adolescent and young adult males in islands that are undergoing a rapid transition from a traditional to a Western cultural orientation. Prior to Westernization, Micronesian families practiced an "incest taboo" in which pubertal males were extruded from the family homes and taken into all-male houses led by older unmarried males, until the adolescent or young adult was ready for marriage. With increasing urbanization, the "mens' house" institution withered away, but unfortunately, the incest taboo and accompanying extrusion of pubertal males from the family home continued, leading to displacement, alienation, hopelessness, substance abuse, and suicide among this population. A similar picture can be seen in the Native American population in the United States (Berlin, 1987), where the overall suicide rate among adolescents and young adult males is extraordinarily high, but quite variable from tribe to tribe. The increased suicide rates among Native Americans are most frequently found in reservations where the traditional cultural practices have eroded, with high rates of family mobility, instability, unemployment, and alcoholism.

Why the rate differs between whites and African Americans is unclear. This may become a moot point, as the rate among African American males has increased rapidly over the past two decades (CDC, 1990; Shaffer, Gourd, and Hiems, 1994). There appears to be an effect of class on the suicide rate among African Americans, but not among whites. Some possible explanations for this finding include that higher SES African Americans are more assimilated, experience estrangement from the traditional supports of extended family, and feel isolated in a still hostile and racist mainstream American society. Other evidence along these lines comes from a geographic analysis of suicide among youthful African Americans. The suicide rate among African Americans is highest in the urbanized and industrial Northeast and Midwest, and lowest in the rural South, possibly also relating to degree of assimilation and loss of traditional family supports (Shaffer & Fisher, 1981). Another question with respect to the difference between white and African American suicide rates is the extent to which some homicides among African American are really "victim-precipitated suicides" (Wolfgang, 1959), thereby obscuring the "true" suicide rate in this ethnic minority.

Secular Trends

The suicide rates among adolescents and young adults have increased dramatically in the past three decades (Shaffer et al., 1988). Similar secular trends among youth have been reported for substance abuse, depression, and conduct disorder (Klerman, Lavor, Rich, Reich, Endicott, Andreasen, Keller, and Hirschfield 1985; Reich, Cloninger, Van Eerdewegh, Rich, and Mullaney 1988; Robins & Price, 1991; Ryan, Birmaher, Perel, Dahl, Meyer, Al-Shabbout, Iyengar, and Puig-Antich 1992). There are data that tie the increase in the suicide rate to increases in substance use at the time of the suicide, and to secular trends in substance abuse (Brent, Perper, and Allman 1987; Carlson, Rich, Grayson, and Fowler 1991; Rich , Young, and Fowler 1986b). A greater availability of firearms in the homes of youth may also have contributed to an increase in the suicide rate (Boyd, 1983; Boyd & Moscicki, 1986; Brent et al., 1987). Other possible contributors to the secular trend are more broad social changes that have occurred post-World War II, including increased mobility, decreased religious orientation, and an increased divorce rate (McArnaney, 1979). There are data to indicate that increase mobility is associated with an increased risk of emotional and behavioral problems in youth (Simpson & Fowler, 1994). Our own data suggests that family breakup may place males at increased risk for completed suicide (Brent & Moritz, 1993). Perhaps as a concomitant of the above-noted social changes, other attitudinal shifts that may have occurred are an increased tolerance and acceptance of suicide as an option and an increased external locus of control, leading to helplessness and hopelessness (Boor, 1976). One intriguing hypothesis is that of Holinger and Otter (1982), who observed correlations between the size of the adolescent-aged cohort relative to the total population and the adolescent suicide rate. They hypothesized that when the adolescent cohort was relatively large, as in the case with the baby boom generation, then that cohort would be at higher risk for completed suicide. The theoretical reason behind the proposed interrelationship of size and risk for suicide is that when the cohort is larger, there is greater competition among its members for scarce resources (i.e., jobs, entry into advanced education), as well as for attention from parents and teachers, resulting in increased stress and an increased risk of suicide.

Family Environment

We have already noted the association between a non-intact family and completed suicide. Further examination of this relationship reveals that non-intactness is associated with parental psychopathology (Brent, Perper, Moritz, Liotus, Schweers, Balach, and Roth 1994a). We found that nonintactness was still a significant predictor of suicide among males, but not females, even when controlling for differences in parental psychopathology (Brent & Moritz, 1993). However, looking in both genders combined, parental substance abuse and depression were more powerful predictors of suicide than family constellation. In fact, parental psychopathology was associated with suicide even after controlling for these same disorders in the proband (Brent et al., 1994a). This implies, for example, that parental depression places an adolescent at risk for completed suicide above and beyond the increased transmission of risk for that disorder to the offspring. The potential mechanisms whereby parental depression may affect risk for depression and suicide in children are legion. Depressed parents are less engaged with their children, monitor their children less well, are less positive and more hostile, and may engender less secure attachment in their youngsters (Downey & Coyne, 1990; Weissman, Paykel, and Klerman 1972). Parent-child conflict is one of the main precipitants for completed suicide (Brent et al., 1988b, 1993a). Depressed parents may be less sensitive to their child's own affect and difficulties and, hence, may serve as a source for dysfunctional cognitions, such as worthlessness and a sense of poor self-esteem in the offspring (Hammen, 1991). A better understanding of the impact of parental depression on the development of children and adolescents may lay the groundwork for viable preventive strategies.

Aside from parental psychopathology, several other family-environmental factors have been associated with completed suicide. Suicide victims were more likely to have moved within the year before their suicide than demographically matched controls (Brent et al., 1993a). A move during adolescence can be quite stressful, as peer relationships become increasingly important. A move may also signal family instability, although the breaking of family, neighborhood, and school supports in and of themselves may put an already vulnerable adolescent at greater risk for suicide. Finally, as noted above, multiple family moves are associated with increased risk for emotional and behavioral disorders (Simpson & Fowler, 1994).

Two other familial risk factors associated with completed suicide were parent-child discord and abuse. Both risk factors were noted to be more frequent in completers than controls in the year prior to the suicide and over the lifetime of the suicide victims and controls (Brent et al., 1994a). This suggests that both factors exert their deleterious effects over the long run and not just acutely. It has been well-demonstrated that parent-child hostility may very well mediate the link between parental and offspring psychopathology (Rutter & Quinton, 1984). Studies of suicidal children and adolescents also note that parent-child hostility or rejection is common (Asarnow & Carlson, 1988; Kosky, Silburn, and Zubrick 1986; Taylor & Stansfeld, 1984; Williams & Lyons, 1976).

Physical and sexual abuse and exposure to domestic violence have also been associated with an increased risk for depression and suicidal behavior (Green, 1978; Hibbard et al., 1988; Kaufman, 1991; Kosky et al., 1986; Taylor & Stansfeld, 1984) and completed suicide (Brent et al., 1993a; Shafii, Carrigan, Whittinghill, and Derrick 1985). For both discord and abuse, the interaction of the timing of the stressor with the developmental stage of the exposed individual is unknown, so that the mechanisms by which discord and abuse lead to suicidality remain to be clarified. However, one possible mechanism is that abuse, discord, and criticism all conspire to lead to an internal working model that is associated with an unstable attachment and low self-worth, which in turn could lead to dysphoria and suicidality, especially under stress (Bifulco, Brown, and Adler 1991). Parent-child discord was associated with a fivefold increased risk of suicide, even after controlling for risk of parental depression and substance abuse, confirming the central role of this severe stressor as a risk factor for adolescent suicide (Brent et al., 1994a). As noted above, lifetime history of parent-child discord was a more powerful predictor of completed suicide than discord within the previous year, supporting the view that discord exerts its deleterious impact over the long run, rather than acutely.

Additional family-environmental characteristics that may relate to risk for psychopathology and suicide include affectionless control and low degree of parental monitoring. Affectionless control is a family-environment condition associated with high control and coercion and low degree of emotional support that we have observed in many families of depressed adolescents in an ongoing clinical trial (Brent, unpublished data). The absence of cohesion has been reported to be characteristic of families of suicidal children and adolescents (Asarnow, and Carlson 1988; Topol & Reznikoff, 1982; King, Hill, Naylor, Evans, and Shain 1993). Others have also found this to be associated with risk for depression, especially in the absence of familial loading for depression (Fendrich, Warner, and Weissman 1990). On the other hand, we have also observed that a disengaged, uninvolved parental style is also associated with elevated suicidal risk, as confirmed by a recent study (Richardson, Radziszewska, Dent, and Flay 1993).

Exposure to Suicide

One important developmental and epidemiologic question is the extent to which exposure to suicide serves as a model for adolescents so exposed and thereby increases their risk for completed suicide. The evidence suggesting that exposure to suicide increases the risk of subsequent suicidal behavior stems from several sources. First, there are reports of epidemics of suicide and suicidal behavior (Brent, Kerr, Goldstein, Bozigar, Wartella, and Allan 1989; Dizmang, Watson, May, and

Bopp 1974; Robbins & Conroy, 1983; Ward & Fox, 1977). Second, there are reports of media coverage of suicides or fictional stories about suicide leading to an increase in suicides and suicide attempts (Gould & Shaffer, 1986; Phillips, 1974; Phillips, and Paight 1987; Schmidtke & Hafner, 1988). Third, there is evidence of time-space clustering of suicides, specifically in the adolescent and young adult age range (Gould, Wallenstein, and Kleinman 1990). The phenomena of "contagion" seems to be confined to adolescents and young adults, at least in the time-space clustering analyses and in the media studies that examined an effect by age.

Suicidal behavior is known to be familial, although it is unclear if this familial aggregation of suicidality is primarily due to genetic factors (Kety, 1986). In one study, exposure to familial suicide differentiated suicidal from nonsuicidal adolescents more so than did genetic loading (Brent, Kolko, Goldstein, Allan, and Brown 1990). However, a cross-sectional study cannot discern the etiological direction of this relationship. Exposure to nonfamilial suicide did not differentiate between the two groups, nor did this factor differentiate between completers and suicidal inpatients (Brent et al., 1988b).

In light of the probable unreliability of retrospective recall of exposure to suicide, we also performed a prospective study of adolescents exposed to a peer's suicide, expecting to find evidence of contagion and imitation in the social networks of suicide victims. In fact, in comparison to matched, unexposed controls, there was no increase in suicide attempts (Brent, Perper, Moritz, Allman, Schweers, Roth, and Balach 1993e). Instead, we found evidence of an increased risk of depression, suicidal ideation (attributable to the increased incidence of depression), and post-traumatic stress disorder in the exposed youth compared to controls. Those youth who were closest to the suicide victim, talked to the victim within 24 hours of the death, found the body, or had a previous personal or family history of depression or both were at increased risk to develop a depressive disorder (Brent, Perper, Moritz, Allman, Liotus, Schweers, Roth, Balach, and Canobbio 1993f). On 12–18 month follow-up, those exposed youth who became depressed continued to be at increased risk for subsequent depressive recurrences, independent of preexisting personal or family history of depression (Brent, Perper, Moritz, Liotus, Schweers, Canobbio 1994b). However, among the exposed group, the rate of suicide attempts continued to be quite low and no different from the control group.

These data suggest that exposure to suicide is probably not a major risk factor for completed suicide and that imitation is quite unlikely to happen within the suicide victim's inner circle of friends. On the other hand, the friends of suicide victims are quite vulnerable to psychopathology and functional impairment on the basis of "assortive mating," as they have very high rates of preexisting psychiatric disorders. Moreover, we suspect that the friendships among the members of the social networks of suicide victims were unusually close, possibly because of a perceived or actual lack of familial support. The loss of a friend to suicide does appear to have devastating and long-lasting consequences on both incident and even recurrent depressive episodes (Brent et al., 1994b).

Precipitants

We have previously reviewed certain family-environmental stressors that may contribute to risk for completed suicide, namely parental psychopathology, parent-child discord, family mobility, and abuse. Additional stressors that occur within the year before the completed suicide have also been identified, specifically, interpersonal conflict with a boyfriend or girlfriend, disruption of a romantic attachment, and legal or disciplinary problems (Brent et al., 1993a). Interpersonal loss was particularly strongly associated with suicide in those with substance abuse disorders, which replicated findings by Murphy and Robbins (1967); Murphy, Armstrong, Hermele, Fisher, and Ciendenin (1979) and Rich, Ricketts, Fowler, and Young (1988). Legal or disciplinary problems were most closely associated with either substance abuse or conduct disorder, whereas interpersonal conflict was equally associated with all the major psychiatric syndromes (Brent et al., 1993a).

Each of these stressors has developmental implications. First is the developmental meaning of disruption of an attachment to an adolescent substance abuser. In the formulation of Murphy et al. (1979), suicide ensues after a lifetime of drinking in which the final support erodes, whether it is from marriage or work. It is more difficult to imagine this scenario occurring in an adolescent after a relatively short "career" in substance abuse, so that one must formulate a mechanism perhaps based on the impulsivity and lability of the substance abuser's response to interpersonal rejection. With respect to legal and disciplinary issues, one can also invoke the impulsivity and poor problem solving of those with conduct disorders and/or substance abuse disorders, who when faced with a real or imagined catastrophic situation, to explain a suicidal outcome. Alternatively, it may be that the conduct disordered or substance abusing youth may truly have exhausted his or her social resources, or at least perceived that this was the case. Further longitudinal research to elucidate how substance abusing and conduct disordered youth cope with severe stressors is required to better understand their vulnerability to suicide.

Psychopathology

Several psychological autopsy studies of adolescents and young adults have contributed to our knowledge about completed suicide (Brent et al., 1988b, 1993b; Marttunen et al., 1991; Rich et al., 1986b; Runeson, 1989; Shaffer et al., 1988; Shafii, Carrigan, Whittinghill, and Derrick (1985) Shafii, Steltz-Lenarsky, Derrick, Beckner, and Whittinghill (1988). Although the rates of various disorders differ among studies, it is generally agreed that affective disorder, substance abuse, conduct disorder, and personality disorder all contribute to risk for completed suicide. In terms of overall course of illness before suicide ensues, we and others have found an average of 6–7 years from onset of first illness to suicide (Brent et al., 1988b, 1993b; Marttunen et al., 1991). However, this progression may very well differ by disorder.

Affective Disorder

Affective disorder is generally acknowledged to be the most central risk factor for completed suicide in youth (Brent et al., 1988b, 1993b; Marttunen et al., 1991; Shaffer et al., 1988; Shafii et al., 1985, 1988). Major depression appears to convey about a 30-fold risk for completed suicide (Brent et al., 1993b), which is consistent with prospective studies on primary affective disorders (Guze & Robins, 1970). With respect to course, a high proportion of suicide victims with depression kill themselves relatively *early* in the course of their illness—within six months of the first episode (Brent et al., 1993b), also consistent with previous prospective studies (Guze & Robins, 1970). It is possible that because adolescents have not had to cope with many frustrating experiences that they are unprepared for the challenge of coping with depression and, therefore, become hopeless, helpless, and suicidal. This highlights two critical issues: the need to learn how to prevent the first episode of depression and the ability to rapidly identify major depression early in its course, offer vigorous treatment, and restore hope. In summary, these findings suggests that suicide often does not ensue after a lifelong struggle with affective illness, but rather close to the onset of the course of illness.

Are there any clinical correlates of affective illness that may predispose to suicide? We compared suicide victims to controls, all with a lifetime history of affective illness. We found that completers were more likely to have major depressive disorder, bipolar affective disorder, comorbid substance abuse, past history of suicide attempt, family history of depression, handgun available in the home, and history of treatment with a tricyclic antidepressant (Brent, Perper, Moritz, Baugher, Schweers, and Roth, 1994c). We have found a high proportion of suicide victims who died in a mixed bipolar state (Brent et al., 1988b, 1993b). Although one other psychological autopsy study in adolescents that examined the rate of bipolar disorder did not find any cases with this condition (Marttunen et al., 1991), one longitudinal study of early-onset affective illness did find that of seven suicides, two, and possibly three, had bipolar illness (Rao, Weissman, Martin, and Hammong 1993). Early-onset affective illness may be enriched with bipolar affective disorder (Puig-Antich, Guetz, Davies, Kaplan, Davies, Ostrow, Asnis, Twomey, Iyengar, and Ryan 1989). There are now several studies indicating that the combination of affective illness and substance abuse is particularly lethal (Brent et al., 1993b, Brent, Perper, Moritz, Liotus, Schweers, and Roth 1994c; Shafii et al., 1988). In terms of the progression of symptomatology, substance abuse often appears to develop *secondary* to the affective illness, at least in youthful populations of completers (Brent et al., 1993b; Shafii et al., 1988). With regard to the issue of the use of tricyclic antidepressants, two factors emerged. One is that more severely ill completers (presumably at higher risk for suicide) received tricyclics, and the second is that those who received tricyclics frequently used them in overdose. There may be developmental reasons why adolescents do not respond as well to tricyclic antidepressants as do adults (Ryan, 1990), but given this unfavorable cost-benefit ratio, their use as a first-line intervention with depressed, suicidal adolescents seems inadvisable. Several other studies looking across the age span have indicated that tricyclic antidepressants may more often result in attempts and completions compared both to other types of antidepressants (Kapur & Mann, 1992) and to other psychosocial interventions (Brent, Kolko, Wartella, Boylan, Moritz, Baugher, and Zelenak 1993g; Pfeffer, Hurt, Kakuma, Peskin, Siefker, and Nagabhai Raua 1994).

Substance Abuse

Several studies support a key role for substance abuse in the secular trends in completed suicide among youth (Brent et al., 1987; Carlson et al., 1991; Rich et al., 1986b). Unlike depression, where suicide often ensues relatively early in the course, substance abuse ends in suicide after several year's duration, frequently precipitated either by the breakup of a romantic relationship or a legal / disciplinary problem (Brent et al., 1993a). The identification of additional factors that contribute to risk of suicide in those with substance abuse come from a case-control study comparing suicide victims and controls with a lifetime history of substance abuse (Bukstein, Brent, Perper, Moritz, Baugher, Schweer, Roth, and Balach 1993). Completers were more likely to have had active substance abuse at the time of death, comorbid depression, a past suicide attempt, family history of depression and substance abuse, availability of a handgun, and a history of legal/disciplinary problems. These findings suggest several clear approaches to the prevention of suicide among substance abusers, including identification and treatment of the underlying or coexisting depression, achievement of abstinence, addressing issues of familial psychopathology, and removal of guns from the home. There is a suggestion that the severity of the substance abuse may worsen in a crescendo-like fashion before death ensues. Another potent contributor to suicidal risk among substance abusers is that they frequently are intoxicated at the time of the suicide attempt and, at least in males, the risk of suicide by a gun greatly increases when the victim uses alcohol associated with the suicidal episode (Brent et al., 1987; Brent, Perper, Moritz, Baugher, Schweers, and Roth 1993c).

Conduct Disorder

The course of disorder is somewhat similar to that detailed for substance abuse above: a chronic or subchronic course that, as a consequence of a severe precipitant, may result in suicide. Legal/disciplinary problems are, not surprisingly, most closely associated with suicide in conduct disordered youth (Brent et al., 1993a). Conversely, suicide rarely occurs in a conduct disordered youth without a precipitating legal or disciplinary crisis. What is of particular interest is that comorbidity of conduct disorder with affective disorder did not alter the risk for suicide substantially (Brent et al., 1993b) and that, at least in one study, conduct disorder only conveyed an increased risk for completed suicide in males (Brent & Moritz, 1993). However, other studies have found an association between antisocial symptomatology and risk for suicide and suicidal behavior in both genders (Shaffer et al., 1988; Cairns et al., 1988).

Anxiety Disorder

There exists in the literature a description of an anxious, perfectionistic type of adolescent suicide victim (Shaffer, 1974; Shafii et al., 1985). It may be that some of the "adjustment disorder" suicides described by Marttunen et al. (1991) fit into this taxonomy, as did some of our suicides with no apparent psychopathology (Brent et al., 1993d). We also have identified, apart from anxiety disorders, that avoidant personality disorder may be related to attempted and completed suicide (Brent, Johnson, Bartle, Bridge, Rather, Matta, Connolly, and Constantine 1993h, Brent, Perper, Moritz, Baugher, Schweers, and Roth 1994d). In a controlled series of completed suicides, the rates of anxiety disorders are higher than in the controls, but almost all the anxiety disorders in the suicide victims were comorbid with depression (Brent et al., 1993b). Therefore, anxiety disorder or an anxious temperament may heighten the risk for completed suicide, but usually in the context of a major depression. Longitudinal studies of children with predisposition to anxiety disorder are necessary to understand the interrelationship of anxiety, depression, and suicidal behavior.

No Apparent Psychopathology

A subset of suicide victims, even after extensive interviews with numerous informants, did not appear to have evidence of psychopathology. These "normal" suicide victims were compared to the remainder of the suicides, as well as to controls who were also found to have no evidence of psychopathology (Brent et al., 1993d). Compared to the other suicide victims, the suicide victims with no apparent psychopathology had lower familial loading for psychiatric disorders, less treatment contact, and less evidence of previous functional impairment, but were much more likely to have access to a loaded gun. In comparing the suicide victims with no apparent psychopathology to the normal controls, the suicide victims were more likely to have had legal problems, previous suicidal ideation, family history of psychiatric problems, and a loaded gun in the home. This suggest that suicidal ideation should be taken seriously even in the absence of clear psychopathology, especially in the face of legal problems. These findings raise interesting developmental questions: Can one predict those youngsters who will react with extreme consequences to seemingly minor stressors, even in the absence of psychopathology? Perhaps longitudinal studies of stress reactivity and cognitive problem solving will enable us to understand more about these puzzling and concerning outcomes. While awaiting the results of such studies, the convergent evidence supports the removal of firearms, particularly loaded ones, as one preventive strategy.

Previous Suicidality

One consistent finding from the literature is that one of the best predictors of future suicidal behavior is past suicidal behavior (Brent et al., 1993b, g; Pfeffer, Klerman, Hurt, Lesser, Peskin, and Siefker 1991, 1994; Shaffer et al., 1988). How does one person with a psychiatric condition develop in such a way that suicidality is *not* part of the presentation and another person develop such that it is an *integral* part of the presentation? Are the processes whereby suicidal ideation and psychopathology develop independent or interactive? Given the above-noted findings on suicidal individuals without clear Axis I psychopathology, this question seems to be warranted. Moreover, some investigators have posited that suicidal risk is at least in part related to certain personality traits as discussed below. Others have viewed the correlates of suicidality as emerging from family circumstances, genetic predisposition to suicide, exposure to suicide, or other comorbid conditions like substance abuse (Brent & Kolko, 1990; Ryan, Puig-Antich, Ambrosini, Rabinovich, Robinson, Nelson, Iyengar, and Twomey 1987; Shaffer et al., 1988).

Personality Disorder and Tendency to Impulsive Disorder

Clinical researchers have posited that suicidal behavior is most likely to occur when there is a co-occurrence of Axis I and II disorders (Blumenthal & Kupfer, 1986). Biological research in adult samples has established that suicidal behavior is associated with a dysfunction in central serotonergic metabolism and that this dysfunction is actually correlated with a tendency to impulsive violence, rather than suicide specifically (Coccaro et al., 1989). There may be an underlying personality trait of affective lability and impulsivity associated with suicide, at times corresponding to the categorical diagnosis of borderline personality disorder (Coccaro, Siever, Klar, Mauer, Cochrane, Cooper, Mohs, and Davis 1989). Borderline personality disorder has been reported to be frequently associated with attempted and completed suicide (Brent et al., 1993h; Brent et al., 1994d; Marttunen et al., 1991; Runeson & Beskow, 1991). One unanswered question has to do with the onset of personality disorders, or better put, how personality disorders develop. Some posit that borderline disorders are part of an affective disorder spectrum and really represent the "scarring" effect of prolonged episodes of affective illness (Akiskal, Chen, Davis, Puxantian, Kashgarian, and Bolinger 1985). Others view borderline disorder as resulting from numerous traumas in close attachment relationships, such as physical or sexual abuse, which of course may also contribute to risk for affective illness (Herman, Perry, and vander Kolk 1989). Still others see such affective lability as emerging from biochemically mediated central nervous system changes, although certainly not ruling out that these differences may have resulted from severe exogenous stress. Longitudinal studies of children at risk for affective illness and of those traumatized by physical or sexual abuse are likely to yield important information about the developmental progression of Axis II disorders.

As noted above, we also observed that avoidant disorder was associated with suicide and suicidal behavior (Brent et al., 1993h; Brent et al., 1994d). This may be in keeping with the contributions of constitutional anxiety to suicidal risk (Shaffer, 1974; Shafii et al., 1985). Longitudinal studies of behaviorally inhibited youth are likely to be informative as to the contribution of this temperamental quality to suicidal risk.

Biological Risk Factors

There has been little in the way of investigation of the biology of suicidality in children and adolescents. Certain biological changes have been noted in suicidal, depressed adolescents that seem to distinguish them from normal controls and from other depressed youngsters. First, Ryan, Puig-antich, Rabinovich, Ambrosini, Robinson, Nelson, and Novacenko (1988) reported that depressed, suicidal adolescents showed a blunted growth hormone response to desipramine, which were presumably indicated of a defect in noradrenergic neurotransmission. Second, Dahl, Ryan, Puig-Antich, Nelson, Dachille, Cunningham, Trubnick, and Klepper report a failure to suppress cortisol secretion before sleep-onset and reduced REM latency in hospitalized, depressed, suicidal adolescent inpatients (1990, 1991). Robbins & Alessi (1985) also reported that suicide attempting adolescent inpatients were more likely to be dexamethasone nonsuppressors. In children with depression, blunted cortisol and prolactin responses to L-5HTP were associated with depression, but not suicidality or impulsive violence (Ryan et al., 1992). Work is now ongoing to understand what role the serotonergic system plays in youthful suicide. An analysis of postmortem studies that included young adults aged 18–25 found similar effect sizes with respect to decreased number of imipramine-binding sites and increased number of beta-adrenergic receptors, making it relatively likely that similar biological alterations occur in more juvenile suicides (Mann, De Meo, Keilp, and McBride 1989). Such work is currently ongoing.

Case Studies

In closing, we describe four vignettes of completed suicide in adolescents that illustrate some of the different developmental trajectories that all tragically converged on the outcome of suicide. These represent composites rather than any particular individual.

Case 1

Ned, a 14-year-old, white male, who was the youngest of three siblings, lived with both natural parents. Both parents worked. His father was a maintenance worker at a warehouse after losing a steel mill job and his mother was an office manager of a small accounting firm. Ned was an average student through the seventh grade, played sports and music, and had good interpersonal relationships. He was part of a large extended family and had good relationships with his two siblings. Ned had not started any heterosexual dating, but he did attend school and church dances. During the summer prior to eighth grade, Ned began to want more freedom and assert some independence beyond that allowed by the family rules. The more Ned tested the rules, the more restrictive his family environment became. After school began, Ned's parents began to receive deficiency notices regarding his academic work. The parents' response was to regiment homework time and to restrict TV and video-game times. Within eight weeks of the beginning of eighth grade, Ned stopped sport and music participation and quit his paper route. On the first semester report card, his grades were poor and many incomplete assignments were found in the home. Ned's reaction at home was to become irritable and oppositional. By Christmas, it was noted that his clothes were too big and that he did not seem to enjoy Christmas as much as usual. The family continued to be concerned about Ned's academic decline and discussed the possibility of changing schools. Ned returned to his regular school after Christmas vacation. After the first day of school, he came home and, while the house was empty, went to his room and killed himself with a handgun.

In this case, the precipitant for the suicide was a conflict between parent and child. The supportive, but authoritarian environment was a good fit for Ned until he entered adolescence. The stress of ongoing parental disapproval and school failure may have also precipitated a depressive episode, associated with anhedonia, weight loss, irritability, and withdrawal from previously valued activities. This case illustrates how suicide often ensues early in the development of a depressive episode, and the importance of developmental and family systems perspective.

Case 2

Janet, a 19-year-old white female, lived with her natural parents and two younger brothers in Pittsburgh. Both parents worked, and they were middle to upper middle class. Janet was an average academic student in high school and an active participant in extracurricular activities until the end of the 11th grade, when she developed a severe case of mononucleosis. During Janet's 12th grade year, her academic work declined sharply, participation in extracurricular activities ceased, and Janet lost weight and often stayed home sleeping. Her family attributed the weight loss, hypersomnia, and her declining academic performance to the aftermath of mononucleosis. During 12th grade, Janet was twice caught in school participating in drug transactions. Despite two school-mandated psychiatric assessments, the family supported their daughter in her contention that she had been set up. Accompanying her decline in school function was a change in her peer group to a more conduct disordered, substance abusing group. Janet did not graduate because of academic failure and enrolled in summer school to complete graduation requirements. Friends reported that during the summer following high school, Janet was often euphoric, was hypersexual, drove recklessly and had several automobile accidents, and was using street drugs. It was also reported that she had pressured speech and an increase in motor activity. Following a rejection by her current boyfriend, Janet spent a night visiting all her friends. She then returned home and overdosed on a mixture of medication available in the home and was found dead the next morning.

This case illustrates how derailment from an expected developmental trajectory can lead to suicide. Subsequent to Janet having developed mononucleosis, she fell behind in school, lost interest in extracurricular activities, and began associating with a much more deviant peer group. It was unclear if Janet also developed a bipolar affective illness or if her behavior could be attributable to use of stimulants. However, as is often characteristic of substance abusing people, an interpersonal rejection by her boyfriend was the precipitant that led to her suicide.

Case 3

Brad, a 15-year-old white male, lived with his younger brother and natural parents in rural Allegheny County. His father was a self-employed electrician and his mother was a homemaker. Brad was described by his parents as an easy and compliant child. He was an outstanding academic student and participated in a large number of extracurricular and academic activities. Brad aspired to attend one of the service academies, but had taken an entrance examination and had not done well. He was noted to be "particular" about the way his hair and clothing appeared and was perfectionistic in his approach to all aspects of his daily life. He was also noted sometimes to overreact to minor criticism or teasing from his younger brother or peers. One night while his parents were out for the evening, Brad took his father's car without permission. The next day after school, Brad's mother informed him that she knew of his misbehavior and that he would have to speak to his father when he arrived home. Brad went to his room and immediately took his life with his own hunting rifle.

This case illustrates a suicide in a high-functioning, somewhat rigid, and perfectionistic individual, precipitated by a seemingly minor incident. In the background was his concern that his lifelong goal of attending a service academy would not be realized. The suicides of these individuals are almost impossible to predict or prevent, except perhaps by keeping firearms out of the home.

Case 4

Jeff, a 19-year-old male, was the third of four children from his mother and father's original marriage and was living in his mother's former home with several friends in suburban Pittsburgh. The three other siblings lived with their mother and stepfather. Jeff's mother was a bartender and his stepfather was a construction worker. Jeff had previously lived with his mother and other siblings and also had lived with his maternal grandmother. Jeff had rare contact with his biological father. After his parents divorced when he was 9 years old, in early middle school, Jeff began to exhibit symptoms of conduct disorder and was taken to a community mental health center for assessment and treatment, but recommendations for treatment were not followed. In early high school years, Jeff was failing academically, frequently truant, and was suspected of abusing alcohol. Jeff finished high school at an alternative setting that accommodated low-functioning students. Following high school, Jeff did seasonal work on a landscaping crew. His alcohol abuse continued and increased considerably when he took up residence at his mother's former home. Jeff consistently and frequently engaged in alcohol and marijuana use, with occasional cocaine use. Jeff had a long-term relationship with a girl, who abruptly broke off the relationship and married. Jeff's mother noted a significant decline in his mood following this rejection. Within a few months, the ex-girlfriend left her husband and Jeff hoped to renew the relationship. She again rejected a relationship with him but wanted to remain in his circle of peers. Jeff's peers described repeated discussions of Jeff's suicidality and his revelations of suicide plans. One night after several hours of drug and alcohol use, Jeff went into his bedroom and ended his life with a handgun, which he always had kept next to his bed.

This case illustrates the trajectory to suicide followed by a conduct disordered and substance abusing youth. As is prototypical for this type of suicide, Jeff was exposed to marked family instability, parental divorce, failure to follow treatment recommendations, and low degree of parental monitoring and supervision. Jeff's substance abuse accelerated shortly before his death and was precipitated by the rejection by his erstwhile girlfriend. Of note were the lack of family involvement and support, prior suicidal ideation and threats, and the unfortunate ready availability of a gun, which Jeff used in an intoxicated state.

Conclusions

Suicide in adolescents is a tragedy. A better understanding of how developmental issues and risk for psychopathology interact is likely to yield improved methods of prevention and treatment of conditions leading to completed suicide.

REFERENCES

Akiskal, H. S., Chen, S. E., Davis, G. C., Puzantian, V. R., Kashgarian M., Bolinger, J. M. (1985). Borderline: An adjective in search of a noun. *Journal of Clinical Psychiatry*, 46, 41–48.

Asarnow, J. R., Carlson, G. (1988). Suicide attempts in preadolescent child psychiatry inpatients. *Suicide and Life-Threatening Behavior*, 18, 129–136.

Ballenger, J. C., Goodwin, E. K., Major, L. F., Brown, G. L. (1979). Alcohol and central serotonin metabolism in men. *Archives of General Psychiatry*, 36, 224–227.

Berlin, I. N. (1987). Suicide among American Indian adolescents: An overview. *Suicide and Life-Threatening Behavior*, 17, 218–232.

Bifulco, A., Brown, G. W., Adler, Z. (1991). Early sexual abuse and clinical depression in adult life. *British Journal of Psychiatry*, 159, 115–122.

Blumenthal, S. J., Kupfer, D. J., (1986). Generalizable treatment strategies for suicidal behavior. In J. J. Mann & M. Stanley (Eds.), *Psychobiology of Suicidal Behavior* (Vol. 487, pp. 327–340). New York: New York Academy of Sciences.

Boor, M. (1976). Relationship of internal-external control and United States suicide rates, 1966–1973. *Journal of Clinical Psychology*, 32, 795–797.

Boyd, J. H., (1983). The increasing rate of suicide by firearms. *New England Journal of Medicine*, 308, 872–874.

Boyd, J. H., Moscicki, E. K., (1986). Firearms and youth suicide. *American Journal of Public Health*, 76, 1240–1242.

Brent, D. A., Johnson, B., Bartle, S., Bridge, J., Rather, C., Matta, Jr., Connolly, J., Constantine, D. (1993h). Personality disorder, tendency to impulsive violence, and suicidal behavior in adolescents. *Journal of the American Academy of Child and Adolescent Psychiatry, 32*, 69–75.

Brent, D. A., Johnson, B. A., Perper, J., Connolly, J., Bridge, J., Bartle, S., Rather, C. (1994d). Personality disorder, personality traits, impulsive violence and completed suicide in adolescents. *Journal of the American Academy of Child and Adolescent Psychiatry, 33*, 1080–1086.

Brent, D. A., Kolko, D. J., Goldstein, C. E., Allan, M. J., Brown, R. V., (1990). Suicidality in affectively disordered adolescent inpatients. *Journal of the American Academy of Child and Adolescent Psychiatry, 29*, 586–593.

Brent, D. Perper, J., Allman, C. (1987). Alcohol, firearms, and suicide among youth: Temporal trends in Allegheny county, PA, 1960–1983. *Journal of the American Medical Association, 257*, 3369–3372.

Brent, D. A., Perper, J. A., Goldstein, C. E., Kolko, D. J., Allan, M. J., Allman, C. J., Zelenak, J. P. (1988b). Risk factor for adolescent suicide: A comparison of adolescent suicide victims with suicidal inpatients. *Archives of General Psychiatry, 45*, 581–588.

Brent, D. A., Perper, J. A., Kolko, D. J., Zelenak, J. P. (1988a). The psychological autopsy: Methodological considerations for the study of adolescent suicide. *Journal of the American Academy of Child and Adolescent Psychiatry, 27*, 362–366.

Brent, D. A., Perper, J. A., Moritz, G. Allman, C., Friend, A., Roth, C., Schweers, J., Balach, L., Baugher, M. (1993b). Psychiatric risk factors for adolescent suicide: A case-control study. *Journal of the American Academy of Child and Adolescent Psychiatry, 32*, 521–529.

Brent, D. A., Perper, J., Moritz, G., Allman, C., Liotus, L., Schweers, J., Roth, C., Balach, L., Canobbio, R. (1993f). Bereavement or depression? The impact of the loss of a friend to suicide. *Journal of the American Academy of Child and Adolescent Psychiatry, 32*, 1189–1197.

Brent, D. A., Perper, J. A., Moritz, G., Allman, C., Schweers, J., Roth, C., Balach, L. (1993e). Psychiatric sequelae to the loss of an adolescent peer to suicide. *Journal of the American Academy of Child and Adolescent Psychiatry, 32*, 509–517.

Brent, D. A., Perper, J. A., Moritz, G., Baugher, M., Allman, C. (1993d). Suicide in adolescents with no apparent psychopathology. *Journal of the American Academy of Child and Adolescent Psychiatry, 32*, 494–500.

Brent, D. A., Perper, J. A., Moritz, G., Baugher, M., Roth, C., Balach, L., Schweers, J. (1993a). Stressful life events, psychopathology, and adolescent suicide: A case-control study. *Suicide Life Threatening Behavior, 23*, 179–187.

Brent, D. A., Perper, J. A., Moritz, G., Baugher, M., Schweers, J., Roth, C. (1993c). Firearms and adolescent suicide: A community case-control study. *American Journal of Diseases of Children, 147*, 1066–1071.

Brent, D. A., Perper, J. A., Moritz, G., Baugher, M., Schweers, J., Roth, C. (1994c). Suicide in affectively ill adolescents: A case-control study. *Journal of Affective Disorders, 31*, 193–202.

Brent, D. A., Moritz, G. Gender, conduct disorder, and suicide: An explanation of the gender differences for adolescent suicide, presented at the Rochester Symposium on Developmental Psychopathology, Rochester, NY, October 5&5, 1993.

Brent, D. A., Perper, J. A., Moritz, G., Liotus, L., Schweers, J., Balach, L., Roth, C. (1994a). Familial risk factors for adolecent suicide: A case control study. *Acta Psychiatrica Scand, 89*, 52–58.

Brent, D. A., Perper, J. A., Moritz, G., Liotus, L., Schweers, J., Canobbio, R. (1994b). Major depression or bereavement? A follow-up of youth exposed to sucide. *Journal of the Amercian Academy of Child and Adolescent Psychiatry, 33*, 231–239.

Bukstein, O. G., Brent, D. A., Perper, J. A., Moritz, G., Baugher, M., Schweers, J., Roth, C., Balach, L. (1993). Risk factors for completed suicide among adolescents with a lifetime history of substance abuse: A case-control study. *Acta Psychiatrica Scan*, 88, 403–408.

Cairns, R., Peterson, G., Neckerman, H. J., (1988). Suicidal behavior in aggressive adolescents. *Journal of Clinical Child Psychiatry*, 17, 298–309.

Carlson, G. A., Asarnow, J. R., Orbach, I. (1987). Developmental aspect of suicidal behavior in children. *Journal of the American Academy of Child and Adolescent Psychiatry*, 26, 186–192.

Carlson, G. A., Rich, C. L., Grayson, P., Fowler, R. C. (1991). Secular trends in psychiatric diagnoses of suicide victims. *Journal of Affective Disorders*, 21, 127–132.

Centers for Disease Control, U.S. Department of Health and Human Services. (1985). Suicide Surveillance 1970–1980.

Centers for Disease Control, National Health Center for Health Statistics (1990), Suicide death rates for year 1984–1990.

Coccaro, E., Siever, L., Klar, H. M., Mauer, G., Cocharane, K., Cooper, T. B., Mohs, R. C., Davis, K. L. (1989). Serotonergic studies in patients with affective and personality disorders. *Archives of General Psychiatry*, 46, 587–599.

Costello, E. J., (1989). Developments in child psychiatric epidemiology. *Journal of the American Academy of Child and Adolescent Psychiatry*, 28, 836–841.

Dahl, R. E., Ryan, N. D., Puig-Antich, J., Nelson, B., Dachille, S., Cunningham, S. L., Trubnick, L., Klepper, T. P., (1990). EEG sleep in adolescents with major depression: The role of suicidality and inpatient status. *Journal of Affective Disorders*, 19, 63–75.

Dahl, R. E., Ryan, N. D., Puig-Antich, J., Nguyen, N. A., Al-Shabbout, M., Meyer, V. A., Perel, J. (1991). 24-hour cortisol measure in adolescents with major depression: A controlled study. *Biological Psychiatry*, 30, 25–36.

Dizmang, L. W., Watson, J., May, P. A., Bopp, J. (1974). Adolescent suicide at an Indian reservation. *American Journal of Orthopsychiatry*, 44, 43–49.

Downey, G., Coyne, J. C. (1990). Children of depressed parents: An integrative review. *Psychiatric Bulletin*, 108, 50–76.

Earls, F., Jung, K. G. (1987). Temperment and home environment characteristics as casual factors in the early development of childhood psychopathology. *Journal of the American Academy of Child and Adolecent Psychiatry*, 4, 491–498.

Eccles, J. S., Midgley, C., Wigfield, A., Buchanan, C. M., Reuman, D., Flanagan, C., MacIver, D. (1993). Development during adolecence: The impact of stage-environment fit on young adolescents' experiences in schools and in families. *American Psychol*, 48, 90–101.

Farrington, D. P., (1986). Age and crime. In M. Tonry and N. Morris (Eds.), *Crime and justice* (Vol. 7, pp. 29–99). Chicago, IL: Chicago University Press.

Fendrich, M., Warner, V., Weissman, M. (1990). Family risk factors, parental depression, and psychopathology in offspring. *Developmental Psychology*, 26, 40–50.

Fleming, J. E., Offord, D. R. (1990). Epidemiology of childhood depressive disorders: A critical review. *Journal of the American Academy of Child and Adolescent Psychiatry*, 29, 571–580.

Gjerde, P. F., Block, J., Block, J. H. (1988). Depressive symptoms and personality during late adolescence: Gender differences in the externalization-internalization of symptom expression. *Journal of Abnormal Psychology*, 97, 475–486.

Gould, M. S., Shaffer, D. (1986). The impact of suicide in television movies: Evidence of imitation. *New England Journal of Medicine*, 315, 690–694.

Gould, M. S., Wallenstein, S., Kleinman, M. (1990). Time-space clustering of teenage suicide. *American Journal of Epidemiology*, 131, 71–78.

Green, A. H., (1978). Self-destructive behavior in battered children. *American Journal of Psychiatry, 135*, 579–582.

Guze, S. B., Robins, E. (1970). Suicide and primary affective disorders. *British Journal of Psychiatry, 117*, 437–438.

Hammen, C. (1991). *Depression runs in families. The social context of risk and resilience in children of depressed mothers.* New York: Springer-Verlag.

Hawton, K., Catalan, J. (1987). *Attempted suicide: A practical guide to its nature and management.* New York: Oxford University Press.

Herman, J. L., Perry, J. C., van der Kolk, B. A. (1989). Childhood trauma in borderline personality disorder. *American Journal of Psychiatry, 146*, 490–495.

Hetherington, E. M., Clingempeel, W. G., Anderson, E. R., Deal, J. E., Hagan, M. S., Hollier, E. A., Lindner, M. S. (1992). *Coping with marital transitions. Monographs of the Society for Research in Child Development, 57*, 2–3.

Hibbard, R. A., Brack, C. J., Rauch, S., Orr, D. P. (1988). Abuse, feelings, and health behaviors in a student population. *American Journal of Dis Children, 142*, 326–330.

Holinger, P. C., Offer, D. (1982). Prediction of adolescent suicide: A population model. *American Journal of Psychiatry, 139*, 302–307.

Jaffe, P., Wolfe, D., Wilson, S. K., Zak, L. (1986). Family violence and child adjustment: A comparative analysis of girls' and boys' behavioral symptoms. *American Journal of Psychiatry, 143*, 74–77.

Kapur, S., Mann, J. J. (1992). Antidepressant medications and the relative risk of suicide attempt and suicide. *Journal of the American Medical Association, 268*, 3441–3445.

Kaufman, J. (1991). Depressive disorders in maltreated children. *Journal of the American Academy of Child and Adolescent Psychiatry, 30*, 257–265.

Kazdin, A. E., French, N. H., Unis, A. S., Esveldt-Dawson, K., Sherick, R. B., (1983). Hopelessness, depression, and suicidal intent among psychiatrically disturbed inpatient children. *Journal of Cons Clinical Psychology, 51*, 504–510.

Kety, S. S. (1986). Genetic factors in suicide: Family, twin, and adoption studies. In J. J. Mann & M. Stanley (Eds.), *Psychobiology of suicidal behavior* (Vol. 487) pp. 327–340). New York: New York Academy of Sciences

King, C. A., Hill, E. M., Naylor, M., Evans, T., Shain, B. (1993). Alcohol consumption in relation to other predictors of suicidality among adolescent inpatient girls. *Journal of the American Academy of Child and Adolescent Psychiatry, 32*, 82–88.

Klerman, G. L., Lavori, P. W., Rich, J., Reich, T., Endicott, J., Andreasen, N. C., Keller, M. S., Hirschfeld, R. M. A. (1985). Birth-cohort trends in rates of major depressive disorder among relatives of patients with affective disorder. *Archives of General Psychiatry, 42*, 689–693.

Kobak, R. R., Sudler, N., Gamble, W. (1991). Attachment and depressive symptoms during adolescence: A developmental pathway analysis. *Development and Psychopathology, 3*, 461–474.

Koocher, G. (1974). Talking with children about death. *American Journal of Orthopsychiatry, 44*, 409–413.

Kosky, R., Silburn, S., Zubrick, S. R., (1986). Symptomatic depression and suicidal ideation: A comparative study with 628 children. *Journal of Nerv Mental Disease, 174*, 523–528.

Lewinsohn, P. M., Rohde, P., Seeley, J. R. (1993). Psychosocial characteristics of adolescents with a history of suicide attempt. *Journal of the American Academy of Child and Adolescent Psychiatry, 32*, 60–68.

Maccoby, E. E., Jacklin, C. N. (1980). Sex differences in aggression: A rejoinder and reprise. *Child Development, 51*, 964–980.

Mann, J. J., DeMeo, M. D., Keilp, J. G., McBride, P. A. (1989). Biological correlates of suicidal behavior in youth. In C. Pfeffer (Ed.), *Suicide among youth* (pp. 185–202). Washington, DC: APA Press.

Marttunen, J. J., Aro, H. M., Henriksson, M. M., Lonnquist, J. K. (1991). Mental disorders in adolescent suicide DSM-III-R axes I and II diagnoses in suicides among 13- to 19-year-olds in Finland. *Archives of General Psychiatry, 48,* 834–839.

McAnarney, E. R., (1979). Adolescent and young adult suicide in the United States: A reflection of societal unrest? *Adolescence, 14,* 765—774.

Murphy, G. E., Armstrong, J. W., Hermele, S. L., Fisher, J.R., Ciendenin, W. W. (1979). Suicide and alcoholism: Interpersonal loss confirmed as a predictor. *Archives of General Psychiatry, 36,* 65–69.

Murphy, G. E., Robins, E. (1967). Social factors in suicide. *Journal of the American Medical Association, 199,* 81–86.

Offord, D. R., Boyle, M. H., Szatmari, P., Rae-Grant, N. I., Links, P. S., Cadman, D. T., Byles, J. A., Crawford, J. W., Blum, H. M., Byrne, C., Thomas, H., Woodward, C. A. (1987). Ontario child health study. II. Six-month prevalence of disorder and rates of service utilization. *Archives of General Psychiatry, 44,* 832–836.

Otto, U. (1972). Suicidal acts by children and adolescents: A follow-up study. *Acta Psychiatrica Scand.* (Suppl 233: entire).

Overholser, J. C., Hemstreet, A. H., Spirito, A., Vyse, S. (1989). Suicide awareness program in the schools: Effects of gender and personal experience. *Journal of the American Academy of Child and Adolescent Psychiatry, 28,* 925–930.

Pfeffer, C. R., Conte, H. R., Plutchik, R., Jerrett, I. (1979). Suicidal behavior in latency-age children: An empirical study. *Journal of the American Academy of Child and Adolescent Psychiatry, 18,* 679–692.

Pfeffer, C. R., Hurt, S. W., Kakuma, T., Peskin, J. F., Siefker, C. A., Nagabhairava, S. (1994). Suicidal children grown up: Suicidal episodes and effects of treatment during follow-up. *Journal of the Amercian Academy of Child and Adolecent Psychiatry, 33,* 225–230.

Pfeffer, C. R., Klerman, G. L., Hurt, S. W., Lesser, M., Peskin, J. R., Siefker, C. A. (1991). Suicidal children grown up: Demographic and clinical risk factors for adolescent suicide attempts. *Journal of the American Academy of Child and Adolescent Psychiatry, 30,* 609–616.

Phillips, D. P. (1974). The influence of suggestion on suicide: Substantive and theoretical implications of the Werther effect. *American Soc Review, 39,* 340–354.

Phillips, D. P., Paight, D. J. (1987). The impact of televised movies about suicide: A replicative study. *New England Journal of Medicine, 317,* 809–811.

Puig-Antich, J., Goetz, D., Davies, M., Kaplan, T., Davies, S., Ostrow, L., Asnis, L., Twomey, J., Iyengar, S., Ryan, N. D. (1989). A controlled family history study of prepubertal major depressive disorder. *Archives of General Psychiatry, 46,* 406–418.

Rao, U., Weissman, M. M., Martin, J. A., Hammond, R. W. (1993). Childhood depression and risk of suicide. A preliminary report of a longitudinal study. *Journal of the American Academy of Child and Adolescent Psychiatry, 32,* 21–27.

Reich, T., Cloninger, C. R., Van Eerdewegh, P., Rich, J. P., Mullaney, M. (1988). Secular trends in the familial transmission of alcoholism. *Alcoholism: Clinical and Experimental Research, 12,* 458–464.

Reinherz, H. Z., Giaconia, R. M., Pakiz, B., Silverman, A. B., Frost, A. K., Lefkowitz, E. S. (1993). Psychosocial risks for major depression in late adolescence: A longitudinal community study. *Journal of the American Academy of Child and Adolescent Psychiatry, 32,* 1155–1163.

Remafedi, G., Farrow, J. A., Deisher, R. W. (1991). Risk factors for attempted suicide in gay and bisexual youth. *Pediatrics, 87,* 869–875.

Rich, C. L., Fowler, R. C., Young, D., Blenkush, M. (1988b). San Diego Suicide Study: Comparison of gay to straight males. *Suicide and Life-Threatening Behavior, 16,* 448–457.

Rich, C., Ricketts, J., Fowler, R. C., Young, D. (1988a). Some differences between men and women who commit suicide. *American Journal of Psychiatry, 145,* 718–722.

Rich, C.L., Young, D., Fowler, R. C. (1986). San Diego suicide study. I. Young vs. old subjects. *Archives of General Psychiatry, 43,* 577–582.

Richardson, J. L., Radziszewska, B., Dent, C. W., Flay, B. R. (1993). Relationship between after-school care of adolecents and substance use, risk taking, depressed mood, and academic achievement. *Pediatrics, 92,* 32–38.

Robbins, D. R., Alessi, N. E. (1985). Suicide and the Dexamethasone Suppression Test in adolescence. *Biological Psychiatry, 20,* 107–110.

Robbins, D., Conroy, R. C. (1983). A cluster of adolescent suicide attempts: Is suicide contagious? *Journal of Adolescent Health Care, 3,* 253–255.

Robins, L. N., Price, R. K. (1991). Adult disorders predicted by childhood conduct problems: Results from the NIMH Epidemiologic Catchment Area Project. *Psychiatry, 54,* 116–132.

Rubenstein, D. H. (1983). Epidemic suicide among Micronesian adolescents. *Social Science and Medicine, 17,* 657–665.

Runeson, B. (1988). Mental disorder in youth suicide DSM-III-R axes I and II. *Acta Psychiatrica Scand, 79,* 490–497.

Runeson, B., & Beskow, J. (1991). Borderline personality disorder in young Swedish suicides. *Journal of Nerv Mental Disorders, 179,* 153–156.

Rutter, M. & Quinton, D. (1984). Parental psychiatric disorder: Effects on children. *Psycholog Medicine, 14,* 853–880.

Ryan, N. D. (1990). Pharmacotherapy of adolescent major depression: Beyond TCA's. *Psychopharmacology Bulletin, 26,* 76–79.

Ryan, N. D., Birmaher, B., Perel, J.M., Dahl, R. E., Meyer, V., Al-Shabbout, M., Iyengar, S., & Puig-Antich, J. (1992). Neuroendocrine response to L-5-hydroxytryptophan challenge in prepubertal major depression. *Archives of General Psychiatry, 49,* 843–851.

Ryan, N. D., Puig-Antich, J., Ambrosini, P., Rabinovich, H., Robinson, D., Nelson, B., Iyengar, S., & Twomey, J. (1987). The clinical picture of major depression in children and adolescents *Archive of General Psychiatry, 44,* 854–861.

Ryan, N. D., Puig-Antich, J., Rabinovich, H., Ambrosini, P., Robinson, D., Nelson, B., & Novacenko, H. (1988). Growth hormone reponse to desmethylimipramine in depressed and suicidal adolescents. *Journal of Affective Disorders, 15,* 323–337.

Schmidtke, A., & Hafner, H. (1988). The Werther effect after television films: New evidence for an old hypothesis. *Psychol Medicine, 18,* 665–676.

Shaffer, D. (1974). Suicide in childhood and early adolescence. *Journal of Child Psychol Psychiatry, 15,* 275–291.

Shaffer, D. (1993, May). Shouts and murmurs: Political Science. *The New Yorker, 23,* 116.

Shaffer, D., & Fisher, P. (1981). The epidemiology of suicide in children and young adolescents. *Journal of the Amercian Academy of Child Psychiatry, 20,* 545–565.

Shaffer, D., Garland, A., Gould, M., Fisher, P. & Trautman, P. (1988). Preventing teenage suicide. A critical review. *Journal of the American Academy of Child and Adolescent Psychiatry, 27,* 675–687.

Shaffer, D., Gourd, M., & Hicks, L. C. (1994). Worsening Suicide Rate in Black Teenagers. *American Journal of Psychiatry, 151,* 1810-1812.

Shaffer, D., Vieland, V., Garland, A., Rojas, M., Underwood, M. & Busner, C. (1990). Adolescent suicide attempters' response to suicide-prevention programs. *Journal of the American Medical Association, 264,* 3151–3155.

Shafii, M., Carrigan, S., Whittinghill, J. R. & Derrick, A. M., (1985). Psychological autopsy of completed suicide in children and adolescents. *American Journal of Psychiatry, 142,* 1061–1064.

Shafii, M., Steltz-Lenarsky, J., Derrick, A. M., Beckner, C., & Whittinghill, J. R. (1988). Comorbidity of mental disorders in the postmortem diagnosis of completed suicide in children and adolescents. *Journal of Affective Disorders, 15,* 227–233.

Simpson, G. A., Fowler, M. G. (1994). Geographic mobility and children's emotional / behavioral adjustment and school functioning. *Pediatrics, 93,* 303–309.

Taylor, E. A. & Stanfeld, S. A. (1984). Children who poison themselves. I. A clinical comparison with psychiatric controls. *British Journal of Psychiatry, 145,* 127–135.

Topol, P. & Reznikoff, M. (1982). Perceived peer and family relationships, hopelessness and locus of control as factors in adolecent suicide attempts. *Suicide and Life Threatening Behavior, 12,* 141–150.

U.S. Department of Health and Human Services. (1989). *Report of the Secretary's Task Force on Youth Suicide: Vol. II, Prevention and interventions in youth suicide.*

Velez, C. N., & Cohen, P. (1988). Suicidal behavior and ideation in a community sample of children: Maternal and youth reports. *Journal of the American Academy of Child and Adolescent Psychiatry, 27,* 349–356.

Ward, J. A., & Fox, J. (1977). A suicide epidemic on an Indian reserve. *Canadian Psychiatric Association Journal, 22,* 423–426.

Weissman, M. M., Paykel, E. S., & Klerman, G. L. (1972). The depressed woman as a mother. *Soc Psychiatry, 7,* 98–108.

Williams, C. L., & Lyons, C. M. (1976). Family interaction and adolescent suicidal behaviour: A preliminary investigation. Australian and New Zealand Journal of Psychiatry, 10, 243–252.

Wolfgang, M. E. (1959). Suicide by means of victim-precipitated homicide. *Journal of Clinical and Experimental Psychopathology, 20,* 335–349.

IX African American Youths' Ecocultural Challenges and Psychosocial Opportunities: An Alternative Analysis of Problem Behavior Outcomes

Margaret Beale Spencer & Davido Dupree

For American youth of color, one's racial status significantly alters the normative developmental pathways and ecological character generally associated with adolescence as a stage of life (Spencer & Dornbusch, 1990; Spencer & Markstrom-Adams, 1990). Similarly, chronic poverty exacerbates the normative difficulties often linked with the adolescent period irrespective of gender, ethnicity, and color. Thus, for chronically impoverished African American youth, notably males, both cultural and ecological factors, synergistically and significantly, may compromise and influence the adaptive capacities of adolescents. Suggested is that the ecocultural character of developmental processes may be very different for some of society's most vulnerable young people. Ecocultural character refers to the embedded ecological niche and interactively linked cultural patterns, practices, traditions, values, and beliefs that reciprocally influence individual developmental processes and social interactions among members. More specifically, behavioral pathways of chronically impoverished youth in general, African American youth specifically, may demonstrate diverse adaptive responses more so than same ethnicity but more economically advantaged youth.

For chronically impoverished African American youth, the adolescent transition from middle childhood is fraught with risks that extend beyond those generally associated with the developmental stage. A state of heightened risk is particularly evident for African American boys (e.g., Cunningham, 1994; Cunningham & Spencer, in press; Spencer, 1995). The unusual level of challenge may be due to a variety of reasons. The most frequently alluded to explanations refer to (1) developmental themes such as the stage's significant biological changes and independence-focused intergenerational tensions; (2) cognitive shifts that are linked to new "meaning-making processes" and social understandings that require the interpretation of both objective and subjective phenomena; (3) heightened social and emotional needs that correlate with new ways of relating to adults and corresponding

Funding for the project was made available from the Spencer Foundation, Ford Foundation, the Commonwealth Fund, and the W. T. Grant Foundation.

An earlier version of the manuscript was based upon a symposium presentation for the Biennial Meetings of the Society for Research in Child Development, New Orleans, LA, March, 1993.

peer alliances that afford special meaning to the term "peer culture" (e.g., male peer groups may be burdened with a priori labeling as gangs); and (4) complex identity processes and corresponding self-focused concerns that may be linked to group stigma and unavoidable visibility and, particularly for African American and Hispanic youths, an imposed identification with and classification as a disfavored group member (see Matute-Bianchi, 1986). It is not surprising, then, that unique ecocultural characteristics and experiences often result in cultural dissonance or expressed developmental specific ambivalence about group membership (Spencer, 1985). The phenomenon appears early in development, is assumed to be linked to cognitive maturation (Spencer & Markstrom-Adams, 1990), and has been observed for a variety of "minority-status" individuals (Ogbu, 1985). Cultural dissonance is evident, in fact, for minority-status individuals in a nation as homogeneous appearing as Japan (see DeVos, 1980; DeVos & Wagatsuma, 1972), although the effects appear attenuated when individuals immigrate elsewhere to places where less stigma is attached to group membership (see Spencer, Kim, & Marshall (1987). Considering the impact of membership in a favored or disfavored group as a chronic source of risk may provide an opportunity for better theory construction. Consistent with social psychology (e.g., symbolic interactionism) and developmental theories (e.g., social cognition and social understanding), introducing social constructs, perceptual mechanisms (e.g., phenomenological processes), and context characteristics (e.g., residing in persistently underserved communities) improves our ability to explain patterns of coping and adaptation. More directly, an interactive and multidimensional analysis can specify (1) sources of challenge: risk and stress; (2) adaptive and maladaptive coping methods and strategies; and (3) resulting coping outcomes: products, abilities, skills, and health status. Identifying particular patterns of adverse or productive "coping behaviors" or adaptive skills and abilities have implications for realizing diverse life-course opportunities versus expressed vulnerability.

An ecoculturally inclusive and more comprehensive perspective introduces significant challenges for traditionally trained developmental psychologists and developmental psychopathologists since ecological and cultural considerations have been infrequently introduced into normative developmental theorizing and designs of empirical research. A group's status and associated societal views about its cultural patterns have bidirectional implications for individual members' treatment in particular contexts. Coping outcomes, adverse and productive, can potentially affect macrolevel beliefs and microlevel experiences. However, ecocultural considerations and interpretative frameworks, in general, remain outside of traditional theorizing about developmental processes. Developmental psychopathology has not been spared the oversight.

As noted by Sroufe and Rutter (1984), developmental psychopathologists are interested in the origins and time course of a given disorder, its varying manifestations with development, its precursors and sequelae, and its relation to nondisordered patterns of behavior (p. 18). They may be as concerned with a group manifesting precursors of a disordered behavior pattern although not developing the actual disorder (Sroufe & Rutter, 1984). The opportunity for the field of developmental psychopathology to take the lead in integrating ecocultural factors into its interpretative frameworks of adaptive and maladaptive behavioral tendencies has important potential, although in principle it remains untapped. More specifically, in the case of poor youngsters and minority youth, there continues to be a monocular focus on disordered behaviors, both real and expected. The long-term behavioral science interest in minority status and assumed connections to psychopathology has seldom considered its linkages to the character and quality of contextual conditions. The assumption of negative, direct, and patterned responses to caste status has basked in nearly a half century of attention (e.g., Kardiner & Ovesey, 1951; Pettigrew, 1962) (see reviews by Gutherie, 1976; Slaughter & McWorter, 1985). Unfortunately, the attentiveness, in general, has lacked a much needed multifaceted, culturally sensitive, multidimensional, and in fact, developmentally linked focus. From the developmental psychology literature, the field of psychopathology has benefited from significant research efforts in a variety of areas that include, among others, stress and competence (e.g., Mulholland, Watt, Philpott, & Sarlin, 1991), stress and normal development during (or across) normative periods (e.g., Garrison, Schoebach, Schluchter, & Kaplan, 1987), stress and coping (e.g., Compas, 1987), and childhood stress and adulthood projections (e.g., Werner, 1989). Unfortunately, even for the latter study that included ethnic Hawaiians, little information concerning specific ecocultural experiences of Hawaiians is considered. For the majority of these studies, if samples were ethnically diverse, ethnicity was not linked to the child's cultural and ecological niche. In the case of visible minorities, particularly during developmentally vulnerable periods such as adolescence, the oversight sidesteps important opportunities for understanding the prediction of resiliency and vulnerability.

Indeed the linkages between group status and individual treatment is salient independent of how adolescents actually present themselves or, at least initially, the specific quality of their activities (e.g., Chestang, 1972; West, 1993). The broader and undergirding issues of stigma (e.g., Gutherie, 1976; Isaacs, 1968; Ito, 1967), between-group conflict (e.g., Bobo, 1988), and their relationship to social experience have been treated thoroughly only in the social psychology literature. There have been a few exceptions of "new" empirical analyses in the developmental and educational psychology literature (e.g., Holliday, 1985); they address particular theoretical contributions (e.g., Boykin, 1985; Boykin & Ellison, 1995; Cross, 1985; Ogbu, 1985; Spencer, 1985, 1995), differential educational experiences (Ladson-Billings, 1994; Irvine, 1990), diverse identity-linked apprenticeship opportunities (Sullivan, 1989), culturally sensitive and ecologically sensible analyses of the family (Billingsley, 1992; Dupree, 1994; Staples & Johnson, 1993), and varying physical characteristics of neighborhoods (Burton et al., in press; Hambrick-Dixon, 1989, 1990; Spencer et al., in press a, in press b). In general, reference to more cultural and contextually informed models are not included in reviews of developmental science (e.g., Parke, Ornstein, Rieser, & Zahn-Waxler, 1994) and would appear not to represent the Zeitgeist for a century of "mainstream" developmental psychology.

Developmental psychopathology, as a field, would do well to improve current explanations of outcome variability by focusing more attention on ecocultural constructs. Particularly for the psychologically vulnerable adolescent, specific ecocultural experiences that are associated with caste status or chronic poverty have particular associations with (1) the nature of stress experienced (Chestang, 1972; West, 1993); (2) the specific quality of youths' meaning-making processes (e.g., Fordham & Ogbu, 1986); (3) the resulting attitudes, expectations, coping and adaptive ploys required (e.g., Cunningham, 1993; Majors & Billson, 1992; Matute-Bianchi, 1986); and (4) the associated complex, multidimensional identity processes of adolescence that represent several domains of functioning. For example, domains may be sex-role germane (e.g., Cunningham, 1993), have cultural relevance (e.g., Spencer, 1985; Spencer, Swanson, & Cunningham, 1991; Swanson & Spencer, 1991), maintain an efficacy focus (e.g., Connell, Spencer, & Aber, 1994; Hare & Castenell, 1985; Spencer, Dobbs, & Swanson, 1988), introduce peer linkages (see Cunningham, 1995), or explore home-related "self-processes" (e.g., Spencer, 1988).

In response to the myriad oversights noted, we introduce an identity-focused cultural ecological (ICE) perspective. The perspective represents programmatic research and theorizing about the relationship between several critical human development themes: (1) risk and stress, (2) coping methods and strategies, and (3) coping products. Together the overlapping themes represent ongoing dynamic processes that commence over time as spiraling lifelong processes from conception to death. The specific contents of each theme vary at different developmental periods as a function of the diverse developmental tasks associated with the period. Figure 1 represents the spiraling characterization of the hypothesized set of overlapping themes during adolescence.

There are specific issues associated with the overlapping themes that are linked to early adolescence. For example, early physical maturation may cause significant social stresses for both boys and girls, although for different reasons (e.g., Dupree et al., under review; Spencer et al., in press). For girls in mother-headed households, increased risk may be associated with girls having less opportunity to receive feedback about femininity from emotionally bonded father figures (i.e., "safe men"). For boys, the extra stress may be associated with physical coordination that may not "fit" with the unexpected physical size of early maturation. On the other hand, late adolescent stress may be associated with youth's awareness of inadequate self efficacy or limited academic prowess. Particularly for males, academic prowess is linked with sex-role and instrumentality issues. An early adolescent peer-focused coping strategy may help adaptive processes, albeit only in the short term. However—of salience for the long term, particularly so for males—realizing the importance of translating academic prowess into future work and schooling opportunities and life course instrumentality becomes an important source of stress.

The notion of a lifelong set of spiraling interactive themes is not new. In fact, Brewster Smith (1968) alluded to such a core when describing everyday competence processes.

Figure 1. A phenomenological variant of ecological systems theory (Spencer, 1995).

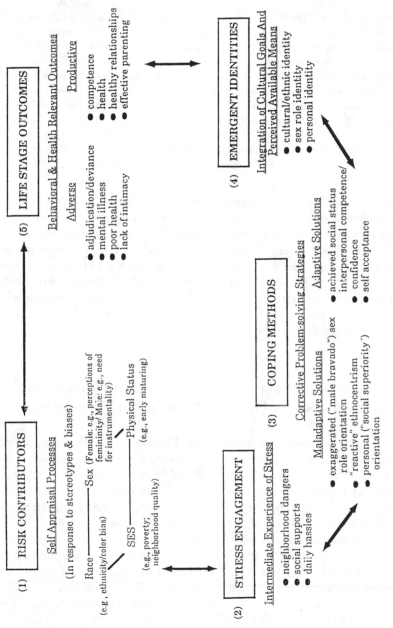

Note. The five domains represent three dynamic themes (i.e., risk and stress contributors, coping strategies and coping products: abilities, skills, adverse outcomes, and health status); the interactive themes occur as overlapping processes from conception until death.

There is a core of interrelated personal attributes which in some way plays a crucial role in the person's effectiveness in interaction with the environment. The cluster of attributes that I am looking for is one that would make a decisive difference in the cumulative direction of the outcomes of a person's interaction. Underlying my search for such a key cluster is a view of causation in personal and social development as inherently circular or spiral, rather than linear in terms of neatly insoluble causes and effects. As the very concept of interaction implies, developmental progress or deficit is typically a matter of benign circles or vicious ones . . . Launched on the right trajectory, the person is likely to accumulate successes that strengthen the effectiveness of his orientation toward the world while at the same time he acquires the knowledge and skill that make his further success more probable . . . Off to a bad start, on the other hand, he soon encounters failures that make him hesitant to try. What to others are challenges appear to him as threats; he becomes preoccupied with defense of his small claims on life at the expense of energies to invest in constructive coping. And he falls increasingly behind his fellows in acquiring the knowledge and skills that are needed for success on those occasions when he does try." (Smith, 1968, p. 277)

For Smith the goal for concerned professionals is to isolate a core of "interrelated personal attributes" of importance for insuring effective coping and successful interaction with the environment. Issues associated with chronic stress and coping are familiar in the developmental psychology and developmental psychopathology literatures. However, omitted is an examination of the specific and chronic stresses and adaptive opportunities associated with ecocultural challenges, particularly those associated with minority group status. The overlapping and interactive relationships among risk and stress, coping strategies, and coping outcomes, we believe, represent the core alluded to by Smith (1968).

Theorizing About Minority Youth: The Role of Unique Ecocultural Conditions for Diverse Challenges and Opportunities

As noted, a century review of developmental science (see Parke et al., 1994) suggests an omission of the ethnic-specific challenges generally confronted by minority youth. The recent publication sponsored by the American Psychological Association, A Century of Developmental Psychology, not only fails to integrate the experiences of nonwhites in its century review of the field but simply alludes to the oversight with statements about what is needed in the future. There is no delineation of the consequences of the oversight for developmental science application more generally. As reviewed in the literature on African American and other minority-status youth that began over a decade ago, there are examples of new theorizing and empirical research published and available on minority youth (see Gibbs, Huang, & Associates, 1989; McAdoo & McAdoo, 1985; Spencer, Brookins, & Allen 1985; Spencer & McLoyd, 1990). The persistent reluctance to accept the contribution of minority-focused research to developmental science reinforces a pattern that omits non-Caucasian youth from a normal human development framework. The dilemma contributes to patterns of simplistic, noncontextual, unidimensional, and monocultural analyses of human development. The situation, in and of itself, presents one of the most difficult challenges currently facing the field, since a lack of understanding about minority-youth development further compromises the quality of research efforts, the possibility of more inclusive theory, the introduction of better policy decisions, and the application of more specific and responsive prevention and intervention strategies.

The conceptual model introduced in this paper draws attention to overlapping themes of an identity-focused cultural ecological perspective. It falls under the category of classical ecological psychology theory (e.g., Barker & Wright, 1949, 1954) and more contemporary ecological systems theorizing (Bronfenbrenner, 1989; Bronfenbrenner & Crouter, 1980). Figure 2 suggests the dynamism associated with a culturally and ecologically inclusive theoretical perspective recently introduced (Spencer, 1995).

The theoretical perspective explicitly speculates about the challenges and predictors of both positive and vulnerable outcomes. Across the life course, the spiraling relationships between risk and stress contributors, coping methods, coping products, and their overlap are conceptually and behaviorally manifested in diverse ways at different developmental periods. For example, sources of early adolescent stress may be peer related and those of later adolescence may be work focused. As such, the specific risks identified, coping methods applied, and coping products accrued as life-stage outcomes will vary by developmental stage and ecocultural factors. Figure 3 introduces theoretically based specificity to the more generic identity-focused cultural ecological perspective suggested in Figure 2. As illustrated in Figure 3, the first theme, "risk factors and stress contributions," may be due to self-appraisals that are context linked (i.e., numbers 1 and 2 depicted in Figure 2). The second theme, "coping strategies, methods, and processes," are represented by factor 3 (i.e., coping methods of Figure 2), and factor four (i.e., emergent identity processes) previously illustrated in Figure 2. The third theme of Figure 3, "coping products, abilities, and skills," includes the fifth factor (illustrated previously in Figure 2) and represents life-stage outcomes that may be adverse or productive coping products, abilities, skills, or health status. As illustrated as an age-specific spiral in Figure 1, these factors or domains represent three dynamic themes that are spiraling (across the life course): interactive, dynamic and overlapping. The behaviors produced are developmental and ecocultural specific. The specific and illustrative organization of variables in Figure 3 demonstrates the identification of specific measures to be used in a set of exploratory or hypothesis-testing analyses.

Importantly, as indicated, the overlapping domains occur recursively and over time. For example, as an illustration of Figure 2, emergent identity issues (as coping strategies) for an adolescent might include sex-role identity themes. Thus, for an early adolescent male, a critical psychosocial sex-role theme might be peer-linked stress related to physical prowess. For late adolescence, risks and stress may be more closely associated with self-efficacy-linked apprenticeship experiences that precede

Figure 2. Spiraling and interactive processes across the life course from conception to death: adolescent segment.

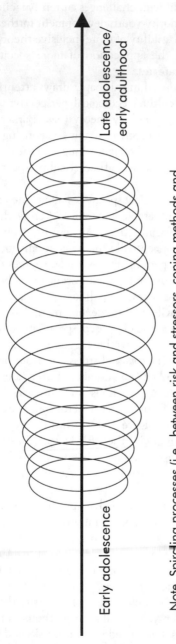

Early adolescence

Late adolescence/ early adulthood

Note. Spiraling processes (i.e., between risk and stressors, coping methods and strategies, and coping products) occur in multiple settings and at multiple levels (e.g., micro- to macro-levels) across the life course. Included are effects caused by historical, sociocultural, psychological and biological changes. Shown here is the adolescent segment only.

Figure 3. A Phenomenological Variant of an Ecological Systems Theory of Human Development (Spencer, 1995).

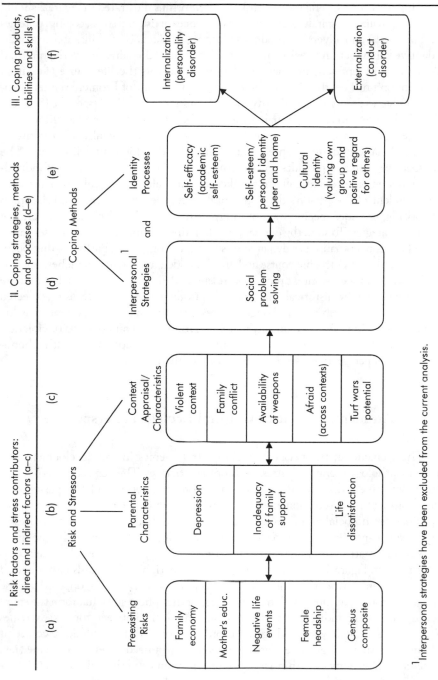

self-sufficiency and appropriate responsibilities as an adult. Coping strategies and methods implemented may be linked to various factors, such as the magnitude of existing risks. The several broadly linked relationships illustrated in Figure 2 should be conceptualized as spiraling and linked experiences of risk and stress, coping strategies and methods, and adverse or productive coping products (i.e., skills, abilities, mal-adaptive behaviors, and health outcomes). The linked domains are overlapping and dynamic, and represent interactive relationships across the life course (i.e., from conception until death). As a phenomenological variant of Bronfenbrenner's eco-logical systems theory and an identity-focused cultural ecological perspective (Spencer, 1995), the alternative conceptualization described introduces more specific and, we believe, valid ways of including the daily experiences and generally stable contexts of traditionally underrepresented groups into the developmental literature. The perspective proposed provides an opportunity to delineate the diversity of individ-ual symbolic processes as culturally linked to particular ecological characteristics. Independent of the person's developmental stage, the framework, like Eriksonian theorizing, provides specific ways of interpreting psychosocial processes as critical coping strategies. To sum, the overlapping domains and interactive processes intro-duced are conceptualized as dynamic sets of relationships and pathways that occur continuously across the life course and are embedded in ecocultural niches. Specifi-cally, person-process-context processes are embedded in ecocultural niches that are influenced, too, by historical changes and episodic exigencies such as unexpected unemployment, harsh policy changes, recessions (that may be experienced as life course depressions for economically marginal groups), and random acts of violence. Psychosocial and inherent social cognitive processes, consistent with pheno-menological perspectives, serve a central role in the theory.

An Identity-Focused Cultural Ecological Perspective

The elements of the perspective are central themes in the development and psychopathology literature—stress and coping processes. The perspective is contex-tually sensitive and, as noted, consistent with Bronfenbrenner's (1989) person-pro-cess-context view. And parallel with Eriksonian (1968) theorizing, the perspective specifies psychosocial processes as central coping methods, strategies and processes that predict specific coping products: behavioral outcomes, coping skills / abilities, and health status (see Figure 2). The size variability of the spiral's spheres, as illustrated in Figure 1, may vary between its midsection and towards either end. If viewed as an entire life course (i.e., conception to death), as opposed to the adolescent segment illustrated, the variability of the sphere sizes that form the spiral across time would indicate greater behavioral variability between the two extremes (e.g., between infancy and old age). Accordingly, observed as a total life course, it should be assumed that there is greater universality of needs at either extreme (i.e., greater homogeneity of within-group needs for infants and the elderly) and significant variability within life stages (e.g., adolescence, middle and late adulthood) depicted as the midsection for a life-course-wide set of spiraling relationships. Thus, evident would be smaller spheres making up the spiral in the beginning (infancy) and end (elderly) with much larger spheres representing the midsection.

The theoretical approach introduced enhances the interpretation of human behavior independent of developmental stage, gender, color, context, socioeconomic status, and ethnicity. For example, the perspective should afford a better understanding of developmental processes for chronically impoverished, abused, and health-compromised children. There would be little argument about the greater difficulties accompanying developmental transitions for youths from families with few resources. Such youths are confronted with normative challenges and unvarying expectations for the positive completion of stage-linked developmental tasks in addition to unique ecocultural social presses (Havighurst, 1953, 1977). A phenomenological variant of Bronfenbrenner's ecological systems theory, as an alternative and more culturally sensitive theoretical approach, attempts an integration of an individual's subjective experience of risk (e.g., economic survival concerns) with an exposure to contexts that may be chronically impoverished, differentially supportive, privy to unacknowledged ethnic bias, and generally high risk (see Spencer, 1995; Spencer, Cole, Dupree, Glymph, & Pierre, 1993).

Although particularly useful as an explanatory vehicle for minority-status individuals, the theoretical perspective has more general utility. As indicated in Figure 2, the confluence of factors depicted, assumed to occur recursively across the life course, should foster a better understanding of individual ontogeny under conditions of great uncertainty such as chronic poverty and ethnicity-linked bias. An ICE perspective should increase the likelihood for an accurate accounting of problematic trajectories as well as more traditional developmental pathways and outcomes for youth who share ecocultural niches.

Economic and sociological macrolevel analyses such as Charles Murray's *Losing Ground* (1984) and William Julius Wilson's *The Truly Disadvantaged* (1987) continue to rekindle concerns about the plight of poor people. Particularly for the field of developmental psychopathology, pressing forward beyond macrolevel-focused analysis is of particular importance; articulating linkages and processes between macrolevel structural conditions (e.g., beliefs, attitudes, informal practices) to individual-level characteristics and psychosocial responses afford the consideration of additional and more ecoculturally meaningful and developmental-stage-specific constructs. The availability of social supports (e.g., psychological characteristics of parents or other significant adults) would influence the impact of risk and stress that is experienced. In response, coping methods, strategies and processes required will vary. These include short-term adaptive solutions constructed (e.g., the "bravado" style of inner-city underemployed males) and the associated identity processes engendered (e.g., sex-role identity and self-efficacy). Conceptualizing the overlapping domains as spiraling sets of processes provides an opportunity to explore diversity within groups, between individuals, and aids in the specification of particular hypotheses. Of note is that the theoretical approach introduced links stress and coping to context and culture in specific ways. In summary, the perspective and theoretical assumptions provide an exploration of conceptual relationships and outcomes that extend beyond macrolevel analyses to an examination of linkages between macrolevel structures and individual characteristics, strategies for coping, developmentally relevant parental "buffering" efforts, and perception of context (e.g., perceptions of threat or safety, bias or opportunity, violence or support).

Contemporary Views About Stress and Coping

Anthony's (1974) analysis of stress and coping also links the context as a source of stress with individual coping strategies. His work is unique in the stress and coping literature, which generally does not integrate individual coping strategies with context-linked sources of risk and stress such as ethnicity, chronic poverty, and gender. Anthony posits that individuals differ in their reactivity to the environment; more frequently vulnerable children tend towards apprehension and fearfulness. Such children have the potential for "encapsulating themselves protectively" in the environment. Erikson (1968) would suggest that "encapsulating" activities could prevent vulnerable youths from taking advantage of diverse opportunities (e.g., learning and achievement options) that ultimately contribute to an achieved identity and positive adult life-course tendencies. Fordham and Ogbu's (1986) research suggest "missed" opportunities or the misconstrual of a context as culturally unsupportive and as a source of cultural dissonance. According to Anthony, highly vulnerable youths would not necessarily suffer impediments to affective functioning as long as the environment is relatively safe, responsive, and predictable. To illustrate, economically vulnerable youth, like those described by Fordham and Ogbu (1986) may use school as a physically safe place but demonstrate diminished intellectual engagement since, from an ecocultural perspective, it may for some contribute to cultural dissonance vis-à-vis black culture. From an ICE perspective, the short-term coping strategies (e.g., diminished academic engagement) become linked to a particularly patterned identity (e.g., low self-efficacy), and hence, as an outcome (expecially for boys), the strategy may be a less effective coping skill for life-course instrumentality as an adult provider. Eriksonian theorizing is quite specific about the role of successful instrumentality, particularly for males, for positive life course psychosocial functioning and health. In our attempt to understand the predictors of personality and conduct disorders, our broad research strategy has been to provide conceptual linkages between individual variations and stress reactivity. There are multiple correlates and sources of stress: (1) chronic poverty, (2) unavoidable maturational processes that include physical changes, (3) microsystem-level parental support efforts (e.g., psychological well-being, and personality characteristics), (4) context characteristics either perceived or objectively assessed, and (5) youths' own identity processes. The latter concerns are particularly salient for members of identifiable and disfavored groups and are particularly so during adolescence, a period of development when peer groups and social evaluations are unusually important.

The process-person-context model (Bronfenbrenner, 1989) affords an analysis of variations in developmental processes and outcomes as a *joint function* related to the characteristics of the environment and of the person (p. 197). The ecological perspective is particularly helpful when directed to minority youth of impoverished families (see Bronfenbrenner, 1985). Despite the nearly two decades of prominence that Bronfenbrenner's theory has enjoyed, it has been too often ignored in the traditional treatment of poverty either by developmentalists or the more macro-level analyses presented by sociologists, economists, policy analysts, and political scientists. The lack of alternative, broadly contextualized explanatory hypotheses, as indicated, is not evident and thus, not warranted as an explanation. The absence impedes an understanding of discrepancies and inconsistencies in the interpretation of behavioral differences that leaves theory, policy, and practice incomplete at best and ineffective at worst. Current reactive, decontextualized, nondevelopmental, and culturally generic approaches to violence illustrate the point (see Glymph, 1993).

Illustrative research demonstrating the ICE perspective comes from a large longitudinal effort that explores the effects of persistent poverty and the predictors of resilience for a sample of African American youth sampled as part of the Project for the Promotion of Academic Competence (i.e., Project PAC) (Spencer, 1988 a). Illustrated as Figure 3, this paper describes findings from a set of analyses (see Spencer & Dupree, under review) that test specific questions from the broader theoretical framework.

As noted, the theory supports the ICE perspective and reflects the integration of phenomenological and ecological systems theorizing (Spencer, 1995). The research design described by Spencer and Dupree responded to several research questions and included constructs that represent specific components of the theoretical framework. Risk and stress contributors included (1) census and family-level measures (preexisting risks), (2) parental mental health variables, and (3) child perception of context variables. Coping strategies, methods and processes constructs included the identity variables of (1) self-efficacy, (2) self-esteem (home and peer), and (3) internalized / proactive afrocentric cultural identity (i.e., valuing own group *and* positive regard of other ethnicities). Coping outcome variables were adolescent reports of internalizing (i.e., personality disorder) and externalizing (i.e., conduct disorder) problem behaviors; the several questions (1) examined differential predictive pathways for risk and stress contributors, coping strategies, and the coping outcomes, and (2) explored assumptions about the effects of context and gender.

Sample Description and Research Findings

The research presented comes from a large longitudinal study of 562 African American urban youth (Spencer, 1988 a). In this section, we present a description of findings reported by Spencer and Dupree (under review) that sought to test aspects of the broader theoretical framework presented as Figure 2 for a particular set of variables illustrated as Figure 3. The specific set of analyses described examined predictors of adolescent personality disorders as measured by adolescent self-ratings on the Youth Self Report (internalizing and externalizing behavior) (see Spencer and Dupree). Neighborhood and family background indices, parental psychological-functioning measures, and child-perceived context characteristics as risk and stress contributors were included as risk factors and stress contributors. Identity processes (i.e., self-esteem, self-efficacy, and cultural [internalized/proactive afrocentric] identity) represented measures of coping strategies. Parental psychological status was included as a risk contributor since our conceptual framework suggests that, irrespective of available resources, parental psychological attributes contribute to protective factors that effectively buffer or enhance societal-level risk factors, neighborhood factors, child-coping mechanisms, identity processes, and behavior disorder outcomes.

The sixth, seventh and eighth grade adolescents were randomly selected from four middle school populations; the consent rate by school ranged between 55% and 80%. The specific description of the sample demographics are described elsewhere (Spencer and Dupree, under review) and the youths, in general, are characterized as extremely impoverished.

Consistent with the overall theoretical model described as Figure 2, socioeconomic status (SES) was a salient issue for this sample, who had young mothers (70% were either in their 20s or 30s) and, in general, were from families with incomes between $2,500 and $10,999. There were significant or trend-level sex differences for subsets of variables for each of the three themes described in Figure 3, Risk factors and stress contributors; Coping strategies, methods, and processes; and Coping products, abilities, and skills. For example, for the first theme Risk factors and stress contributors), girls reported more negative (stressful) life events (trend level), although there was a trend for boys to come from families with lower mean incomes; girls reported perceiving the context as significantly more fearful, although boys perceived a significantly higher "weapons (harboring) context." For the second theme (Coping strategies, methods, and processes), girls obtained significantly higher means for both of the identity processes measures, the identity composite and self-efficacy scores. And for the third theme (Coping products, abilities, and skills), boys obtained a significantly higher mean for externalizing problems. The findings suggested the importance of examining specific questions separately by gender.

A set of regression and cluster analyses reported by Spencer and Dupree (under review) indicate the need to more specifically interpret the ICE perspective, given the pattern of findings. For example, for male adolescents, and as described in Figure 3, a combination of background measures, parental variables, and child-perceived context measures make up the first theme, risk factors and stress contributors. However, for background variables or preexisting risks, it was the negative life events score in addition to age (being young) that accounted for the majority of variance accounted for in the prediction of boys' internalizing (personality disordered) problem behaviors. Parental characteristics (i.e., depression) accounted for little or nothing, although the boys' context appraisal (family conflict) did contribute to a significant overall F change score. However, it was the addition into the regression model of the identity variables (i.e., II. Coping strategies, methods and processes) that virtually doubled the adjusted R^2 due to two identity variables, (low) self-efficacy and self-esteem scores; it is important to note the unexpected finding given prior reviews (e.g., Spencer, 1988 b; Spencer, 1985; Spencer & Markstrom-Adams, 1990) that cultural identity (internalization / proactive afrocentrism did not play a role. As noted by Spencer and Dupree, virtually the same pattern was evident for adolescent girls: negative (high-stress) life events played an important role, parental variables did not add to the regression model, and cultural identity (internalizing / proactive afrocentrism) did *not* contribute to the overall prediction equation, although for both males and females virtually 25% of the variance for the prediction of internalizing (personality disordered) problem behavior was accounted for (see Spencer and Dupree). Similarly, the prediction equations were similar for both male (overall adjusted R^2 was 24.4%) and female (overall adjusted R^2 was 18.7%) adolescents for externalizing (conduct disordered) problem behavior: as with internalizing behaviors, negative (high-stress) life events were most consistently and significantly important, parental measures contributed little to nothing, and cultural identity (internalizing or proactive afrocentric identity) did not contribute to the equation.

The findings from the study by Spencer and Dupree (under review) warrant independent discussion here given the theorized importance of race / ethnicity for the life-course experiences of youth of color. Earlier research has suggested that cultural identity plays an important role in the lives of preschool and middle-childhood youths as they transition into early adolescence even under conditions of unusual stress (see Spencer, Dobbs, & Swanson, 1988). Prior findings suggested a buffering role for cultural identity processes as a coping strategy, along with the proposal that parental cultural socialization plays an important psychologically enhancing function (see Spencer, Swanson, & Cunningham, 1991; Spencer, 1990; Spencer, Kim, & Marshall, 1987). Current findings by Spencer and Dupree for early and middle adolescents suggest a different relationship for parental characteristics and youths' cultural (internalizing / proactive afrocentric) identity. Before integrating their regression findings into a discussion of the theory, their subsequent set of cluster analysis results were suggestive and should be mentioned.

As reported by Spencer and Dupree (under review), in order to consider alternative ways of considering resilience with respect to social context, cultural identity, and conduct disorders, a hierarchical cluster analysis was conducted. To this point, low scores on internalizing and externalizing behaviors were considered indicative of resilience. However, internalizing and externalizing behaviors may actually represent reactive coping, which itself can be mediated by other factors including social context and cultural identity. Resilience may more properly be conceived of in terms of stress engagement. Thus a cluster analysis was conducted to determine whether such an analysis would yield clusters of youth who differed significantly in stress engagement, as measured by an interaction term composed of items indicating whether or not a stressful event occurred as well as the impact of that event.

Scores from five of the measures included in the regression analysis described and listed in Figure 3 were entered into the cluster analysis. The measures included were cultural identity (internalized / proactive afrocentrism), maternal depression, self-esteem (composite score), and the conduct disorder variables—internalizing and externalizing behaviors. In order to conduct an ANOVA comparing the clusters on stress engagement, a 10-cluster solution was accepted because it yielded sufficient variability among the clusters without resulting in clusters composed of less than five subjects. The overall F for a 10-cluster solution was not significant, suggesting that the constellation of variables chosen, particularly parental depression, did not distinguish youth with respect to stress engagement.

Given the findings from the first analysis, it was decided to add a more developmentally appropriate parental characteristic for the adolescent sample, (youth-reported) home and family hassles. Careful scrutiny of the items in this factor suggest that experiences that are perceived of as hassles by the adolescent may actually represent parental monitoring. While maternal depression may be a proxy for the absence of parental socialization or monitoring, the home and family hassles variable represents a more proactive role taken by the parent. Thus a second cluster analysis was conducted including the home and family hassles score.

As reported by Spencer and Dupree (under review), comparisons of the error variance for the second cluster analysis that included the home and family hassles score offered the possibility of a 5, 8, 14, or 24 cluster solution. The 8-factor solution was chosen in order to provide sufficient variability yet still meet criterion for ANOVA comparison. The ANOVA for cluster comparisons obtained an overall $F(7,181)= 4.27$ that was highly significant ($p. < 001$), that is, the clusters did differ significantly in stress engagement. Significant comparisons based on the Tukey's studentized ranged test indicated somewhat patterned findings. For example, cluster 7 had the lowest level of stress engagement and was significantly lower in stress engagement than five of the remaining clusters. The youth in cluster 7 had average means based on t scores of approximately 50 for maternal depression and youth-internalized cultural identity scores. The youth also had below average means for self-esteem as well as internalizing and externalizing behaviors. Most importantly however, cluster 7 indicates above average home and family hassles or parental monitoring. Cluster 1 was the only other cluster with an above average mean for home and family hassles. Cluster 1 was not significantly different from cluster 7 in stress engagement. The remaining clusters with stress engagement scores significantly higher than cluster 7 showed scores for home and family hassles that were average or below average. Thus, the content of cluster 7 and the fact of its significant difference from five of the other clusters suggested alternative ways of thinking about the theory given the pattern of regression and cluster analysis reported.

Discussion

The theoretical perspective introduced proposes linkages between three overlapping themes of an identity-focused cultural ecological perspective. Research was integrated that investigated the contributions of stress and risk factors to adolescents' self-rated internalizing (i.e., personality disordered) and externalizing (i.e., conduct disordered) behaviors (see Spencer & Dupree, under review). Relative to the regression analyses, no gender differences were evident for the prediction of internalizing problem behaviors (i.e, personality disordered). However, for boys, greater vulnerability is associated with age—younger boys report more internalized problem outcomes. The sex differences indicated for conduct disorder (externalizing behaviors) are quite similar to those noted for personality disorder (internalizing behaviors)—being young and reporting high rates of negative life events is associated with more conduct disorder for boys. Both genders are negatively and significantly affected by reports of negative life events.

Regression findings from Spencer and Dupree were considered and hypothesized relationships explored between the three themes described in Figure 3: I. Risk factors and stress contributors, II. Coping strategies, methods, and processes; and III. Coping products, abilities, and skills. The most consistent finding for both genders for either behavior problem (i.e., internalizing or personality disordered and externalizing or conduct disordered) was the significant role played by negative life events. Although appraisal of context characteristics was an important contributor, as was self-efficacy or self-esteem, most importantly, parental depression as an hypothesized important parental characteristic or potential buffering agent was *not* important. Similarly, counter to the expectations of the theoretical framework undergirding the model, cultural identity (proactive afrocentrism) did not contribute to the model for either gender. The lack of a relationship was a significant omission since the issue of identifiability (here, race / ethnicity-linked) is a critical aspect of the theory. However, the cluster analysis findings aided in explaining the lack of a main effect from the regression analysis both for the parental characteristic (depression) and the cultural-identity variable (proactive [internalized]) afrocentric cultural identity): clearly the relationships between themes are *not* linear, although the illustrated relationships in Figures 2 and 3 would suggest linear linkages.

To recap, theorizing has consistently suggested the buffering role of parents in the rearing of children. This has been particularly underscored in the case of minority youth (see Spencer, 1990; Spencer, Kim, & Marshall, 1987; Spencer, Swanson, & Cunningham, 1991); accordingly, that the parental depression score would not contribute to the regression equation might have raised significant concerns about the efficacy of the framework. However, because youth are less parent-focused and more peer-focused during adolescence, perhaps the lack of parental depression as a significant predictor is less important than the findings from the cluster analysis, which suggests nonlinear relationships. When parental monitoring (youth-reported home / family hassles) is included in the cluster analysis, it is this variable along with scores for youths' more proactive (internalized) afrocentric cultural identity that are associated with lower levels of personality disordered (internalization) and conduct disordered (externalization) behavior (i.e., cluster 7). Accordingly, relative to the theoretical framework introduced, the traditional parental variables usually included in the family research literature may not be of great relevance and importance uniformly across the life course of youth development: parental variables like depression may be more important during the early years of development (e.g., infancy through middle childhood), while more parental intrusion variables such as parental monitoring (home / family hassles) are more salient for the independence-seeking middle adolescent.

Certainly the research findings integrated for use as illustrations of the theory and ICE perspective provide only a glimpse of the suggested complex set of relationships between the overlapping themes indicated for the perspective introduced. Continued application of programmatic research findings to the model proposed should be helpful in determining the strength of the relationships (if they exist at all) and their specific behavioral expression for a particular developmental period. The data application to the model did not result in a set of definitive findings but, instead, suggested nonlinear and dynamic relationships specific to *middle adolescence* for a group of impoverished youth of color. The exercise suggested that, as theorized and illustrated in Figures 2 and 3, family characteristics and youth's cultural identity matter and are not independent from behavior problems. The relationships are clearly not linear, however, but are developmentally sensitive and are ecoculturally relevant in ways not generally considered.

REFERENCES

Achenbach, T. M. (1991). *Manual for the Youth Self-Report and 1991 profile*. Burlington: University of Vermont Department of Psychiatry.

Allen, W. R. (1985). Race, income and family dynamics: A study of adolescent male socialization processes and outcomes. In M. B. Spencer, W. R. Allen, and G. K. Brookins (Eds.), *Beginnings: The social and affective development of Black children* (pp. 273–292). Hillsdale, NJ: Erlbaum.

Anthony, E. J. (1974). Introduction: The syndrome of the psychologically vulnerable child. In E. J. Anthony and C. Koupernik (Eds.), vol. 3, *The child in his family: Children at psychiatric risk* (pp. 3–10). New York: Wiley and Son.

Anthony, E. J. (1987). Risk, vulnerability, and resilience: An overview. In E. J. Anthony and B. J. Cohler (Eds.), *The invulnerable child* (pp. 3–48). New York: Guilford Press.

Barker, R. G., & Wright, H. F. (1954). *Midwest and its children: The psychological ecology of an American town*. Evanston, IL: Row & Peterson.

Barker, R. G. & Wright, H. F. (1949). Psychological ecology and the problem of psychosocial development. *Child Development, 20*, 131–143.

Baron, R. M. & Kenney, D. A. (1986). The moderator-mediator variable distinction in social psychological research: Conceptual, strategic, and statistical considerations. *Journal of Personality and Social Psychology, 51*(6), 1173–1182.

Beck, A.T. (1987). *Beck depression inventory manual*. San Francisco: The Psychological Corporation.

Billingsley, A. (1968). *Black families in white America*. New York City: Simon & Schuster.

Billingsley, A. (1992). *Climbing Jacob's ladder*. New York: Simon and Schuster.

Bloom, B. (1964). *Stability and change in human characteristics*. New York: Wiley.

Bobo, L. (1988). Group conflict, prejudice, and the paradox of contemporary racial attitudes. In P. Katz & D. Taylor (Eds.), *Eliminating racism* (pp. 85–114). New York: Plenum.

Bowman, H., & Viveros-Long, A. (1981). *Balancing jobs and family life*. Philadelphia: Temple University Press.

Boyd-Franklin, N. (1989). *Black families in therapy: A multisystems approach*. New York: Guilford Press.

Boykin, A. W. (1978). Psychological behavioral verve in academic task performance: A pretheoretical consideration. *Journal of Negro Education, 47*, 343–354.

Boykin, A. W. (1986). The triple quandry and the schooling of Afro-American children. In U. Neisser (Ed.), *The school achievement of minority children* (pp. 57–92). Hillsdale, NJ: Erlbaum.

Boykin, A. W. & Ellison, C. M. (1995). The multiple ecologies of black youth socialization: An afrographic analysis. In R. L. Taylor (Ed.), *African-American youth: Prospectives on their status in the United States* (pp. 93–128). Westport, Conn.: Praeger.

Bronfenbrenner, U. (1985). Summary. In M. B. Spencer, G. K. Brookins, and W. R. Allen (Eds.), *Beginnings: The social and affective development of black children* (pp. 67–73). Hillsdale, NJ: Erlbaum.

Bronfenbrenner, U. (1989). Ecological systems theory. In R. Vasta (Ed.), *Annals of child development* (pp. 87–248). Greenwich, CT: JAI Press.

Bronfenbrenner, U., & Crouter, A. C. (1983). The evolution of environmental models in developmental research. In P.H. Mussen (Ed.), *Handbook of child psychology: Vol. 1, History, theory, and methods* (pp. 357–414). New York: Wiley.

Brookins, G. K. (1988). Making the honor role: A black parent's perspective on private education. In T. T. Slaughter and D. J. Johnson (Eds.), *Visible now: Blacks in private school* (pp.12–20). New York: Greenwood Press.

Burton, L. M., Price-Spratlen, T., & Spencer, M. B. (in press). Alternative approaches to conceptualizing neighborhoods: Implications for the study of development among ethnically diverse children. In J. Brooks-Gunn and G. Duncan (Eds.), *Neighborhood, poverty and youth outcomes*. New York: Sage.

Chestang, L. W. (1972). Character development in a hostile environment. *Occasional Paper No. 3 (Series)* Chicago: University of Chicago Press, p. 1–12.

Cicchetti, D. (1993). Developmental psychopathology: Reactions, reflections, projections, *Developmental Review, 13*(4), 471–502.

Clark, R. (1983). *Family life and school achievement: Why poor black children succeed and fail*. Chicago: University of Chicago Press.

Compas, B. E. (1987). Coping with stress during childhood and adolescence. *Psychological Bulletin, 101*(3), 393–403.

Connell, J. P., Spencer, M. B., & Aber, J. L. (1994). Educational risk and resilience in African-American youth: Context, self, action, and outcomes in school. *Child Development, 65*, 493–506.

Conte, V. A., & Salamon, M. J. (1982). An objective approach to the measurement and use of life satisfaction. *Measurement and Evaluation in Guidance, 15*, 194–200.

Cross, W. E., Jr., (1985). Rediscovering the distinction between personal identity and reference group orientation. In M. B. Spencer, G. K. Brookins, and W. R. Allen (Eds.), *Beginnings: The social and affective development of black children* (pp. 155–171). NY: Erlbaum.

Cunningham, M. (1993). Sex role influences on African Americans: A literature review. *Journal of African American Male Studies, 1*(1), 30–37.

Cunningham, M. (1994). *Expressions of manhood: Predictors of educational achievement and African American Adolescent males.* Unpublished doctoral dissertation, Emory University, Atlanta.

Cunningham, M. (1995, March). *The influence of contextual peer-based perceptions on African American males' gender role development.* Symposium presentation to the Black Caucus of the Meetings of the Society for Research in Child Development, Indianapolis.

Cunningham, M., & Spencer, M. B. (in press). The black male experiences measure. In R. L. Jones (Ed.), *Handbook of tests and measurements of black populations.* Hampton, VA: Cobb & Henry Press.

DeVos, G. (1980). Ethnic adaptation and minority status. *Journal of Cross Cultural Psychology, 11*(1), 101–124.

DeVos, G. (1982). Ethnic pluralism: Conflict and accommodation. In G. DeVos and L. Romanucci-Ross (Eds.), *Ethnic identity: cultural continuities and change* (pp. 5–41). Chicago: University of Chicago Press.

Devos, G., & Wagatsuma, H. (1972). Family life and delinquency: Some perspectives from Japanese research. In W. P. Lebra (Ed.), *Transcultural research in mental health: II. Mental health research in Asia and the Pacific.* Honolulu: University of Hawaii Press.

Deyhle, D. (1986). Breakdancing and breaking out: Anglos, Utes, and Navajos in a border reservation high school. *Anthropology and Education Quarterly, 17,* 111–127.

Dupree, D. (1994). *The effect of experience with violence on the affective and cognitive functioning of African-American adolescents.* Unpublished doctoral dissertation, Emory University, Atlanta.

Dupree, D., Spencer, M. B., Cunningham, M., & Swanson, D. P. (under review). The association of physical maturation with family hassles in African American adolescent males. *Journal of Comparitive Family Studies* (Special issue).

Erikson, E. H. (1986). *Identity: Youth and crisis.* New York: Norton.

Fordham, S., & Ogbu, J. U. (1986). Black students' school success: Coping with the "burden of 'acting white.' " *Urban Review, 18*(3), 176–206.

Garmezy, N., & Masten, A. S. (1986). Stress, competence, and resilience: common frontiers for therapists and psychopathologist, *Behavior Therapy, 17,* 50;0–521.

Garrison, C. Z., Schoenbach, V. J., Schuchter, M., & Kaplan, B. (1987). Life events in early adolescence. *Journal of the American Academy of Child and Adolescent Psychiatry, 26*(6), 865–872.

Gibbs, J. T. (1988). *Young Black and male in America: An endangered species.* Dover, MA: Auburn House Publishers.

Gibbs, J. T., Huang, L. N., *et al,* eds. (1989). *Children of Color.* San Francisco: Jossey-Bass.

Glymph, A. (1993). *Assessing youths' perception of their neighborhood; Development of the student perception of neighborhood scales.* Unpublished master's thesis, Emory University, Atlanta.

W. T. Grant Foundation (1988). *Commission on work, family and citizenship. The forgotten half: Pathways to success for America's youth and young families.* Final Report on youth and America's future. Washington, DC: William T. Grant Foundation.

Gutherie, R. (1976). *Even the rat was white: A historical view of psychology.* New York: Harper & Row.

Hacker, A. (1992). *Two nations: Black and white, separate, hostile, unequal.* New York: Scribner's Sons.

Hambrick-Dixon, P. J. (1989). The effects of experimentally imposed noise on the task performance of black children attending day care centers near elevated subway trains. *Developmental Psychology, 22*(2), 259–264.

Hare, B. R. and Castenell, L. A. (1985). No place to run, no place to hide: Comparative status and future prospects of black boys. In M. B. Spencer, G. K. Brookins, & W. R. Allen (Eds.), *Beginnings: The social and affective development of black children* (pp. 185–200). Hillsdale, NJ: Erlbaum.

Havighurst, R. J. (1953). *Human development and education.* New York: McKay.

Havighurst, R. J. (1972). *Developmental tasks and education* (3rd ed.). New York: David McKay.

Holliday, B. G. (1985). Towards a model of teacher-child transactional processes affecting black children's academic achievement. In M. B. Spencer, G. R. Brookins, and W. R. Allen (Eds.) *Beginnings: The social and affective development of black children* (pp. 117–131). Hillsdale, NJ: Erlbaum.

Irvine, J. J. (1988). An analysis of the problem of disappearing black educators. *The Elementary School Journal, 88*(5), 503–513.

Irvine, J.J. (1990). *Black students and school failure.* New York: Greenwood Press.

Isaacs, H. R. (1968) Group identity and political change: The role of color and physical characteristics. In J. H. Franklin (Ed.), *Color and race* (pp. 75–97). Boston: Houghton Mifflin.

Ito, H. (1967). Japan's outcastes in the United States. In G. A. DeVos & H. Wagatsuma (Eds.), *Japan's invisible race.* Berkeley: University of California Press.

Kardiner, A., & Ovesey, L. (1951). *The mark of oppression: Explorations in the personality of the American negro.* Cleveland: World Publishing.

Ladson-Billings, G. (1994). *The dream keepers.* San Francisco: Jossey-Bass.

Majors, R., and Billson, J. M. (1992). *Cool pose: The delimmas of black manhood in America.* New York: Lexington Books.

Markstrom-Adams, C., and Spencer, M. B. (in press). Interventions with minority cultures. In S. Archer (Ed.), *Interventions for adolescent identity development.* Beverly Hills, CA: Sage.

Masten, A., Best, K., & Garmezy, H. (1990). Resilience and development: Contributions from the study of children who overcome adversity, *Development and Psychopathology, 2,* 425–444.

Masten, A., Garmezy, N., Tellegen, A., Pellegrini, D., Larkin, K., & Larsen, A. (1988). Competence and stress in school children: The moderating effects of individual and family qualities. *Journal of Child Psychiatry and Psychology and Allied Disciplines, 29*(6), 745–764.

Matute-Bianchi, M. E. (1986). Ethnic identity and patterns of school success and failures among Mexican descendent and Japanese-American students in a California high school: An ethnographic analysis. *American Journal of Education, 95,* 233–255.

McAdoo, H. P., & McAdoo, J. (1985). *Black children.* Beverly Hills, CA.: Sage.

Mead, G. H. (1934). *Mind, self and society.* Chicago: University of Chicago Press.

Mulholland, D. J., Watt, N. F., Philpott, A., & Darlin, N. (1991). Academic performance in children of divorce: Psychological resilience and vulnerability. *Psychiatry, 54,* 268–280.

Murray, C. (1984). *Losing ground: American social policy, 1950–1980.* New York: Basic Books.

Ogbu, J. U. (1985). A cultural ecology of competence among inner-city blacks. In M. B. Spencer, G. K. Brookins, and W. R. Allen (Eds.), *Beginnings: The social and affective development of black children* (pp. 45–66). Hillsdale, NJ: Erlbaum.

Parham, T. A., & Helms, J. E. (1981). The influence of black student's racial identity attitudes on preferences for counselor's race. *Journal of Counseling Psychology, 28,* 250–257.

Parke, R. D., Ornstein, P. A., Rieser, J. J., & Zahn-Waxler, C. (1994). The past as prologue: An overview of a century of developmental psychology. In R. D. Parke et al (Eds.), *A century of developmental psychology* (pp. 1–70). Washington, DC: American Psychological Association.

Pettigrew, T. F. (1964). *A profile of the Negro American.* Princeton: Van Nostrand.

Pierce, C. M. (1989). Unity in diversity: Thirty-three years of stress. In G. L. Berry & J. K. Asamen (Eds.), *Black students: Psychosocial issues and academic achievement* (pp. 296–312). Newberry Park, CA: Sage.

Richters, J., & Cicchetti, D. (1993). Mark Twain meets *DSM-III-R.* Conduct disorder, development, and the concept of harmful dysfunction. *Development and Psychopathology,* 5–29.

Shade, B. J. (1989). *Culture, style, and the educative process.* Springfield, IL: Charles C. Thomas.

Simmons, R. G., Rosenberg, F., & Rosenberg, M. (1973). Disturbance in the self-image at adolescence. *American Sociological Review, 38,* 553–568.

Slaughter, D. T. & Johnson, D. J. (1988). *Visible now: Blacks in private schools.* New York: Greenwood Press.

Slaughter, D. T. & McWorter, G. A. (1985). Social origins and early features of the scientific study of black American families and children. In M. B. Spencer, G. K. Brookins, and W. R. Allen (Eds.) *Beginnings: The social and affective development of black children* (pp. 5–14). Hillsdale, NJ: Erlbaum.

Spencer, M. B. (1983). Children's cultural values and parental child rearing strategies. *Developmental Review, 4,* 351–370.

Spencer, M. B. (1985). Cultural cognition and social cognition as identity factors in black children's personal-social growth. In M.B. Spencer, G. K. Brookins, and W. R. Allen (Eds), *Beginnings: The social and affective development of black children* (pp. 215–230). Hillsdale, NJ: Erlbaum.

Spencer, M. B. (1986). Risk and resilience: How black children cope with stress. *Journal of Social Sciences, 7,* 22–26.

Spencer, M. B. (1988a). Persistent poverty of African American adolescents. (Ford and Spencer Foundations funded research proposal.)

Spencer, M. B. (1988b). Self-concept development. In D. T. Slaughter (Ed.), *Perspectives in black child development: New directions in child development* (pp. 59–72). San Francisco: Jossey-Bass.

Spencer, M. B. (1990). Parental values transmission. In J. B. Stewart and H. Cheatham (Eds.), *Interdisciplinary perspectives on black families* (pp. 111–130). New Brunswick NJ: Transactions Press.

Spencer, M. B. (1995). Old issues and new theorizing about African American youth. In R. L. Taylor (Ed.), *Black youth: Prospectives on their status in the United States* (pp. 37–70) Westport, CT: Praeger.

Spencer, M. B., Aber, J. L., Connell, J. P., Cole, S. P., & Halpern-Felsher, B. (a in press). Neighborhood and family influences on young urban adolescents' academic efficacy and behavior problems: A multi-sample, multi-site analysis. In J. Brooks-Gunn & G. Duncan (Eds.), *Neighborhood, poverty and youth outcomes.* New York: Sage.

Spencer, M.B., Cole, S.P., Dupree, D., Glymph, A., & Pierre, P. (1993). Self-efficacy among urban African American early adolescents: Exploring issues of risk, vulnerability, and resilience. *Development and Psychopathology, 5,* 719–739.

Spencer, M. B. & Dupree, D. M. (under review). Ecocultural challenges and problem outcomes among African American youth (submitted paper).

Spencer, M. B., Dobbs, B., & Swanson, D. P. (1988). Afro-American adolescents: Adaptational processes and socioeconomic diversity in behavioral outcomes. *Journal of Adolescence, 11*, 117–137.

Spencer, M. B., & Dornbusch, S. (1990). American minority adolescents. In S. Feldman & G. Elliot (Eds.), *At the threshold: The developing adolescent* (pp. 123–146). Cambridge, MA: Harvard University Press.

Spencer, M. B., Dupree, D., Swanson, D. P., & Cunningham, M. (c in press). The influence of physical maturation and hassles of African American adolescents' learning behaviors [Special issue]. *Journal of Comparitive Family Studies*.

Spencer, M. B., Kim, S., & Marshall, S. (1987). Double stratification and psychological risk: Adaptational processes and school experiences of Black children. *Journal of Negro Education, 56*(1), 77–86.

Spencer, M. B., & Markstrom-Adams, C. (1990). Identity processes among racial and ethnic minority children in America. *Child Development, 61*, 290–310.

Spencer, M. B., McDermott, P., Burton, L., & Cole., S. (b in press). An alternative approach for assessing neighborhood effects on early adolescent achievement and problem behavior. In J. Brooks-Gunn & G. Duncan (Eds.), *Neighborhood, poverty and youth outcomes*. New York: Sage.

Spencer, M. B., & McLoyd, V. (1990). Minority child development [Special issue]. *Child Development, 61*(2).

Spencer, M. B., Swanson, D. P. & Cunningham, M. (1991). Ethnicity, ethnic identity, and competence formation: Adolescent transitions and cultural transformations. *Journal of Negro Education, 60*(3).

Sroufe, L. A., & Rutter, M. (1984). The domain of developmental psychopathology. *Child Development, 55*, 17–29.

Staples, R., & Johnson, L. B. (1993). *Black families at the cross-roads: Challenges and prospects.* San Francisco: Jossey-Bass.

Stewart, J. B., & Cheatham, H. (Eds.)(1990). *Interdisciplinary perspectives on black families.* New Brunswick, NJ: Transactions Press.

Sullivan, M. L. (1989). *Getting paid: Youth, crime and work in the inner city.* Ithaca NY: Cornell University Press.

Swanson, D. P. & Spencer, M. B. (1991). Youth policy, poverty, and African-Americans: Implications for resilience. *Education and Urban Society, 24*(1), 148–161.

Terkel, S. (1992). *Race: How blacks and whites think and feel about the American obsession.* New York: New Press.

Washington, V. (1980). Teachers in integrated classrooms: Profiles of attitudes, perceptions, and behavior. *Elementary School Journal, 80*(4), 192–201.

Washington, V. (1982). Racial differences in teacher perceptions of first and fourth grade pupils on selected characteristics. *Journal of Negro Education, 51*(1), 60–72.

Weiss, R. V., Spencer, M. B., Schaeffer, B. A., & McDermott, P. A. (under review) (1995). Construct validity of the child behavior checklist and youth self-report with urban African-American adolescents.

Werner, E. E. (1989). High-risk children in young adulthood: A longitudinal study from birth to 32 years. *American Journal of Orthopsychiatry, 59*(1), 72–81.

West, C. (1993). *Race matters.* Boston: Beacon Press.

White, R. (1959). Motivation reconsidered: The concept of competence. *Psychological Review, 66*, 297–333.

Wilkinson, D. Y., & Taylor, R. L. (1977). *The black male in America: Perspectives on his status in contemporary society*. Chicago: Nelson-Hall.

Wilson, W. J. (1987). *The truly disadvantaged: The inner city, the underclass, and public policy*. Chicago: University of Chicago Press.

Wingard, D. L. (1987). *Social, behavioral and biological factors influencing the sex differential in longevity*. Background paper prepared for the National Institute on Aging. Washington, DC: U.S. Government Printing Office.

X Definitions of Competence During Adolescence: Lessons From Puerto Rican Adolecent Mothers

Cynthia T. García Coll & Heidie A. Vázquez García

Adolescent pregnancy and childbearing is not a new phenomenon. Cross-cultural and historical data suggest that adolescent pregnancy, childbearing, and parenthood was and continues to be a viable life course for many young women and their families around the world. However, as societies in general, and North American society specifically, redefine educational, occupational, and life-span goals for both men and women, the normative aspects of teenage pregnancy, childbearing, and parenting change. In the face of new familial, societal, and occupational demands, the concept of adolescent motherhood has taken on a very different, mainly negative, connotation than it has had in the past.

How do the "normative" behavioral patterns of early childbearing and parenting become maladaptive? Is it accurate to consider these life processes as expressions of incompetence, maladaptation, or even expressions of psychopathology across cultures and contexts? In other words, are there universal definitions of competence in the area of female sexuality and reproduction during the adolescent period? If so, how do we explain the persistence of these patterns across history, societies and cultures?

If we do not consider teenage childbearing, and parenting a universally maladaptive response, then what are the major sources of influence that change the meaning of these events from signs of competence to incompetence? Is there wide variation on how individual adolescents and their families cope and adapt to these life events and their ultimate impact on their life course?

We contend that teenage pregnancy, childbearing, and parenting need to be reconceptualized, not as universal expressions of incompetence but as adaptive or maladaptive responses given certain motivational, contextual, and life-span considerations. The same behavioral responses (getting pregnant, deciding to have and keep the child, and becoming a very young parent) can have very different antecedents and consequences depending on the interaction between the individual, the context, and the expectations for life-span development of that particular society.

This conceptualization of adolescent parenthood leads to fundamental questions about how to define competence and adaptation cross-culturally and how to conceive the salient developmental tasks for the adolescents and their families as a function of contextual demands. Specifically, it emphasizes the importance of environmental demands—both historical and concurrent, on the group as well as on the individual—as a major consideration in our analysis of competence during adolescence.

This conceptualization also reinforces the notion of how different pathways can lead to similar developmental outcomes (i.e., an adolescent can become a parent as a result of either adaptive or maladaptive processes or both), a basic principle of various contemporary models of normative development as well as of developmental psychopathology (Bertanlanffy, 1968; Sadler & Catrone, 1983). Finally, this conceptualization has profound implications for public policy and for understanding the needs and providing services to bicultural individuals and those who belong to minority groups in the United States.

Our main thesis is that the consequences of teenage pregnancy, childbearing, and parenting for the mother as well as the child are a product of both intrinsic (personal) and extrinsic (contextual) factors. In this chapter, we will examine the extant literature on the antecedents and consequences of adolescent pregnancy, childbearing, and parenting, which for the most part, characterize these events as signs of maladaptation or incompetence. Subsequently, we will summarize findings from several studies conducted with adolescent mothers in traditional, low socioeconomic status (SES), Puerto Rican culture. These studies emphasize the role of contextual circumstances on defining the antecedents and consequences of adolescent pregnancy for both mother and child. They also suggest that adolescent pregnancy is considered to be an adaptive rather than a maladaptive life process within traditional, low SES Puerto Rican culture. Finally, we will examine the implications of this reconceptualization on social policy and on clinical interventions with adolescents in general and adolescent mothers in particular.

Adolescent Pregnancy: Why Is it a Problem?

The majority of the extant literature on teenage pregnancy, childbearing, and parenting identifies the negative antecedents and outcomes of adolescent pregnancy (Nord, Moore, Morrison, Brown, & Myers, 1992; Stevens-Simon & White, 1991; Zuckerman, Walker, Frank, Chase, & Hamburg, 1984). Two major trends can be identified within this literature: One trend focuses on the underlying reasons why adolescents get pregnant and the behaviors and risks associated with said reasons. The primary argument presented throughtout this literature is that adolescents, unlike their older counterparts, get pregnant for the "wrong reasons." The second trend identifies the negative consequences for the teenage mother and her child in a variety of areas. Again, the primary argument here is that adolescent pregnancy, childbearing, and parenting have negative consequences for both mother and child, and therefore they are maladaptive.

The section that follows reviews the antecedents of adolescent pregnancy that have been identified in the literature, including personal characteristics and individual motivational factors.

Antecedents of Adolescent Pregnancy

Cognitive Limitations and Feelings of Invulnerability

The literature on adolescent pregnancy suggests that there are a variety of reasons why adolescents get pregnant. One of the reasons proposed is that adolescents get pregnant because they are egocentric thinkers (Mercer, 1983; Redl, 1972). Some suggest that adolescents' preoccupation with themselves limits their ability to conceptualize the viewpoints of others, thereby restricting their ability to think about how their decisions will affect others (Elkind, 1974). As a result, the adolescent perceives herself as unique, special, and invincible (Franklin, 1988).

In addition, Elkind (1967) suggests that this feeling of uniqueness also stems from a limited time perspective of the "here and now," which is also characteristic of adolescence. The reluctance or inability to consider future occurrences and consequences renders the adolescent captive in the present. Thus, the limited perception of time on the part of the adolescent, coupled with the sense of uniqueness and preoccupation with the self, leads to perceptions of invulnerability as well as misperceptions of chance and probability (Norris, 1988).

This false sense of invulnerability and immunity to all negative consequences permeates into the realm of sexual activity, becoming the adolescent's armor against pregnancy. The ability to anticipate sexual encounters, to plan and use effective methods of birth control, and to anticipate the consequences of impulsive sexual behaviors is not characteristic of early adolescent development but rather develops later on in adolescence as the shift from concrete to formal operational thought takes place (Piaget, 1952). For example, Sommer and colleagues (1993) suggest that deficits in cognitive readiness predispose adolescent mothers to stress and stressful parenting experiences and that the level of cognitive readiness is more a function of the developmental stage of the adolescent versus the unique characteristics of the adolescent. Thus, research seems to suggest that the younger the adolescent at the time of her pregnancy, the more immature and limited her thought processes and consequently her behavior will be regarding her pregnancy and its effects on herself as well as others.

Developmental and Emotional Needs

A number of studies report that some adolescents become pregnant because of the underlying developmental, motivational, or emotional needs that having a child during adolescence fulfills. The need to establish independence from their parents, to become autonomous from their families, and to be considered an adult with adult responsibilities have been suggested as factors that motivate teenage girls to get pregnant (Buchholz & Gol, 1986). Research suggests that parenthood for some adolescents is a validation of their maturity, allowing them access to a desirable "adult role" (Stevens-Simon & White, 1991). Thus, becoming a parent during this developmental stage may be seen as fulfilling the needs for separation and individuation that are part of the normative processes of adolescence.

Similarly, there are numerous implications in the literature that suggest that adolescents get pregnant because of their needs for affiliation and love. For example, Musick (1993) suggests that some teenagers get pregnant in order to fill an emotional void in the home such as parents getting a divorce, the need to be cared for, or to fill an inner sense of emptiness. Ladner (1971) states that sexual activity and early pregnancy for some girls provides a feeling of utility and a sense of identity. Other research suggests that adolescents get pregnant as an acting-out behavior in response to a conflict or hidden wish toward her parents (Young, 1954). However, these needs are usually not completely or temporarily fulfilled by the baby and thus might contribute to the dissatisfaction or lower maternal self-concept found among adolescent mothers (Stevens-Simon & White, 1991).

Sensation Seeking and Risk Taking

Research has also identified adolescents' desire and need to take risks as another potential factor leading to early pregnancy. Strauss and Clarke (1992) state that "risk taking is the behavioral manifestation of the transition from immature to mature thinking" (p. 71). As previously discussed, this shift in thinking occurs as the adolescent moves from concrete operational thought to abstract or formal operational thought. An adolescent who is in the concrete operational stage will tend to focus more on the excitement, the fantasy, and the pleasure gained from taking risks than the repercussions associated with said risks.

Thus, sensation-seeking behavior is also identified with high-risk behavior, in that risks are usually taken on the part of the adolescent in order to fulfill their need for stimulation and novel experiences (White & Johnson, 1988). Other research has reported that unconventionality is also related to sensation seeking, possibly accounting for the higher levels of impulsivity in adolescents and the lack of contraceptive use among teenagers (Zuckerman, Bone, Neary, Mangelsdorff, & Brustman, 1972). In other words, explorations with unprotected sex, multiple partners, or just premarital sex with one partner, might all be related to the need for stimulation and the questioning of conventional rules that are very much a part of the normative processes of adolescent development.

However, while experimenting and risk-taking behaviors help adolescents to mature (in the process, they eventually leave behind their narcissistic feelings of power and invulnerability), some researchers argue that pregnancy during adolescence can slow this process and can compromise the adolescent's development in general as well as that of her child (Sahler & McAnarney, 1981). Elster and McAnarney (1980) suggest that the cognitive, emotional, and psychosocial development of the adolescent in addition to the child's physical and emotional well-being may be jeopardized by pregnancy during the adolescent years.

Low Self-Concept

It has been suggested that adolescents who have lower self-concept are more prone to getting pregnant (Vernon, Green, & Frothingham, 1983). However, very few empirical studies examine the relationship between self-concept and early sexual activity and childbearing, and their findings are contradictory. Some studies suggest that pregnant teenagers have lower self-esteem and more feelings of incompetence, low motivation, poor future orientation, and inability to cope when compared to nonpregnant teenagers (Brown, 1977; Paik, 1992). Similarly, Zongker (1977) reported that pregnant adolescents scored lower on several dimensions of self-esteem, such as "self-identity" and "family and social relationships." Yet while these studies indicate that some adolescents who become pregnant have low self-concepts, there is no empirical evidence to support the hypothesis that adolescents with low self-concepts may be more prone to getting pregnant (Paik, 1992; Vernon et al., 1983) since we do not know if the low self-concept is an antecedent or a consequence of pregnancy. In fact, in one study by Robbins, Kaplan, and Martin (1985), the findings suggest that adolescents with low self-esteem and high feelings of powerlessness were at *less* risk for nonmarital pregnancy.

Moreover, recent studies are suggesting that there are differences across ethnic / racial groups in the level of self- concept among pregnant teenagers. Paik (1992) found that, among pregnant teenagers, African American girls had a higher self-concept in all areas—identity; behavior; and moral, family, and social dimensions—than did the Anglo girls in the sample. This finding can either be interpreted to mean that the self-concept of African American adolescents affects less or is less affected by their pregnancy compared to pregnant Anglo teenagers or that teen pregnancy is more accepted in the African American community than it is in the Anglo community, thereby allowing the adolescent to maintain a higher level of self-concept. Thus, when examining the role of self-concept in teenage pregnancy, it is important not to make universal attributions that low self-concept characterizes all pregnancy-prone adolescents. Low self-concept may be an important factor for pregnancy in some individuals, but not others.

Peer Influences and Role Models

Peer influences and role models may also contribute to the adolescent's motivation to get pregnant. Researchers have indicated that sexual behavior and pregnancy among adolescents may be a result of the perceived social norms of that age group (Hardy & Zabin, 1991). Reiss (1967) states that during the years of early sexual contact, young people relate very strongly to their peer group and, thus, might behave more similarly to their perceptions of their peers' norms rather than their own personal standard of behavior. Even if they do not want to be sexually active nor want to have a child, they still end up pregnant because they are following the standards of the peer group.

In addition to peer group influences, studies are increasingly suggesting that the family context greatly influence the sexual attitudes and behaviors of adolescents (Hogan & Kitagawa, 1985; McLanahan & Bumpass, 1988; Zelnick, Kanter, & Ford, 1981). For example, Marini (1984) suggests that for some African-American teenagers having a baby is more the customary behavior in that it conforms to the behaviors of role models in their community (i.e., mothers or other signinicant female figures). Therefore, teenage pregnancy in this context is not only acceptable but is expected and the supportive behavior of significant others such as the adolescent's mother reinforces the pattern as a customary rather than deviant behavior. In a study of 30 African American teenage mothers between the ages of 15 and 18, Williams (1991) found that the presence of consistent adult role models (i.e., mothers, cousins, aunts, grandmothers) who themselves were unwed adolescent mothers, was very important for study subjects, as it reinforced the message that adolescent pregnancy was an accepted life course. What this information ultimately suggests is that there are patterns within the family as well as the immediate community that influence the behavior of a teenager with regards to sexual activity and pregnancy and that this behavior may be considered either normative or deviant in different contexts and cultures.

Contraception Knowledge and Use

Finally, the lack of contraception knowledge and use and limited access to sex education and contraceptive methods also plays a part in adolescent pregnancy. Statistics show that while more teenagers are using contraception, the majority are using a less effective method or no method at all (Torres & Singh, 1986). In addition, there are clear ethnic differences in the rate of contraception use. According to COSSMHO (1987), only 23% of sexually active Hispanic adolescents were using contraception at first intercourse compared to 36% of African American adolescents and 55% of Anglo adolescents.

In comparison to other industrialized countries, the United States has a higher rate of pregnancies (but not a higher rate of sexual activity) perhaps due to the fact that it also has less available contraception for teenagers. According to Jones and her colleagues (1986), only the United States has an increasing rate of teenage pregnancies when compared to industrialized countries. The U.S. teenage abortion rate alone is at least as high as the combined teenage abortion and birthrates of the other five countries in the world (Canada, England and Wales, France, Netherlands, and Sweden) that are comparable to the United States in industrial development, stage of economic growth, and the essential cultural tradition of northwestern Europe (Jones, Forrest, Goldman, Henshaw, Lincoln, Rosoff, Westoff, & Wulf, 1986). Only Chile, Cuba, and Puerto Rico and certain Eastern European countries (Hungary and Bulgaria) had teenage birth rates that exceeded that of the

United States (Jones, et al., 1986). Since all industrialized countries have comparable rates of teenage sexual activity, the differences in access to and use of contraception services might implicated in the high rate of teenage pregnancies in the United States.

Review

In short, there are a variety of reasons why teenagers get pregnant. However, most of the documented reasons in the literature suggest that teenage mothers are having babies at this age for reasons other than those of reaching maturity, getting married, and the natural consequence of having children. The reasons for which adolescents get pregnant, though, as varied as they are, are in part a result of normative processes of the adolescent period, and therefore, teenage pregnancy might represent in these cases an inappropriate response to age-appropiate developmental needs.

Consequences of Adolescent Pregnancy

The second trend in the literature on adolescent pregnancy emphasizes the negative consequences of adolescent pregnancy and childbearing primarily on the biological, social, and psychological outcomes for both mother and child. The following sections outline the risks associated with early pregnancy and childrearing and review the findings of specific studies.

Obstetric Risks

It has been suggested that "very young women . . . are biologically too immature for effective childbearing" (Lowe, 1979). Support for this view is derived from the various population statistics and empirical studies demonstrating that obstetric risks, such as pregnancy-induced hypertension, toxemia, anemia and malnutrition, cephalopelvic disproportion, abruptia placentae, contracted pelvis, glucose intolerance, vaginal bleeding, premature labor, and perinatal mortality are commonly linked to teenage pregnancies (Battaglia, Frazier, & Hellegers, 1963; Braen & Forbush, 1975; Dott & Fort, 1976; Koetsawang, 1990; McAnarney, 1975; Osofsky & Osofsky, 1971; Stevens-Simon, Roghmann & McAnarney, 1991; Stevens-Simon & White, 1991). Furthermore, research suggests that these risks are more prevalent among younger adolescents (15 years of age or younger) than older adolescents (McAnarney, 1987), but because many of the studies group pregnant adolescents of all ages together, this finding is not very widely known.

Other studies have indicated, however, that if given adequate prenatal care, the supposed physiological immaturity of adolescent mothers is not manifested (Mednick, Baker, & Sutton-Smith, 1979). Slap and Schwartz (1989) found that the number of prenatal visits was the most important factor associated with the delivery of a low birth weight infant by an adolescent mother; the greater the number of prenatal visits the higher the birth weight of the baby. Thus, rather than biological, the risks associated with teenage pregnancy (with the exception of those 15 years old and younger), are in great part due to lifestyle (e.g., poor nutrition) and lack of preventative care.

Social Risks

Social factors have been increasingly identified as sources of risk for adolescent mothers. Reduced educational and occupational attainments, lower income, increased public dependency, higher marital separation, divorce, and remarriage have been identified as factors contributing to the negative outcomes for adolescent mothers and their children (Klerman & Karas, 1982). Longitudinal studies report that educational deficits are the major obstacles for the adolescent mother who attempts to become an economically self-sufficient adult (Furstenberg, Brooks-Gunn, & Morgan, 1987). Furthermore, research suggests that when the adolescent mother is unable to attain her educational goals and fails to become economically stable, her support networks (i.e., her partner or her family) are negatively affected. Studies report that life stressors such as the ones mentioned above reduce the number of social supports and resources available to the adolescent mother, potentially contributing to less well-developed coping skills and strategies (Gotlib & Asarnow, 1979), low self-esteem, and depression (Brown & Harris, 1978; Colletta, 1983; Vernon et al., 1983).

In fact, several researchers are suggesting that pregnancy during adolescence is more of a social disadvantage than a biological one (Morris, 1981). In a study examining the effects of perceived social support of adolescent mothers on infant outcome, Turner, Grindstaff, and Phillips (1990) reported that family support was by far the most powerful predictor of infant birth weight. Furthermore, family support for adolescent mothers from lower socioeconomic backgrounds seemed to have direct and independent effects from variations in life stress, whereas family support appeared to buffer the impact of stressful experiences in higher SES adolescent mothers. Similar findings were reported by Gale and colleagues (1989), who found that pregnant teenagers living under stable socioeconomic conditions, receiving economic and social support from their families and the community, had infants with higher birth weights and better overall fetal outcome than teenagers who did not have social, cultural, and economic supports. In other words, social support from both the mother's partner and her family as well as cultural (community) support are important not only for improved infant outcome but for the mother's outcome as well.

Psychological Risks

Negative psychological consequences, specifically psychopathology and particularly depression, have also been associated with adolescent childbearing. While the onset of adolescence in and of itself can bring upon depression, research indicates that depression rates increase for women who have children compared to those who do not (Zuckerman, Amaro, & Beardslee, 1987). Furthermore, the younger the age of the mother at the time of her first birth, the more likely she is to become depressed (Colletta, 1983). Zuckerman and colleagues state that becoming an adolescent parent may interrupt the completion of developmental tasks, consequently predisposing the adolescent mother to be more vulnerable to depression. Other studies have reported that there is an increased incidence of suicides among pregnant adolescents (Gabrielson, Klerman, Currie, Tyler, & Jekel, 1970), suggesting that because depression in adolescents goes for the most part unrecognized, its occurrence and consequences are not examined as closely as in other populations (Inamdar, Siomopoulos, Osborn, & Bianchi, 1979).

However, while it has been suggested that adolescent mothers are at higher risk for depression (Colletta, 1983), not all research suggests that pregnant adolescents are more prone to psychopathology than other adolescents. In a longitudinal study of life events and depressive symptoms in adolescents, Ge and associates (1994) found that increasing age in adolescents, particularly in girls, appeared to predispose them to depressive symptoms and eventually depressive moods. In addition, they found gender differences that suggested that girls appear to be more reactive in the face of environmental adversities and stressors when compared to boys the same age. Furthermore, they reported that depressive moods in adolescents were moderated by parental warmth and support. In other words, the mental health status of adolescents, especially girls, can be altered by the quantity and quality of social support. What is important about this research is that it suggests that *all* adolescents are susceptible to depressive symptoms by the mere fact of their developmental stage. Whether or not depression is a catalyst for teenage pregnancy or a consequence of it, depends on the perception of pregnancy in adolescence as a stressful event.

Caretaking Risks

Research also indicates that adolescent mothers might not display adequate parenting behaviors with their children. Adolescent mothers have been found to be more impatient and insensitive to their babies and tend to have more punitive attitudes toward their infants (Jones, Green, & Krauss, 1980; Osofsky & Osofsky, 1970; Ragozin, Basham, Crnic, Greenberg, & Robinson, 1982). They also display less realistic developmental expectations and child rearing attitudes (Field, 1980), as well as less verbal and more physical interactive styles of child rearing when compared to older mothers (McAnarney, Lawrence, & Aten, 1979; Osofsky &

Osofsky, 1970; Sandler, Vietze, & O'Connor, 1981). This literature suggest that, because of their own developmental needs, adolescent mothers are less able to respond adequately to their children's needs.

However, recent studies have documented wide variations among adolescent mothers. For example, Flanagan and colleagues (1994) found that there was great variability in mothering styles of adolescent mothers as measured by the Parent Infant Interaction Scale. Similarly, McAnarney and associates (1986) suggested that the younger the adolescent mother, the less accepting, less accessible, and less flexible her mothering behaviors and attitudes were likely to be. Specifically, they found that younger adolescent mothers were less likely to reinforce verbal communication in their infants compared to older adolescent mothers (McAnarney et al., 1986). Levine, García Coll, and Oh (1985) also found individual differences during face-to-face interactions among adolescent mothers to be a function of their own ego development.

In addition, other studies have reported wide variations within adolescent mother's mothering styles as a function of culture. Field, Widmayer, Stoller, and de Cubas (1986) found that Cuban adolescent mothers were more indulgent and provided more social stimulation for their infants, whereas the African American teenage mothers in the sample were more restrictive and less stimulating. Similarly in a study with a sample of low SES African American, Dominican, and Puerto Rican teen and adult mothers, Wasserman and colleauges (1990) found that Dominican teen mothers scored higher than Puerto Rican teen mothers on the dimensions of aggravation and strictness regarding discipline. In short, individual differences between adolescent mothers do have an effect on their caretaking abilities and behaviors.

Infant Health and Developmental Status

The extant research has stated that not only are there negative consequences for the adolescent mother, but they are also there for her child. Most reports suggest that infants born to adolescent mothers are at risk for poor health, developmental, and behavioral outcome. Low birth weight (Gold, 1969; Hayes, 1987); prematurity (Dott & Fort, 1976; Ryan & Schneider, 1978); neurological deficits (Dott & Fort, 1976; Ryan & Schneider, 1978); and fetal, neonatal, and infant mortality (Battaglia, Frazier, & Hellengers, 1963) are more prevalent in children born to adolescent mothers. Furthermore, studies suggest that when compared to children of older mothers, children born to adolescent mothers score lower on the Stanford-Binet IQ Test and in reading grade level and are also more dependent (Oppel & Royston, 1971; Broman, 1981; Moore, 1986). Other research reports that these children are also more prone to behavioral problems as well as are more likely to repeat a grade than children born to adult mothers (Furstenberg, Brooks-Gunn, & Morgan, 1987).

Review

The conceptualization that childbearing and parenting during adolescence reflect a maladaptive response from the individual is common to all of these explanatory reasons and documented consequences. However, there are wide individual differences in both antecedents and consequences and several areas of research point out contextual rather than individual, intrinsic influences as major moderators of outcome. Moreover, most of the studies indicating that adolescent pregnancy presents increased risks for both mother and child have not examined other factors, such as health habits, prenatal care, family and social support, and background variables such as race, ethnicity, and socioeconomic status, which might explain a significant percent of the variance in the outcome for both mother and child. In fact, some recent studies indicate that the above-mentioned variables are far better predictors of children's status than is maternal age per se (Baldwin & Cain, 1980; Darabi, Graham, Namerow, Philliber, & Varga, 1984; García Coll, Vohr, Hoffman, & Oh, 1986). Stoiber and Houghton (1993) state that the deficit paradigm utilized in many studies involving adolescent mothers and their children has precluded the examination of other competing factors, such as race, class, and ethnicity, which may be contributing to within-group differences as well as between-group differences. Therefore, examining mother and infant outcomes based solely on the age of the mother is not enough to determine the competency level of the adolescent mother. Rather, it is crucial that we understand the individual as well as the contextual factors that moderate the antecedents and consequences of childbearing during the teenage years.

While it is clear that the characterization of adolescent childbearing and parenting as a maladaptive process is accurate of a large percentage of adolescents that seek or need our public health, mental health, and social services, it is also evident that this conceptualization does not apply to all adolescents in our communities nor does it capture the phenomena of childbearing and parenting during adolescence both historically and cross-culturally. In the sections that follow, we will delineate some historical data regarding adolescent sexuality, marriage, and childbearing, as well as present the case of teenager mothers within traditional, low SES Puerto Rican culture and the consequences of pregnancy among them, to elucidate further the importance of context in our definitions of competence.

Teenage Pregnancy as an Adaptive Life Strategy: Historical and Cross-Cultural Perspectives

While current research in the United States for the most part concludes that adolescent pregnancy, childbearing, and parenting are high-risk conditions for both parent and child, historical and cross-cultural analyses suggest that adolescent sexual activity and pregnancy in the past, as well as presently, have been considered normative, expected, and almost universal occurrences (Konner & Shostak, 1986). Researchers have speculated that the onset of menarche in colonial America was around age 15 or 16 (Thompson, 1986). If the period of subfecundity lasted around two or three years (Talwar, 1965; Trussel & Steckel, 1978), adolescent girls in colonial America would be having their first child at around age 17 or 18. Demo-graphic studies indeed suggest that during the colonial era, girls in New England married as early as 15 or 16 (Demos, 1970; Greven, 1970; Lockridge, 1970). Therefore, it would follow that they would be having their first child in their late teens and that this occurrence was considered part of the transition from adoles-cence to adulthood.

Contrary to expectations, however, not all adolescents that became pregnant during colonial times were married. Smith and Hindus (1975) report that during the 18th and 19th centuries, premarital pregnancies were higher among adolescents than women in their 20s. For example, between 1721 and 1800, premarital pregnancies in girls between the ages of 15 and 19, was 40.9% compared to women aged 20–24, 23.8% and to women aged 25–29, 21.6% (Smith & Hindus, 1975). While initially condemned, premarital pregnancies were later tolerated as long as the community did not perceive the child as an economic burden (Vinovskis, 1988). In sum, adolescent pregnancy and marriage were not considered abnormal or maladaptive processes during colonial times in America and there was actually a high percentage of premarital pregnancies. In fact, early marriage and childbearing were considered favorably because they signified relative independence and implied financial stabil-ity (Stone, 1977).

Likewise, if we consider current worldwide statistics, adolescent childbearing and parenting is still more the norm than the exception. According to Cuadro de Pueblos (1980), there are large numbers of 15–19 year olds around the world who are married, with some of the highest percentages in South America (to name a few, Columbia, 25%; and Guatemala, 28%), Asia (Afghanistan, 49%; Bangladesh, 72%; and India, 56%), and Africa (Guinea, 82%; Niger, 81%; Mali, 79%; and Nigeria, 72%). Trent (1990) states that the younger the adolescent at her marriage, the higher the fertility rate. Early marriage in these countries represents sanctioned sexual activity and results in higher rates of early childbearing and overall fertility: Both consequences are usually perceived as adaptive responses, given high infant mortality rates and shorter life expectancy. While early childbearing and higher fertility rates might be perceived in these countries as adaptive responses, we need to consider that, given the growth of the human population as a whole and the consequent depletion of natural resources, these behaviors might not be at the present time as adaptive as they once were.

Illegitimacy has been both historically and cross-culturally more of an anomaly than births to adolescents. Cultural and ethnic norms play a large role in this phenomenon. For example, early marriages among Hispanics can account for the high rate of adolescent pregnancy among Hispanics in this country. It has been documented that in the United States, Hispanics between the ages of 15 and 19 are far more likely to have been married than their African American or Anglo counterparts (COSSMHO, 1987). Therefore, it seems that adolescent pregnancy, childbearing, and parenting, both historically and cross-culturally, are more the normative responses than the exceptions, as long as they are within a marriage.

Puerto Rican Adolescents: A Case Study

Our work in Puerto Rico provides further support to the notion that within social, cultural, and economic circumstances where these life events are considered normative, the antecedents and consequences of adolescent childbearing and parenting are quite adaptive. Although teenage pregnancy and childbearing might seem to have been the norm in the past among Puerto Ricans (i.e., two generations ago) and not today, the statistics reveal quite a different picture. Among Puerto Ricans living in the United States, 22% of the total births are to mothers under 20 years of age, as compared to 10% for Anglo-American women (Ventura, 1994). Among Puerto Ricans living on the island, the percentage of live births by adolescents aged 15–19 in 1986 was 23.6%, second to women aged 20–24 with a rate of 39.1% (Boletín Estadístico, 1988).

Furthermore, statistics show that in 1990, 42% of Puerto Rican teenage mothers in the continental United States in 1990 were 17 years old or younger compared to 31% of non-Hispanic white teenage mothers (Ventura, 1994). These figures indicate that, whether on the U.S. mainland or on the island, adolescent pregnancy among Puerto Ricans continues at considerably higher rates, and especially at younger ages, than those observed in other communities. Thus, even within the U.S. society in which contraception is relatively more widely advocated among teenagers, Puerto Rican women still procreate at early ages. Our work in Puerto Rico suggests that this trend is supported not only by the high social value placed on motherhood and by the lack of alternatives provided or perceived by young Puerto Rican women, but also by the general attitudes of Puerto Rican society toward women's sexual behavior and contraception (García Coll, 1989).

In a series of studies with Puerto Rican adolescent and adult mothers living in Puerto Rico, we have found that within a traditional, low SES population, the antecedents and consequences of pregnancy, childbearing, and parenting are not as negative as those observed in other populations of teenage mothers. For example, in an early study with adolescent and older mothers in Florida and Puerto Rico, we found the expected increase in obstetric risk associated with teenage pregnancy in the Florida sample but not in the sample from Puerto Rico (Lester, García Coll, & Seposki, 1982). The total sample consisted of 303 infants: 148 born in Gainesville, Florida (38 born to mothers less than 18 years of age), and 155 born in San Juan, Puerto Rico (42 born to mothers younger than 18 years of age). All infants were born to low SES mothers in both places, as indicated by their participation in state-supported programs providing prenatal and postnatal medical care. Immediately after birth, the medical records of the mothers and infants were coded with an obsteric complications score based on the system developed by Prechtl (1968). Of all four groups (adolescent and adult mothers in Florida and Puerto Rico), teenage mothers in Florida had the highest maternal and fetal nonoptimal conditions (Lester, García Coll, & Sepkoski, 1982). No differences were found in obstetric and perinatal outcome between teenage and older mothers in Puerto Rico.

In addition, we have not found negative consequences of early childbearing on the infants' development by one year of age in the Puerto Rican sample. Twenty-eight infants born to adolescent mothers and 27 infants born to older mothers were administered the Bayley Scales of Infant Development by an examiner unaware of the infant's or mother's characteristics. After the examination, the examiner filled out the Infant Behavioral Record from the Bayley Scales (Bayley, 1969) and administered the Carey Infant Temperament Questionnaire (1970) to the mother. It is important to note that 26 mothers in each group were married, again a very different context than in the United States where most adolescent mothers are single. Mental and motor developmental status as well as temperamental ratings by both examiners and mother revealed no significant differences between infants born to adolescent mothers and those born to older mothers (García Coll, Sepkoski, & Lester, 1982). These findings were in spite of the fact that families of adolescent mothers were more often from lower socioeconomic backgrounds than the families of older mothers. These findings are also in contrast with our own findings from a longitudinal study with Caucasian adolescent and adult mothers in Rhode Island, where by eight months of age, infants of adolescent mothers were already showing lower developmental scores (García Coll, Vohr, Hoffman, & Oh, 1986).

Likewise, in contrast with the experience of Caucasian adolescent mothers in Rhode Island (and other samples studied in the extant literature), adolescent mothers in Puerto Rico apparently do not experience more life stress during pregnancy and actually report more child-care support than older mothers (García Coll, 1989; García Coll, Escobar, Cebollero, & Valcarcel, 1989). In the first study, thirty Puerto Rican mothers delivering their babies in San Juan, Puerto Rico (fifteen between 13 and 17 years of age), and thirty Caucasian, Anglo mothers delivering their babies in Providence, Rhode Island (thirteen between the ages of 14 and 17 years), were interviewed shortly after the birth of their babies. All mothers were primiparous from low socioeconomic backgrounds at both sites. We asked all mothers about their life stress during the last year using a modified version of Cochrane and Robertson's (1973) instrument and about their expected help with child care, using an adaptation of Crockenberg's questionnaire (Crockenberg, 1981, 1987). Adolescent mothers from Providence reported the highest frequency of life events within the last year. In contrast, adolescent mothers in Puerto Rico did not experience more life stress than older mothers, but they also reported the highest frequency of help from their child care support network. These findings suggest that adolescent mothers in Puerto Rico not only experienced less stress during pregnancy, but they also expected more help with their baby on a daily basis, increasing their perception of a constant supportive network.

Similar findings were obtained in a recent study of adolescent and older mothers in San Juan, Puerto Rico (García Coll et al., 1989). The sample consisted of three groups: young adolescents (n=20) with a mean age of 16 years (range 14–17), older adolescents (n=20) with a mean age of 18.5 years (range 18,19620), and adult women (n=20) with a mean age of 25 years (range 21–31). The groups were similar in marital status (most of them were married), maternal occupation (most of them were homemakers), income (most of them were from low SES backgrounds), percentage of planned pregnancies (most pregnancies were planned), and previous caretaking experience (most mothers had some caretaking experience). As in our previous study, no differences were found between the groups in life stress or the expected frequency of child-care support among the groups. These findings suggest that teenage mothers in Puerto Rico experience a relatively positive sociocultural context toward teenage childbearing. This interpretation is supported by other findings from the same study, such as teenage mothers reporting a higher frequency of adolescent parenting among relatives, friends, or neighbors or all three, as well as no group differences in how relatives, friends, and so forth, reacted to the news of the pregnancy (most reacted positively).

Similarly, a number of studies with Puerto Ricans conducted in the United States have shown that the concepts of familism, intense family networks, and support systems aid the adolescent in her childrearing (Ortíz & Váquez Nuttall, 1987). In this context, adolescent parenting is seen as a normal process when it is within a marriage and when the right support mechanisms are in place. The extended family network among Puerto Ricans not only serves to support the adolescent mother in her caretaking and childrearing responsibilities, but can also help diminish other risks (e.g., educational and economic) associated with teenage childbearing.

This supportive environment is further demonstrated by the attitudes expressed by both adolescent and adult mothers toward early childbearing. Puerto Rican mothers regardless of their age report more positive beliefs around teenage child-bearing than do non-Puerto Ricans (see García Coll, 1989, previously mentioned). For example, both adolescent and adult mothers in Puerto Rico (in comparison to adolescent and older mothers in Rhode Island) reported that (1) the best age for a woman to have her child is earlier in her life, (2) a teenage mother would naturally become a good, responsible mother on her own, and, (3) a child of a teenage mother would not have any problems if the mother and child lived with either her family or husband or both (García Coll, 1989).

Finally, the psychological consequences of adolescent childbearing seem to also be different among this population. In addition to the life stress and social support scales, the three groups of mothers described earlier in the García et al. (1989) study (young adolescents, older adolescents, and adult mothers), filled out the Beck's Depression Inventory (Beck, Ward, Mendelson, Mock, & Erbaugh, 1961) and the Maternal Self-Esteem Inventory (Shea & Tronick, 1982). No group differences were found in maternal self-esteem: All of the groups' mean scores were reflective of relatively high maternal self-esteem. Higher depressive symptomatology was related to marriage after pregnancy for younger adolescents and unplanned pregnancies for older adolescents and to higher levels of stress for both groups of adolescents. None of these associations were observed for adult mothers (García Coll et al., 1989). Thus, the higher incidence of depression observed among U.S. pregnant teen populations (i.e., Beardslee, Zuckerman, Amaro & McAllister, 1988; Brown, Adams, & Kellam, 1981; Coletta, 1983) is not necessarily a universal phenomenon and seems to be associated with the circumstances surrounding pregnancy.

In short, teenage pregnancy and childbearing within traditional Puerto Rican culture present a very different array of antecedents and consequences than those found in the U.S. populations. In this context (i.e., traditional Puerto Rican, culture, low SES) the adolescent is probably married, has probably planned the pregnancy, and is surrounded by other women who are or were teenage mothers themselves, as well as an extended family who supports the young couple in their parenting tasks.

A resolution: Adolescent Mothers in Context

Keeping the historical and cross-cultural perspectives as well as the current theories and research findings in mind, the question then becomes, How can we consider adolescent childbearing as an expression of competence if most of the psychological and sociological literature suggest that a large percentage of teenagers in the world who are parents are incompetent? In order for us to parsimonously explain childbearing and parenting during adolescence as the possible expressions of both or either maladaptive or adaptive responses of the individual, we need to place the analysis within two theoretical frameworks, those of organization and life span. Within an organizational approach (Ciccetti & Schneider-Rosen, 1986; Sroufe, 1979; Werner, 1948), individual development proceeds as a series of qualitative reorganizations across behavioral systems that result from increasing sophistication of individual developmental abilities and a hierarchical integration of these skills. Competence is a central theme within organizational models (Cicchetti, Toth, & Bush, 1988), as it results from successful resolution of the salient tasks for any given developmental period. Normal development, then, is not defined in relation to the mean. Rather, it is defined more as a series of interlocking social, emotional, and cognitive competencies that facilitate an individual's broad adaptation to his or her environment (Cicchetti & Schneider-Rosen, 1986). Such a framework allows movement away from the universal "incompetence model" of adolescent pregnancy, childbearing, and parenting because competence is defined as adaptation to specific environmental circumstances. Therefore, competence in teenage mothers may be judged in relation to the immediate context or the demands of culturally and historically defined environmental circumstances rather than norm-driven indices of competence derived primarily from the experience of white, middle-class Anglo adolescents.

Life-span conceptualizations of development (i.e., Lerner, 1989) offer additional means of adapting present mainstream theoretical models to make them relevant to developmental processes in adolescent mothers. In essence, life-span approaches detail a developmental-contextual view of the bases of constancy and change over time. The adolescent's individual biology and the contexts in which she interacts are mutually influential such that developmental change represents the product of the interaction of varying biological, psychological, and social factors. Life-span models also suggest the critical importance of cohort effects in understanding developmental influences at any one point in time, as changes in society across the years may serve as important markers for specific developmental attributes that may be salient at one time but not another. Following this type of analysis, adolescent pregnancy might be adaptive for some cohorts and not others, given some life-span and contextual considerations.

Following these theoretical principles, what we would like to propose is that adolescent pregnancy, childbearing, and parenting can be considered either maladaptive or adaptive or it can even be considered both. Adolescent pregnancy is maladaptive if its primary antecedents are a history of physical, emotional, or sexual abuse; low self-concept; cognitive limitations; peer pressure; or the product of high-risk behaviors, or of all of these factors. It can also be maladaptive if for the most part the consequences are negative for either mother, father, or child, that is, if it results in any or all consequences of parental depression, isolation, low educational and economical attainment, poor parenting, or developmental problems for the child.

Alternatively, adolescent pregnancy can be considered adaptive if it is in late adolescence within a supportive context that conceptualizes it or accepts it as part of the normal developmental tasks of adolescents. It also can be considered adaptive if the mother is supported in child care and childrearing by her partner, her parents, and other family members and if she can continue her education or acquire viable economic means of support. Increasing and strengthening her sense of self through social and economic supports can aid the adolescent mother with her increased parental and adult responsibilities. In this case, contextual influences can provide special supports for mother, father, and child to increase the likelihood of adaptations in other areas of development so that the adolescent parents can be successful in completing other developmental tasks. For example, certain family contexts can provide support in parenting tasks; alternatively, existing school-based programs can provide the services necessary for the adolescent mother to have a healthy pregnancy and delivery and provide child care and contraceptive services enabling her to continue her education and master other appropriate developmental tasks. Having a child during the adolescent period, given certain contextual conditions, can contribute to the mother's own growth just as it can in later stages of development.

A Caveat: Adolescent Mothers Caught Between Worlds

Before concluding, one caveat needs to be made about the information presented in this paper. While we suggest that adolescent childbearing and child rearing in Puerto Rican adolescents can be an adaptive strategy, it is important to note that this is not the case for all. Factors such as migration, acculturation, and poverty can serve to debilitate the support systems and coping strategies of adolescent mothers. For example, the process of migration can provoke disruptions not only in the family unit but also in other social support systems, leading to generational conflicts, role confusions, and feelings of not belonging to one place or another (Zavala-Martínez, 1994). Furthermore, the process of migration is usually accompanied and exacerbated by factors such as discrimination, racism, and poverty, which contribute to the debilitatation of the traditional support systems of Puerto Ricans. This suggests that, while traditional Puerto Rican culture accommodates adolescent pregnancy and offers support to both mother and child, there are external stressors such as migration and poverty that may erode the effective coping strategies and support systems available to pregnant adolescents, thereby rendering these adolescents susceptible to negative consequences.

Minority adolescents can be caught between different worlds that can have diametrically opposed definitions of competence in the areas of sexuality, pregnancy, and childbearing. Boykin and Toms (1985) provide us with a framework to understand the unique developmental tasks faced by minority families, adolescents, and children. For example, African-American individuals have an African-American culture whose orientations are antithetical to mainstream cultural understandings. We can expand this to say that, in general, minority adolescents are caught between worlds that—and we would like to emphasize that this is the case in teenage pregnancy and childbearing—sometimes give them opposite definitions of competence: the dominant American culture that believes in delaying pregnancy until marriage and completion of educational goals, traditional third world values that celebrate children and equate motherhood with womanhood, and minority culture in the United States for which educational and economic attainments do not seem to be viable goals given the pervasive experience of racism, discrimination, and distrust in and within the system.

Implications for Social Policy and Clinical Interventions

Very clear implications can be derived of this contextual, organismic analysis of teenage mothers. First of all, it is clear that teenage pregnancy and childbearing is a complex problem of modern societies that needs a multifactorial analysis. The multifactorial analysis needs to take into consideration historical and culturally relevant definitions of human competence as well as be especially sensitive to members of minority groups that are receiving mixed messages about competence during adolescence.

Our analysis suggests that there are multiple pathways to teenage pregnancy and childbearing and their consequences are also multiple. The direct implications of this analysis are that there is no one program, service delivery mode, or prevention strategy that will work for all adolescent mothers; alternatively, some of these programs and strategies might work for some subgroups of adolescents but not others. The multiple pathways that lead to teenage pregnancy and the multiplicity of outcomes given the various contextual conditions suggest that social policies, as well as prevention and clinical programs, need to be responsive to these complexities. Simple answers to complex problems usually bring about either temporary solutions or no solutions at all. It might be that our failure as a society to decrease the number of teenage pregnancies and teenage mothers on public assistance reflects the use of unidimentional approaches to a very complex problem.

At the policy level, in the United States sexuality needs to be recognized as an integral part of adolescent development. Jones and colleagues (1986) suggest that the role of the U.S. government with regards to adolescent pregnancy has to be one that not only recognizes the needs of adolescents with regards to contraception (i.e., accessibility, availability, and cost), but also makes a commitment to increase public understanding, perception, and acceptability of adolescent sexual activity. The belief that sex education and the availability of contraception and abortions only encourage teenage sexual activity is unfounded. In fact, statistics from other developed countries suggest the contrary; the greater availability of contraceptives, sex education, and abortion services to adolescents, the lower the teenage pregnancy rates (Jones et al., 1986).

Teenagers in the United States are as sexually active as those in other industrialized countries (e.g., Sweden, Netherlands, Canada, England and Wales, and France), and yet the United States has a much higher teenage pregnancy rate. Unlike the "just say no" position of the United States, not one of the industrialized countries previously mentioned discourages teenage sexual activity (Jones et al., 1986). Instead, these countries adopt all or a combination of the following strategies to reduce adolescent pregnancy: (1) active contraception campaigns directed at both boys and girls; (2) free or low-cost abortions; (3) comprehensive sex and contraceptive education in schools; and (4) greater health care and welfare services (Jones et al., 1986). These alternative approaches do not condemn the adolescent for having sex and do not preach to the adolescent about controlling sexual urges. Rather they provide supports in the form of education and contraception that allow adolescents to make their own informed decisions regarding their sexual lives while providing support and guidance relating to growth and development. The attitude that Sweden has taken regarding adolescent sexual activity is reflected in the following quote: "Responsible premarital sex is acceptable; sex is not to be regarded as an isolated event" (Jones et al., 1986).

Public policy needs to address the issue of teenage pregnancy in the context of increasing interest in cross-gender relationships as a normative developmental process rather than an isolated event. Moreover, until policy offers and provides real access to equally attractive alternatives (i.e., careers, moving out of poverty) that emphasize the benefits of postponing childbearing until later in life, the "problem" of teenage childbearing will not go away.

At the individual clinical level, there has to be an in-depth assessment of the context in which the adolescent operates that will serve to guide the specific treatment modality and content. This assessment needs to examine several variables, including: (1) the acceptance of teenage pregnancy and childbearing in general, and in this particular instance, by the mother herself, by her partner, and by both families; (2) the prevalence and acceptance of adolescent pregnancy and childbearing in the immediate and the larger societal context; (3) the people who are the role models of the immediate and larger community; (4) other extrinsic influences as immediate contextual forces like educational and occupational opportunities and other alternative pathways to success in adulthood; and (5) other viable alternatives that may potentially exist for these young women and men. Only when these issues are addressed and studied can we take into consideration intrinsic factors like cognitive, social, and emotional competencies and needs that contribute to the individual's choices and actions.

It is evident that sexuality and reproductive urges are an important part of adolescent development. In the past, societies incorporated these physiological changes and provided the social structures to accommodate these needs through early marriage and childbearing. While some contemporary societies maintain these events (i.e., adolescent pregnancy and childbearing) in high regard, in our present society these events are not conceptualized as an expression of competence and therefore present a very high cost to society and to the individuals involved. As we have curtailed fertility in older women through educational and occupational opportunities, we need to develop *real* incentives for young women to postpone the onset of their childbearing careers. Perhaps an appreciation of the complexity of the phenomenon is the first step in the right direction.

REFERENCES

Baldwin, W., & Cain, V. S. (1980). The Children of Teenage Parents. *Family Planning Perspectives*, 12, 34–43.

Battaglia, F. C., Frazier, T. M., & Hellegers, A. E. (1963). Obstetric and pediatric complications of juvenile pregnancy. *Pediatrics*, 32, 902–908.Bayley, N.(1969). *Manual for the Bayley Scales of Infant Development*. New York: Psychological Corporation.

Beck, A. T., Ward, C. H., Mendelson, M., Mock, J., & Erbaugh, H. (1961). An inventory for measuring depression. *Archives of General Psychiatry, 4*, 561–571.

Bertanlanffy, L. von. (1968). *Organismic psychology and systems theory*. Barre, MA: Clark University Press.

Boletin Estadistico (1988). Nacimientos por sector de servicios Puerto Rico, 1986. Departamento de Salud Y A. F. A. S. S. Secretaria auxiliar de Administracion, Division estadisticas e informes. Año 3, serie D-1, num. 1, 1 de junio de 1988.

Boykin, A. W., & Toms, F. (1985). Black child socialization: A conceptual framework. In H. McAdoo & J. McAdoo (Eds.), Black children: Social, educational and parental environments. Newbury Park, CA: Sage.

Braen, N. N., & Forbush, J. B. (1975). School-age parenthood - A national overview. Journal of School Health, 45, 257–271.

Broman, S. H. (1981). Long-term development of children born to teenagers. In K. Scott, T. Field, & E. G. Robertson (Eds.), Teenage parents and their offspring. New York: Grune & Stratton.

Brown, G. W., & Harris, T. (1978). Social origins of depression. London: Tavistock.

Brown, N. W. (1977). Personality characteristics of black adolescents. Adolescence, 12, 45.

Buchholz, E. S., & Gol, B. (1986). More than playing house: A developmental perspective on the strengths in teenage motherhood. American Orthopsychiatric Association, 347–359.

Carey, W. B. (1970). A simplified method for measuring infant temperament. Pediatrics, 77, 188–194.

Cichetti, D., & Schneider-Rosen, K. (1986). An organizational approach to childhood depression. In M. Rutter, C. E. Izard, & P. B. Read (Eds.), Depression in young people: Developmental and clinical perspectives (pp. 71–134). New York: Guilford Press.

Cicchetti, D., Toth, S. L., & Bush, M. (1988). Developmental psycholpathology and incompetence in childhood: Suggestions for intervention. In B. Lahey & A. Kazdin (Eds.), Advances in Clinical Child Psychology (pp. 1–71). New York: Plenum Press.

Cochrane, R., & Robertson, A. (1973). The life events inventory: A measure of the relative severity of psycho-social stressors. Journal of Psychosomatic Research, 17, 135–139.

Colletta, N. D. (1983). At risk for depression: A study of young mothers. Journal of Genetic Psychology, 142, 301–310.

COSSMHO (1987). Adolescents, pregnancy, and Hispanic communities: Statistical review. The National Coalition of Hispanic Health and Human Services Organizations: Washington, DC.

Crockenberg, S. (1981). Infant irritability, mother responsiveness, and social support influences on the security of infant-mother attachment. Child Development, 52, 857–865.

Crockenberg, S. (1987). Support for adolescent mohter during the post-natal period: Theory and research. In Z. Boukydis (Ed.), Research on support for parents and infants in the postnatal period. Norwood, NJ: Ablex.

Cuadro de Pueblos (1980). La mujer en sociedad. England: Stephen Austin.

Darabi, K. F., Graham, E. H., Namerow, P. B., Philliber, S. G., & Varga, P. (1984). The effect of maternal age on the well- being of children. Journal of Marriage and the Family, Nov., 933–936.

Demos, J. (1970). A Little Commonwealth: Family Life in Plymouth Colony. New York: Oxford University Press.

Dott, A. B., & Fort, A. T. (1976). Medical and social factors affecting early teenage pregnancy. Anerican Journal of Obstetric Gynecology, 125, 532–536.

Elkind, D. (1967). Egocentrism in adolescence. Child Development, 38(4), 1025–1034.

Elkind, D. (1974). Children and adolescents: Interpretive essays on Jean Piaget. London: Oxford University Press.

Elster, A. B., & McAnarney, E. R. (1980). Medical and psychosocial risks of pregnancy and childbearing during adolescence. Pediatric Annals, 9(3), 89–94.

Field, T. M. (1980). Interactions of preterm and term infants with their lower and middle-class teenage and adult mothers. In T. Field, S. Goldberg, D. Stern, & A. Sustek (Eds.), *High-risk Infants and Children* (pp. 113–132). New York: Academic Press.

Field, T. M., Widmayer, S., Stoller, S., & de Cubas, M. (1986). School-age parenthood in different ethnic groups and family constellation: Effect on infant development. In J. B. Lancaster & B. A. Hamburg (Eds.), *School-age pregnancy and parenthood* (pp. 263–272). New York: Aldine de Gruyter.

Flanagan, P. J., Coppa, D. F., Riggs, S. G., & Alario, A. J. (1994). Communication behaviors of ingants of teen mothers. *Journal of Adolescent Health, 15,* 169–175.

Franklin, D. L. (1988). Race, class, and adolescent pregnancy: An ecological analysis. *American Journal of Orthopsychiatry, 58*(3), 339–354.

Furstenberg, F. F., Brooks-Gunn, J., & Morgan, S. P. (1987). Adolescent mothers and their children in later life. *Family Planning Perspective, 19,* 142–151.

Gabrielson, I. W., Klerman, L. V., Currie, J. B., Tyler, N. C., & Jekel, J. F. (1970). Suicide attempts in a population pregnant as teenagers. *American Journal of Public Health, 60*(12), 2289–2301.

Gale, R., Seidman, D. S., Dollberg, S., Armon, Y., & Stevenson, D. K. (1989). Is teenage pregnancy a neonatal risk factor? *Journal of Adolescent Health Care, 10,* 404–408.

García Coll, C. T. (1989). The consequences of teenage childbearing in traditional Puerto Rican culture. In K. Nugent, B. M. Lester, & T. B. Brazelton (Eds.), *Cultural context of infancy* (pp. 111–132). New Jersey: Ablex.

García Coll, C. T., Escobar, M., Cebollero, P., & Valcárcel, M. (1989). Adolescent pregnancy and childbearing; Psychosocial consequences during the postpartum period. In C. T. García Coll & M. L. Mattei (Eds.), *The Psychosocial Development of Puerto Rican Women,* New York: Praeger.

García Coll, C. T., Vohr, B. R., Hoffman, J., & Oh, W. (1986). Maternal and environmental factors affecting developmental outcome of infants of adolescent mothers. *Developmental and Behavioral Pediatrics, 7*(4), 230–236.

Ge, X., Lorenz, F. O., Conger, R. D., Elder, G. H., & Simons, R. L. (1994). Trajectories of stressful life events and depressive symptoms during adolescence. *Developmental Psychology, 30*(4), 467–483.

Gold, E. M. (1969). Interconception nutrition. *Journal of American Diet Association, 55,* 28.

Gotlib, I. H., & Asarnow, R. F. (1979). Interpersonal and impersonal problem solving skills in mildly and clinically depressed university studyents. *Journal of Consulting and Clinical Psychology, 47,* 86–95.

Greven, P. J., Jr. (1970). *Four Generations: Population, Land, and Family in Colonial Andover, Massachusetts, Ithaca.* New York: Cornell University Press.

Hardy, J. B., & Zabin, L. S. (1991). *Adolescent pregnancy in an urban environment.* Washington, DC: Urban Institute Press.

Hayes, C. D. (Ed.) (1987). *Risking the future: Adolescent sexuality, pregnancy and childbearing, Vol. I.* Washington, DC: National Academy Press

Hogan, D. P., & Kitagawa, E. M. (1985). The impact of social status, family structure, and neighborhood on the fertility of black adolescents. *American Journal of Sociology, 90*(4), 825–855.

Inamdar, S. C., Siomopoulos, G., Osborn, M., & Bianchi, E. C. (1979). Phenomenology associated with depressed moods in adolescents. *Amercan Journal of Psychiatry, 136*(2), 156–159.

Jones, E. F., Forrest, J. D., Goldman, N., Henshaw, S., Lincoln, R., Rosoff, J. I., Westoff, C. F., & Wulf, D. (1986). *Teenage pregnancy in industrialized countries.* New Haven: Yale University Press.

Jones, F. A., Green, V., & Krauss, D. R. (1980). Maternal responsiveness of primiparous mothers during the postpartum period: Age differences. *Pediatrics, 65,* 579–584.

Klerman, L. V., & Karas, J. A. (1982, September 25). Pregnancy and parenting among Hispanic adolescents: Health and social issues. Talk prepared for delivery at session "Critical Issues in Hispanic Adolescent Para-Natal Care," part of The conference on Critical Health Issues Facing Mainland Puero Ricans, sponsored by the Boston Area Health Education Center, Boston.

Koetsawang, S. (1990). Adolescent reproductive health. *Health Care in Women and Children in Developing Countries* (pp. 491–503). Oakland, CA: Third Party Publishing.

Konner, M., & Shostak, M. (1986). Adolescent pregnancy and childbearing: An anthropological perspective. In J. B. Lancaster & B. A. Hamburg (Eds.), *School-age pregnancy and parenthood* (pp. 325–345). New York: Aldine de Gruyter.

Ladner, J. (1971). *Tomorrow's tomorrow: The black women.* Garden City, NY: Doubleday.

Lerner, R. M. (1989). Individual development and the family system: A life-span perspective. In K. Kreppner & R. M. Lerner (Eds.), *Family systems and life-span development,* (pp. 15–31). Hillsdale, NJ: Erlbaum.

Lester, B. M., García Coll, C. T., & Sepkoski, C. (1982). Teenage pregnancy and neonatal behavior: Effects in Puerto Rico and Florida. *Journal of Youth and Adolescence, 2*(5), 385–401.

Levine, L., García Coll, C. T., & Oh, W. (1985). Determinants of mother-infant interaction in adolescent mothers. *Pediatrics, 75,* 1–23.

Lockridge, K. A. (1974). *Literacy in Colonial New England: An Enquiry into the Social Context of Literacy in the Early Modern West.* New York: Norton.

Lowe, C. J. *Fertility and contraception in America. Adolescent and pre-adolescent pregnancy hearings before the Select Committee on Population,* 95th Cong., 2d Sess., 570 (1979).

Marini, M. M. (1984). Age and sequencing norms in the transition to adulthood. *Social Forces, 63,* 1.

McAnarney, E. (1975). Adolescent pregnancy. *Clinical Pediatrics, 14,* 19–22.

McAnarney, E. R. (1987). Young maternal age and adverse neonatal outcome. *American Journal of Disabled Children, 141,* 1053–1059.

McAnarney, E. R., Lawrence, R. A., & Aten, M. J. (1979). Premature parenthood: A preliminary report of adolescent mother-infant interaction. *Pediatric Research, 13,* 328.

McAnarney, E. R., Lawrence, R. A., Ricciuti, H. N., Polley, J., & Szilagyi, M. (1986). Interactions of adolescent mothers and their 1-year-old children. *Pediatrics, 78*(4), 585–589.

McLanahan, S., & Bumpass, L. (1988). Intergenerational consequences of family disruption. *American Journal of Sociology, 94*(1), 130–152.

Mednick, B. R., Baker, R. L., & Sutton-Smith, B. (1979). Teenage pregnancy and perinatal morality. *Journal of Youth Adolescence, 8*(3), 343–357.

Mercer, R. T. (1983). Assessing and counseling teenage mothers during the perinatal period. *Nursing Clinics of North America, 18,* 293–301.

Moore, K. A. (1986). Children of teen parents: Heterogeneity of outcomes. Final report to the National Institute of Child Health and Human Development. Washington, DC: Urban Institute.

Morris, N. M. (1981). The biological advantages and disadvantages of teenage pregnancy. *American Journal of Public Health, 14,* 412–427.

Musick, J. S. (1993). *Young, Poor, and Pregnant.* New Haven: Yale University Press.

Nord, C. W., More, K. A., Morrison, D. R., Brown, B., Myers, D. E. (1992). Consequences of teen-age parenting. *Journal of School Health, 62*(7), 310–318.

Norris, A. E. (1988). Cognitive analysis of contraceptive behavior. *IMAGE: Journal of Nursing Scholarship, 20*, 135– 140.

Oppel, W., & Royston, A. (1971). Teenage births: Some social, psychological, and physical sequelae. *American Journal of Public Health, 61*(4), 751–756.

Ortíz, C. G., & Vazquez Nutall, E. (1987). Adolescent pregnancy: Effects of family support, education, and religion on the decision to carry or to terminate among Puerto Rican teenagers. *Adolescence, 22*, 897–917.

Osofsky, J. J., & Osofsky, J. D. (1971). Adolescents as mothers: Results of a program for low-income pregnant teenagers with some emphasis upon infants' development. In S. Chess & A. Thomas (Eds.), *Annual progress in child psychiatry and child development.* New York: Brunner-Mazel.

Paik, S. J. (1992). Self concept of pregnant teenagers. *Journal of Health & Social Policy, 3*(3), 93–111.

Piaget, J. (1952). *The origins of intelligence in children.* New York: Norton.

Prechtl, H. F. R. (1968). Neurological findings in newborn infants after pre-and paranatal complications. In J. Jonxis, H. Visser, & J. Troelstra (Eds.), *Aspects of prematurity and dysmaturity.* Leyden: Stenfert Kroese.

Ragozin, A. S., Basham, R. B., Crnic, K. A., Greenberg, M. T., & Robinson, N. M. (1982). Effects of maternal age on parenting role. *Developmental Psychology, 18*, 627–634.

Redl, F. (1972). Pre-Adolescents - What makes them tick? In D. Hamachek (Ed.), *Human Dynamics in Psychology and Education: Selected Readings (2nd ed., pp. 353–366).* Boston: Allyn & Bacon.

Reiss, I. (1976). *The social context of sexual permissiveness.* New York: Holt, Rinehart & Winston.

Robbins, C., Kaplan, H. B., & Martin, S. S. (1985). Antecedents of pregnancy among unmarried adolescents. *Journal of Marriage and the Family, 47*(3), 567–583.

Ryan, G. M., & Schneider, J. M. (1978). Teenage obstetric complications. *Clinical Obstetric Gynecology, 21.*(4), 1191– 1197.

Sadler, L. S., & Catrone, C. (1983). The adolescent parent: A dual developmental crisis. *Journal of Adolescent Health Care, 4*, 100–105.

Sahler, O. H. Z., & McAnarney, E. R. (1981). *The child from three to eighteen.* St. Louis, MO: CV Mosby Co.

Sandler, H., Vietze, P., & O'Connor, S. (1981). Obstetric and neonatal outcomes following intervention with pregnant teenagers. In K. Scotgt, T. M. Field, & E. Robertson (Eds.), *Teenage parents and their offspring* (pp. 249–254). New York: Grune & Stratton.

Shea, E., & Tronick, E. Z. (1982). Maternal self-esteem as affected by infant health and family support. Paper presented at the American Psychological Association, Washington, DC.

Slap, G. B., & Schwartz, J. S. (1989). Risk factors for low birth weight to adolescent mothers. *Journal of Adolescent Health Care, 10*, 267–274.

Smith, D. S., & Hindus, M. S. (1975). Premarital pregnancy in America, 1640–1971: An overview and interpretation, *Journal of Interdisciplinary History, 5*, 537–570.

Sommer, K. Whitman, T. L., Borkowski, J. G., Schellenbach, C., Maxwell, S., & Keogh, D. (1993). Cognitive readiness and adolescent parenting. *Developmental Psychology, 29*(2), 389–3 98.

Sroufe, L. (1979). The coherence of individual development: Early care, attachment and subsequent developmental issues. *American Psychologist, 34*(10), 834–841.

Stevens-Simon, C., & White, M. M. (1991). Adolescent pregnancy. *Pediatric Annals, 20*(6), 322–331.

Stevens-Simon, C., Roghmann, K. H., & McAnarney, E. R. (1991). Early vaginal bleeding, late prenatal care, and misdating in adolescent pregnancy. *Pediatrics, 87,* 838–840.

Stoiber, K. C., & Houghton, T. G. (1993). The relationship of adolescent mothers' expectations, knowledge, and beliefs to their young children's coping behavior. *Infant Mental Health Journal, 14*(1), 61–79.

Stone, L. (1977). *The Family, Sex, and Marriage in England, 1500–1800.* New York: Harper & Row.

Strauss, S. S., & Clarke, B. A. (1992). Decision-making patterns in adolescent mothers. *IMAGE: Journal of Nursing Scholarship, 24*(1), 69–74.

Talwar, P. P. (1965). Adolescenct sterility in an Indian population. *Human Biology, 37,* 256–261.

Thompson, R. (1986). *Sex in Middlesex: Popular mores in a Massachusetts County, 1649–1699.* Amherst, MA: University of Massachusetts Press.

Torres, A., & Singh, S. (1986). Contraceptive practice among Hispanic Adolescents. *Family Planning Perspectives, 18*(4), 193–194.

Trent, K. (1990). Teenage childnearing: Structural determinants in developing countries. *Journal of Biosocial Science, 22,* 281–292.

Trussell, J., & Steckel, R. (1978). The age of slaves at menarche and their first birth. *Journal of Interdisciplinary History, 8,* 477–505.

Turner, R. J., Grindstaff, C. F., & Phillips, N. (1990). Social support and outcome in teenage pregnancy. *Journal of Health and Social Behavior, 31,* 43–57.

Ventura, S. J. (1994). Demographic and health characteristics of Puerto Rican mothers and their babies, 1990. In G. Lamberty & C. T. García Coll (Eds.), *Puerto Rican women and children: Issues in health, growth and development*(pp. 71–84). New York: Plenum Press.

Vernon, M. E. L., Green, J. A., & Frothingham, T. E. (1983). Teenage pregnancy: A prospective study of self-esteem and other sociodemographic factors. *Pediatrics, 72*(5), 632–635.

Vinovskis, M. A. (1988). *An "epidemic" of adolescent pregnancy?* New York: Oxford University Press.

Wasserman, G. A., Rauh, V. A., Brunelli, S. A., García-Castro, M., & Necos, B. (1990). Psychosocial attributes and life experiences of disadvantaged minority mothers: Age and ethnic variations. *Child Development, 61,* 566–580.

Werner, H. (1948). *The comparative psychology of mental development.* New York: Harper & Row.

White, H. R., & Johnson, V. (1988). Risk taking as a predictor of adolescent sexual activity and use of contraception. *Journal of Adolescent Research, 3*(3–4), 317–331.

Williams, C. W. (1991). *Black teenage mothers: Pregnancy and childrearing from their perspective.* Lexington, MA:

Young, L. (1954). *Out of wedlock.* New York: McGraw-Hill.

Zavala-Martinez, I. (1994). Entremundos: The psychological dialectics of Puerto Rican migration and its implications for health. In G. Lamberty & C. T. García Coll (Eds.), *Puerto Rican women and children: Issues in health, growth and development* (pp. 29–38). New York: Plenum Press.

Zelnik, M., Kanter, J. F., & Ford, K. (1981). *Sex and Pregnancy in Adolescence.* Beverly Hills, CA: Sage.

Zongker, C. E. (1977). The self-concept of pregnant adolescent girls. *Adolescence, 12,* 48.

Zuckerman, B. S., Amaro, H., & Beardslee, W. (1987). Mental health of adolescent mothers: The implications of depression and drug use. *Developmental and Behavioral Pediatrics, 8*(2), 111–116.

Zuckerman, B. S., Walker, D. K., Frank, D. A., Chase, C., & Hamburg, B. (1984). Adolescent pregnancy: Biobehavioral determinants of outcome. *The Journal of Pediatrics, 105*(6), 857–863.

Zuckerman, M., Bone, R. N., Neary, R., Mangelsdorff, D., & Brustman, B. (1972). What is the sensation seeker? Personality trait and experience correlates of the sensation-seeking scales. *Journal of Consulting and Clinical Psychology, 32,* 420–426.

XI Competence and Coping in Adolescence: Developmental Transitions and Diversity

Daniel P. Keating & Darla J. MacLean

A key tenet of the emerging field of developmental psychopathology is that mutual benefits accrue to the conjoint study of normalcy and deviance from a developmental perspective. Not only does an understanding of normal development enlighten us with respect to the origins and pathways of any type of deviance, but an understanding of non-normative developmental pathways also affords new opportunities to discover otherwise hard-to-discern developmental processes (Cicchetti, 1993; Sroufe & Rutter, 1984). The work reported in this and other Rochester Symposia provides ample evidence for this claim. The value of contrasting normal with deviant developmental pathways has been shown in a better grasp of fundamental developmental processes, of critical and sensitive periods in development, of the essentially transactional nature of development, and of the multiple and mutually causal interactions among biological, psychological, and social factors in development.

There has also been a recognition within the perspective of developmental psychopathology that this contrast between normal and abnormal development has parallels in the more general study of comparative development, such as that between cultures or between different groups within a society (Cicchetti, 1993). Similar opportunities and similar methodological cautions attend both types of comparative research strategies.

Developmental Diversity and Developmental Psychopathology

In focusing on these richly rewarding strategies of group comparisons, we may tend to adopt implicitly a view that there is some essential, unifying construct of

The first author is Royal Bank Fellow and Director, Human Development Program, Canadian Institute for Advanced Research, to whom he expresses appreciation for their generous support. Comments or questions should be addressed to the first author at the following address: Centre for Applied Cognitive Science, Ontario Institute for Studies in Education, 252 Bloor Street West, Toronto, Ontario M5S 1V6; Telephone / Voicemail: 416 923-6641 (x2482); FAX: 416 926-4708; Internet: DKEATING@OISE.ON.CA.

Support for research reported here was provided to both authors from the Social Science and Humanities Research Council and the Natural Sciences and Engineering Research Council (Canada); Northern Telecom; and the National Institute of Health (US).

normal, healthy, or functional development against which these various comparisons are made. Although this assumption is not in keeping with the broader organismic and organizational perspective from which developmental psychopathology arises, it is an easy assumption to make when our principal attention is focused on particular issues in pathology or on particular aspects of group differences.

A potentially useful antidote to this misleading assumption is a better grasp of the sources and pathways of diversity within "normal" development. There are two complementary ways to regard this approach. First, the study of the origins and pathways to diversity within normative populations can serve as an additional arena of comparison to non-normative groups. Are the processes and factors that yield diversity within normal, healthy development similar to or different from the processes and factors associated with developmental diversity that we identify as deviant? In other words, two potentially rich sources for the study of developmental diversity—between groups and within groups—afford comparisons between patterns of differentiation.

The second way to approach this issue is to seek an integration of the between-group and within-group comparative strategies. If we take the larger issue to be an understanding of the nature of human developmental diversity, then the study of normative diversity and of developmental psychopathology reflect complementary approaches to the same question. In many ways, we have a relatively more impoverished developmental understanding of the nature of normative diversity than of psychopathology or deviance. There are a variety of historical reasons why this is so.

Theoretically, our understanding of normative diversity is embedded within a tradition of individual differences and trait models, which tended toward a fixed rather than a developmental and transactional perspective (Keating, in press). Indeed, the very term "individual differences" tends to bias the search toward internal and inherent characteristics of the person. In contrast, the notion of developmental diversity remains open to multiple sources of influences and their interactions and also focuses the question of diversity toward a search for *how* these multiple influences operate across time to yield phenotypically observed population diversity (Keating, 1996b).

Opening the question in this way does generate a substantial increase in complexity, and this complexity poses daunting methodological challenges:

> Both [correlational and experimental] methods favor main effects over interactions, whereas the development of all aspects of life . . . involves an interaction of heredity with environment. But dissecting that interaction will require a level of detail and precision not now available. (Green, 1992, p. 331)

Dealing with the methodological challenges to our understanding of diversity is thus a key to going beyond the traditional constraints, which emphasized main effects, simple interactions, and exclusively linear models. Much of what we were forced to discard as "noise" in past approaches become accessible through analyses of complex patterns that reflect coherent developmental dynamics. A brief overview illuminates the central methodological concerns and gives us some basis for cautious optimism (Keating, 1996a).

Meeting the Methodological Challenge

One problem in pursuing a comprehensive research strategy to understand diversity is its seemingly overwhelming complexity. The number of potentially important factors and their interactions expands exponentially as we take into account higher order interactions. If we examine the multiplicative interactions of those factors across time, the picture becomes even more complex. We quickly approach a virtually infinite set of possible causal arrangements, even when we begin with a small number of relevant factors. On the basis of statistical fit, at least some of these models will be indistinguishable (Glymour, Scheines, Spirtes, & Kelly, 1987).

On the other hand, we have recognized for some time the value of combining analytic methods to strengthen our inferential capabilities. Cronbach (1957) suggested that the complementary strengths and weaknesses of experimental and correlational approaches could effectively address otherwise intractable problems in understanding human functioning. Experimental approaches are high in internal validity but often fail to generalize beyond the experimental setting, especially when contextual factors are strongly influential. Correlational approaches can be high in external validity, demonstrating what aspects of human functioning are robustly associated with other aspects, but are weak in their ability to specify the sources of those strong associations.

Methodological developments that have contributed to our ability to make sound inferences continue to rewrite history. The development and sophistication of multivariate analytic techniques has been a major step forward. Campbell and Stanley's (1963) famous treatment of experimental and quasi-experimental designs expanded dramatically the ways that rigorous research on real phenomena in their own context could be done. Cook and Campbell (1979) elaborated on this work, focusing on construct validity as a key to identifying the underlying sources of observed covariance. Cole and Means (1981) identified critical research methods for comparative studies, especially when differences between populations are the focus.

Some have argued that these methods have made issues unnecessarily complex. Indeed, the belief that phenomena shown in the laboratory can easily be generalized to the highly contextualized world outside the laboratory is a common simplifying assumption, but also a dangerous one. Similarly, beliefs that covariance analyses reveal underlying structures are usually based on a preexisting assumption about the nature of those structures. But the complexity that these simplifying assumptions avoid almost surely resides in the real phenomena, not in the techniques used to study them. The task of assembling a sensible and coherent story of human functioning, which does not distort the real complexity, remains the central challenge. The integration of different methodologies seems an important step in the right direction.

How might we go about this integration? Contemporary models of developmental psychopathology point us in the right direction, in two important respects (Cicchetti, 1993; Sroufe & Rutter, 1984). The first is that, in addition to experimental and correlational approaches, we need to take seriously the developmental, or more broadly, historical methods that have been of great value in other fields that confront contextual complexity. The second is that the discipline of trying to solve actual "applied developmental" problems focuses attention on the interrelatedness of development, rather than on pet theories or magic bullets (Keating & Rosen, 1990; Morrison, Lord, & Keating, 1984–1989).

Development as History

Darwin's work on evolution provides an interesting case study of this kind of integration, on an equally complex problem (Gould, 1986). A key part of this approach is the recognition that historical observation can be as robust a method as experimental intervention. Indeed, Gould argues cogently that Darwin's breakthrough was as much methodological as theoretical and that the two were inseparable in establishing evolutionary biology.

How does an historical approach help to resolve complexity? At the simplest level, it provides the conceptual framework, narrative in form, within which the seemingly unrelated complexities form a coherent picture. Further, this coherent picture then becomes a fruitful mechanism for suggesting and contrasting testable hypotheses about the interrelations of many discrete elements.

There are, of course, many "histories" that are relevant to the understanding of human development: the phylogenetic, species-given substrate; the particular genetic composition of the individual; the ontogenetic history of that individual; and the sociohistoric context within which that ontogenesis transpires. Integrating human development so as to encompass all these histories—evolutionary, cultural, ontogenetic—requires that many specific methods be used. Each of these methods needs to be critically examined for its ability to permit robust inferences, but we should not assume that we can define robustness with paradigmatic criteria (Keating, 1996a).

Several areas of methodological progress hold particular promise for contemporary developmentalists. Causal pathway modeling of longitudinal data of many different kinds, based on new techniques arising within the general linear model, is rapidly becoming quite accessible even to nonspecialists (Collins & Horn, 1991; Glymour et al., 1987; von Eye, 1990). Given the expanding opportunities for the secondary analyses of longitudinal data archives (Brooks-Gunn & Chase-Lansdale, 1991), it is possible to imagine multiple waves of analyses conducted through a variety of different lenses, to yield a coherent picture with increasing resolution.

We would also benefit from a better integration of what we can call indicator versus process analyses. The former can offer a clear picture of the distribution of various outcomes in the population, and the life courses and consequences associated with those distributions. Recent advances in identifying the substantial social-class gradients in health outcomes is a good example of this approach (*Daedalus*, Special Issue, 1994). The latter, which incorporates information on developmental processes, offers the opportunity to interpret those patterns taking the several levels of history into account.

Recently, nonlinear dynamic models have sought to capture the history of any system that is self-organizing, as in a recent study of the dynamics of babies learning to walk (Thelen & Ulrich, 1991). For some areas of research on human development, the database does not yet hold sufficient density of detail to permit quantitative estimation of dynamic system functions. Nevertheless, the work on dynamic systems does provide an immediately useful source of theoretical possibilities. Simulation studies can also be informative in many cases, although we must remain attuned to the critical distinction between simulation and reality. Fine-grained analyses of performance based on sound experimental work, especially combined with longitudinal, cross-sequential, and microgenetic designs that enable the observation of acquisitions across time, enable the charting of multiple pathways to the development of expertise and competence (Keating, 1990b).

Although the human development research agenda that we have described is new in some respects, it is important to recognize that the goal of integrating context with development has a distinguished history (Bronfenbrenner, 1975). We can already identify some obstacles in pursuing this agenda with rapidly advancing techniques. The first might be termed the cost of success. Assume for a moment that we could resolve the exponential complexity observed in the study of human development into a coherent developmental framework. The coherent story may then sound much more like "common sense" than "science." For example, important differences in thinking, feeling, coping, and interacting with others appear to be associated with the quality of the attachment relationship in early life. Although it may seem obvious that children do better if they have a close and supportive relationship with at least one primary caretaker, we may be tempted to inquire why we pursue social arrangements that appear to place this relationship at risk, and do relatively little to support it, if its importance is universally obvious.

A second critique is more likely to arise from the scientific community than the public at large: the perceived rigor of innovative methodologies. Historical analyses demonstrate that many scientific breakthroughs owe a substantial debt to innovations in the methods of research (Feyerabend, 1975), and that Darwin, dealing with the equally complex question of biological evolution, was a major methodological innovator (Gould, 1986). The shortest response is that the proof of any method lies in how effective it is in practice, even in such an apparently abstract domain as mathematics (Kline, 1980).

Applied Developmental Psychology

One of the key features of developmental psychopathology's approach is that it expressly avoids the tedious debate between applied and basic research. It does this because the target of study, dysfunctions in development, is already a fundamentally applied concern. In pursuit of its core question, it can appropriate and integrate all manner of research approaches whatever their core questions may be, from neuroscience and psychoneuroimmunology to sociohistorical and cohort analyses.

A similar perspective on the study of developmental diversity as a general phenomenon may be equally productive. Although it is conceivable that we may be merely curious about human diversity, the social uses of the study of diversity more often reveal a more practical or political agenda guiding such investigations. It may prove useful to identify explicitly the agenda guiding the study of developmental diversity, and thus the basis upon which an integration of developmental methods will proceed. One guiding agenda is to understand the developmental sources of human diversity so as to optimize individual and population outcomes in the context of a learning society (Keating, 1995b, 1996b).

Note that one might effectively argue for the integration of historical methodologies as *epistemologically essential* to our full understanding of human development. The value of applied developmental questions, on the other hand, is *heuristic*. Attention to applied developmental questions arises as a preference for the sharper focus that attention to real problems forces upon the investigator; as a preference for socially meaningful as opposed to insulated, purely curiosity-driven research; and as an approach that may advance the prospects of developmental integration more rapidly by requiring attention to the interrelatedness of development.

Developmental Diversity in Adolescence

Prime Time for the Study of Diversity

One of the important insights gleaned from the study of dynamic systems is that the timing of various influences on a complex system is as important as, or sometimes more important than, the magnitude of that influence. In particular, when a system is in a transitional state, the influence of even small events can be magnified many times. Biologically critical periods in early cortical development are a clear example of these transitional-period sensitivities. A corollary insight is that the variability of systems often increases during transition periods as it moves from one reasonably stable state to another reasonably stable state.

We may productively view adolescence, the unifying theme of this volume's contribution to the symposium series, as a transitional period in both senses: a time of increased variability or differentiation, and a phase that contains several critical or sensitive periods.

Increased differentiation has been found in a variety of domains, from cognition to the self system (Harter, 1990; Keating, 1990a). As the adolescent moves from the relatively more protected environments of home, family, and elementary school to the more public and challenging arenas of peer relations, secondary school, and possibly the workplace, individual patterns of performance and personal characteristics undergo more specific definition.

Within this pattern of increased differentiation, there may exist significant critical (or developmentally sensitive) periods. Identity formation (Harter, 1990) is one example, in which the competing tasks of individuation and remaining connected need to be integrated by the adolescent. Another example is the development of critical thinking, which entails not only sufficient complexity in emerging cognitive skills (Keating, 1990a) but also their integration with a sense of psychological and emotional commitment (Keating & Sasse, 1996).

The increasing diversity within normative adolescent populations affords an opportunity to investigate key influences associated with diversity. We can explore both suboptimal development (including psychopathology) along with patterns of difference in healthy functioning. Through such comparisons, we may be able to glean some general lessons not only for the understanding of the processes of diversity, but also for a better understanding of how to deal with many kinds of diversity in our social institutions and practices.

We are confronted with a daunting task: to enhance health and competence in the face of increasing diversity of populations in modern society, which is attributable to many influences, such as increasing ethnic diversity from global population flows or the greater inclusiveness of social institutions (such as schools) through mainstreaming and other means (Keating, 1995b). Understanding the fundamentals of developmental diversity appears to be an essential step in meeting that challenge.

We have been studying adolescent transitions to competence and health within normative populations in several contexts. We provide a brief overview of three investigations in progress, and then draw some general lessons from those studies. First, we examine the relationships among stress, coping, and health / psychological well-being in adolescents and young adults. In particular, we are interested in whether the well-established relationships among these variables begin to be established as early as adolescence, and whether these relationships are similar or different among various groups of adolescents when we examine the effects of gender, race, and social class. Second, we explore patterns of differentiation among a much more select group to see what processes of diversity operate within a high-functioning group. Specifically, we report an investigation of diversity of competence among an identified "gifted" group in early adolescence. Third, we report the findings of several related investigations of gender differences in mathematics and science performance in high school. In contrast to a question of whether gender differences exist (Keating, 1993), we look at the different pathways by which boys and girls pursue or drop their earlier interest and success in these areas. Included in this summary are initial results of a successful intervention to increase the proportion of talented girls who pursue the study of advanced mathematics and science.

Stress, Health, and Coping

A substantial body of literature has documented that psychosocial stress, such as stressful life events, is a major contributor to the development of psychological and physical health problems in children, adolescents, and adults (Compas, 1987a). Self-esteem, perceived control over one's life choices, and social support have been proposed as important resources for coping with stress (e.g., DuBois, Felner, Brand, Adam, & Evans, 1992; Werner & Smith, 1992; Windle, 1992). When these variables are present at sufficiently high levels, they have been reported as serving a protective function for health outcomes. Life events perceived as uncontrollable have a more adverse effect on psychological well-being than those perceived as controllable (Compas, Banez, Malcarne, & Worsham, 1991; Farne, Sebellico, Gnugnolli, & Corallo, 1992).

Self-esteem presents a problem within the current research context, in that it can be viewed as both a moderator and outcome variable (Costello, 1992). In fact, self-esteem may be related to outcome indirectly through mastery beliefs, in that one might expect individuals with high levels of self-worth to anticipate easier mastery of a challenging situation (Seligman, 1990). Compas (1987b) noted, however, that evidence on the relations among stressful life events, social support, locus of control, self-esteem, and psychological well-being is equivocal. Reasons for these inconsistent findings may include the homogeneity of subject populations—with a bias toward white, middle- to upper-class adolescents; relatively small samples; and measurement problems.

In an ongoing investigation, we have sought to assess the role of life stress, locus of control, and social support in predicting psychological well-being in a large and demographically diverse population (Keating, Menna, & Matthews, 1992; Menna & Keating, 1992, 1993; Menna, Keating, & Oatley, 1994). The bulk of stress and health research has been conducted on adult populations. The stress and health model has been more recently applied to children and adolescents (Compas, 1987a). Several of the coping variables that have been identified in the adult population as mediators of the stress and health relationship have been studied in children and middle adolescents.

In our ongoing study, life stress is conceptualized as recent life events, ongoing difficulties, and daily hassles that have occurred within the previous 12 months (Holmes & Rahe, 1967; Sarason, Johnson, & Siegel, 1978). Psychological well-being is conceptualized as the degree to which psychological and physical symptoms are reported by the participants, as the level of positive feelings reported by them, and as their reported self-esteem. We have elected to examine self-esteem as an outcome, given its critical nature in adolescent development.

Note, however, the mutual causality in a dynamic system that likely characterizes all the major variables in this investigation. Persistent high levels of life stress (as major events or daily hassles) may lead to lower levels of psychological health and well-being, culminating in perceptions of lower self-esteem, lack of control, and diminishing social connectedness. The absence of these psychosocial resources may lead to increased vulnerability to stress, creating in effect a "vicious cycle." Of course, one may imagine a contrasting "virtuous cycle" in which healthier outcomes become generative of better self-worth, greater control, and social participation, leading to greater resiliency in the face of stressful life events. Unraveling the core developmental dynamics may thus lead not only to more successfully targeted intervention or prevention but also to the support of more optimal development throughout a diverse population (Csikszentmihalyi, 1990).

We are investigating whether physical and psychological health is negatively associated with life stress, and whether the strength of this association is moderated by locus of control and social support. In particular, we are interested in whether these relationships emerge as early as adolescence and young adulthood, and whether they appear to be similar or different across gender, ethnicity, social class, and age groupings. A key characteristic of the present study is thus its large and demographically diverse sample across dimensions of age, race, sex, and socioeconomic status (SES).

This sample included 383 adolescents and young adults between the age of 15 and 27 years from the Baltimore metropolitan area. Individuals were distributed by age into four groupings: 15–16 years (n=52), 17–18 years (n=133), 20–22 years (n=96), and 25–27 years (n=102). The mean chronological age for each group was 15.3 years (SD=.36), 17.9 years (SD=.53), 21.1 years (SD=2.14), and 25.9 years (SD=1.27). The sampling procedure was designed to generate substantial numbers within the categories of sex (males, n=209; females, n=184), socioeconomic status (middle class, n=260; working class, n=133), and race (African American, n=225; white, n=168).

Volunteer participants were recruited from a city high school, a local university, and community settings (e.g., YMCA, social service agencies). In the high school, research assistants visited homeroom classes, explained the project, and handed out information packets. At the university, solicitation was through advertisements posted around the campus, announcements in classes, and advertisements placed in the school newspaper. In the community, solicitation was through advertisements posted in public settings (e.g., community centers, the YMCA, shopping centers) and placed in the city newspaper. Generating a diverse sample, especially of adolescents not in school, is a major cost. Our relative lack of knowledge of such underrepresented samples is due in part to these costs.

Participants completed a battery of self-report questionnaires (30 minutes on average). Testing sessions took place either at the high school or at a room in the university that had been set up as a comfortable conversation area. The informed consent of each participant (and parental consent for those under 18 years of age) was obtained. Participants were given a small financial compensation ($10) for their cooperation.

Gender, race, and socioeconomic status information was obtained at the initial contact. Gender and race (African American, white) were coded as dichotomous variables. Socioeconomic status was calculated using Nock and Rossi's (1979) index, based on a weighted combination of parent's educational level and occupational status. This was later dichotomized at a break point between working class (skilled labor / clerical) versus higher SES levels.

We report here on six measures from the self-report battery, administered individually to each subject: (1) Life Events Questionnaire (Sarason et al., 1978), (2) Perceived Social Support Scale (Procidano & Heller, 1983), (3) Hopkins Symptom Checklist (Derogatis, Lipman, Rickels, Uhlenroth, & Cole, 1974), (4) Rosenberg's Self-Esteem Scale (Rosenberg, 1965), (5) Pearlin Mastery Scale (Pearlin & Schooler, 1978), and (6) Bradburn's Affect Balance Scale (Bradburn, 1969).

Three of the self-report instruments administered were adapted from existing measures: the Hopkins Symptom Checklist, Perceived Social Support Scale, and the Life Experiences Survey.

Life Stress

To provide a measure of life stress, subjects completed an adaptation of the Life Events Questionnaire (Sarason et al., 1978) for use with an adolescent population. The adapted version consists of 49 specific events plus 2 blank spaces in which subjects can indicate other events that they may have experienced. For an overall life stress score, a total was computed by summing all the "yes" responses. A high score indicates the subject has experienced high life stress in the last year. A low score indicates the subject has experienced low life stress in the past year.

Three measures of physical and psychological health and well-being were administered: (1) Hopkins Symptom Checklist (Derogatis, Lipman, Rickels, Uhlenroth, & Cole, 1974), (2) Bradburn's Affect Balance Scale (Bradburn, 1969), and (3) Rosenberg's Self-Esteem Scale (Rosenberg, 1965).

Prevalence of Symptoms

The Hopkins Symptom Checklist (HSCL) is a self-report measure of a wide variety of physical and psychological symptoms used to assess maladjustment in nonpsychiatric populations. An adaptation of the measure was used to provide information concerning perceptions of personal adjustment at the time of testing. This measure consists of a 13-item list of problems and complaints. Subjects are asked to rate on a five-point scale (1=not at all, 5=extremely) the extent of discomfort each problem or complaint has caused the subject during the past week including the day of testing. A total is computed by summing the ratings on all items. The possible scores on the prevalence of symptoms and the intensity of distress range from 0 (no symptoms) to 65 (maximum prevalence). The alpha reliability for the HSCL scale (.89) was adequate. In some cases, we report results for the full checklist.

Feelings of Well-Being

The Bradburn Affect Balance Scale (1969) is a measure of a person's feelings of well-being. The measure consists of a 10-item list of statements about feelings. Five of the items are positive feeling (Positive Affect Scale [PAS]) and the other five are negative feeling statements (Negative Affect Scale [NAS]). The difference between the scores on the positive (PAS) and negative feelings (NAS) indices, which Bradburn (1969) named the Affect Balance Scale (ABS), serve as an indicator of an individual's current level of feelings of well-being. Participants indicated by circling a yes or no answer to statements about what they had experienced during the past few weeks. Items are of the following type: Bored; On top of the world; That things were going your way; Depressed or very unhappy. A total is computed by summing the scores on the two affect scales (NAS and PAS), yielding a distribution of scores running from −4 to +4. For computing purposes, a constant of +5 is added to each sum, giving a scale with values of +1 to +9. The possible scores on the feelings of well-being range from 1 (not too happy) to 9 (very happy). The alpha reliability for the Bradburn Affect Balance Scale (.60) was adequate.

Self-Esteem

Self-esteem was assessed by Rosenberg's (1965) Self-Esteem Scale. The measure consists of 10 items addressing global positive or negative attitudes toward the self. Items are of the following type: I feel that I have a number of good qualities; All in all, I am inclined to feel that I am a failure. Participants rated on a five-point scale (1=not at all accurate, 5=completely accurate) the extent to which each statement is felt or experienced. A total is computed by summing the ratings on all items. The possible scores on self-acceptance and self-worth range from 1 (low self-esteem) to 50 (high self-esteem). High self-esteem signifies that the individual respects him or herself and considers him or herself worthy. Low self-esteem reflects both lack of self-respect and feelings of inadequacy. The alpha reliability for the Rosenberg Self-Esteem Scale (.85) was adequate.

Two measures of coping resources were administered: (1) Pearlin Mastery Scale (1978), and (2) Perceived Social Support Scale (Procidano & Heller, 1983).

Pearlin Mastery Scale

The Pearlin Mastery Scale (PMS) is a measure of perceived control of life events. The instrument measures the extent to which one regards one's life chances as being under one's own control in contrast to being fatalistically ruled (Pearlin & Schooler, 1978). The measure consists of seven items addressing self-attitude. Items are of the following type: I have little control over the things that happen to me; There is really no way I can solve some of the problems I have; There is little I can do to change many of the important things in my life. Participants were asked to rate on a five-point scale (from 1=not at all accurate to 5=completely accurate) the accuracy of each statement. A total is computed by summing the ratings on all items. The possible scores on self-attitude range from 1 (no control, chance) to 35 (one's own control). High scores signify that the individual perceives him or herself in control of his or her life. The alpha reliability for the Pearlin Mastery Scale (.67) was adequate.

Social Support

To provide a measure of perceived levels of social support from family and friends, participants completed a shortened version of the Perceived Social Support

Scale (Procidano & Heller, 1983). The shortened versions were derived by principal components analysis. This measure consists of a 12-item subscale addressing
perceived social support from family members (PSS-Fa) and a 10-item subscale
addressing perceived social support from friends (PSS-Fr). Most items appear on
both subscales with identical wording, apart from changes in the referent of the
statement (e.g., My parents give me the moral support I need vs. My friends give me
the moral support I need). The original true–false question format was changed so
that subjects are asked to rate on a five-point scale (1=not at all accurate, 5=com-
pletely accurate) the extent to which each statement is felt or experienced. A total
is computed by summing the ratings on all items. The possible scores on the family
and friend support scales range from 0 (no perceived support) to 60 (maximum
perceived support). Procidano and Heller (1983) report adequate reliability and
validity for the scales. The alpha reliabilities for the parent (.92) and for friend
scales (.87) were adequate.

To further assess perceived social support, a measure of perceived social support
from a most significant person in the subject's life was used. This scale is a direct
adaptation of the friends and family support scale developed by Procidano and
Heller (1983). This measure is a 12-item scale addressing perceived social support
from a most significant person (PSS-MSP). The wording is similar to the PSS-Fa
and PSS-Fr items, apart from the changes in the referent of the statement (e.g., I
provide more support than I receive vs. I provide more support to my friends than I
receive from them vs. I provide more support to my parents than I receive). Subjects
were asked to rate on a five-point scale (1=not at all accurate, 5=completely
accurate) the extent to which each statement is felt or experienced. A total score is
computed by summing the ratings on all items. The possible score for perceived
support from a most significant person in one's life ranges from 0 (no perceived
support) to 60 (maximum perceived support).

Although the primary focus of our concern is on the relationships among stress,
coping, and health, we have explored the group differences on these variables. This
is fraught with interpretive pitfalls, however, given that there are 90 potential
effects arising from four-way analyses of variance (ANOVAs) (age x sex x race x
class) of six dependent variables. A number of expected effects do appear: for
example, working-class youth report more symptoms than middle-class youth; girls
report higher levels of social support than boys; middle-class youth report higher
levels of self-esteem than working-class youth. Other findings may conform less to
traditional expectations: African American youth, for example, report higher levels
of parental support and more positive feelings than white youth.

These last findings are good examples of the value of including as many sectors
of the youth population as possible and not generalizing until we have done so. We
sampled so as to obtain a balanced design across our major demographic factors
(although not a randomly representative sample). Thus, the presence of a substantial number of middle-class African American youth in our sample unconfounded
the too often overlooked race / class nexus and yielded patterns of findings on key
psychosocial indicators that are infrequently observed.

The focal results on the relationships among stress, health / well-being, and coping are shown in Tables 1 and 2. In both tables, the results of stepwise regression analyses for the full sample and for various subsamples are reported. The strategy was to predict the three criteria of health or well-being (reported health symptoms, positive feelings, and self-esteem) by first entering stress (using life events as the measure) and then entering coping resources (perceived control and social support) as a second step to see whether they explain additional variance in the criteria.

Although there are many specific findings of interest, a few significant general patterns emerge. For reported symptoms, stress makes a significant and substantial contribution (negatively) for the full sample and for all subgroups. Indeed, the magnitude of effects is quite similar across groups; the apparent difference between whites and African Americans (after step 1, $R^2 = .19$ vs. .12) is not statistically significant.

The effect is also quite substantial in magnitude. Recall that this finding is, in essence, the correlation between their reports of stressful life events over the past year and their reported health symptoms. In all these groups of young people, the level of stress is significantly associated with reported health. Nor is this effect merely a function of the older groups. In Table 2, we see that this effect holds even for the youngest (15-year-old) sample ($R^2 = .20$). When using the full-scale symptom checklist (not available for the youngest group), the association between stress and health is magnified. Thus, even within normative samples of adolescents and youth, there is a substantial association between stressful life events and negative health outcomes.

The addition of coping resources in step 2 of the regression analyses also yields a straightforward picture. For each coping resource, the additional variance explained is reported, as is the semipartial correlation of that variable with the criterion after the effect of stress has been removed. This column is added due to the collinearity of the two coping resource predictors, so as to avoid misinterpretation.

Again focusing on the health symptoms outcome variable, there is a strong and clear effect of perceived control, over and above the effect of stress. At first glance, it appears that social support makes no additional contribution, but this is an artifact of perceived control entering the regression first. The semipartial correlations (in italics in Tables 1 and 2) show that social support is indeed significantly related to reported health symptoms (rs from .19 to .27, all significant at $p < .05$ given the sample sizes). For health symptoms, the total variance explained is quite high overall when all three variables are entered, and this is true of the various subsamples as well.

The similarity among the various groups, demographically and by age, is somewhat more striking than the differences among them. For all groups, there is a substantial negative effect of stressful life events on self-reported health symptoms and a substantial positive contribution (after equating for stress) from perceived control and social support.

This picture contrasts sharply with what we observe for the outcome of self-esteem, in which the role of stressful life events is minimal. This suggests that self-esteem is more stable across transient stressors than are health symptoms, which

Table 1. Prediction of Health and Well-Being:
Stepwise Regression Summary (by group)

Outcome	Stress (Step 1) R^2	Perceived Control (Step 2) R^2	Partial	Social Support (Step 2) R^2	Partial	Total R^2
Health						
All	.14	.24	-.53	.00	-.23	.38
Males	.17	.25	-.55	.00	-.26	.42
Females	.12	.27	-.56	.00	-.24	.39
Middle class	.15	.32	-.57	.00	-.20	.47
Working class	.13	.14	-.39	.01	-.27	.28
White	.19	.25	-.55	.00	-.24	.44
African American	.11	.23	-.51	.00	-.19	.34
Positive Feelings						
All	.07	.16	.47	.05	.37	.29
Males	.11	.22	.49	.09	.48	.41
Females	.05	.14	.38	.04	.32	.23
Middle class	.06	.16	.41	.05	.35	.27
Working class	.12	.16	.42	.06	.41	.34
White	.05	.21	.47	.07	.42	.32
African American	.13	.09	.33	.03	.27	.25
Self-esteem						
All	.03	.27	.53	.04	.38	.34
Males	.04	.32	.58	.03	.37	.39
Females	.02	.23	.49	.07	.39	.31
Middle class	.01	.31	.56	.06	.41	.37
Working class	.07	.21	.47	.02	.31	.29
White	.04	.24	.50	.04	.38	.33
African American	.03	.30	.55	.03	.35	.36

makes theoretical sense. There are strong associations between self-esteem and both coping resources, perceived control and social support, slightly but not significantly higher with perceived control. Self-esteem and perceived control likely have substantial method covariance, but the positive associations among these variables is what we might expect.

Note, however, that the regression patterns are quite similar across groups. This does not mean that the experiences of all groups, or their levels of stress, health, and coping are identical. It does suggest that the pattern of relations among these variables appears to be stable across groups.

The regression patterns for the criterion of feelings appear the most variable across groups. This may be due to the more transient feelings assessed by this scale, which thus may respond more to daily ups and downs than do health symptoms or

Table 2. Prediction of Health and Well-Being:
Stepwise Regression Summary (by age)

Outcome	Stress (Step 1)	Perceived Control (Step 2)		Social Support (Step 2)		Total R^2
	R^2	R^2	Partial	R^2	Partial	
Health						
All	.14	.24	-.53	.00	-.23	.38
15–16	.20	.30	-.61	.08	.14	.59
17–18	.11	.19	-.46	.00	-.09	.30
[53-item:	.22	.20	-.51	.00	-.09	.43]
20–22	.17	.31	-.61	.01	-.35	.49
[53-item:	.21	.23	-.54	.01	-.34	.45]
25–27	.10	.27	-.55	.06	-.45	.43
[53-item:	.15	.33	-.62	.03	-.41	.50]
Positive Feelings						
All	.07	.16	.47	.05	.37	.29
15–16	.18	.15	.42	.00	.16	.33
17–18	.04	.13	.37	.10	.41	.27
20–22	.08	.17	.43	.06	.40	.31
25–27	.10	.20	.47	.06	.42	.35
Self-esteem						
All	.03	.27	.53	.04	.38	.34
17–18	.02	.26	.52	.03	.34	.32
20–22	.03	.24	.49	.10	.50	.37
25–27	.05	.27	.54	.05	.42	.37

Note: "53-item" refers to the full-length health symptom checklist used for these samples.

self-esteem. The effects of stress are statistically significant, given the large sample size, but they are less clear than for the other criteria. The differences across groups in the regression patterns are of only marginal significance, although the positive contributions of perceived control and social support are substantial for all groups.

The general picture that emerges from these analyses is coherent and theoretically sensible. Although there are a number of differences among groups of adolescents and youth in *levels* of reported health, well-being, coping, and stress, the *pattern* of relationships among these variables is quite stable across groups. High levels of stress in the past year are associated with higher levels of reported health symptoms; after equating for stress levels, the positive impact of perceived control and social support on health is found for all groups. For self-esteem, recent stress appears not to have much of an impact, although the positive association of self-esteem with perceived control and social support is consistently found. For current levels of positive feelings, which are likely more transient, the negative contribution of stress is more marginal than for health, but the contributions of coping resources are again substantial.

It is noteworthy that these patterns appear to be so similar across groups that are so different in many ways. It may be that the patterns tap into more fundamental developmental dynamics than do levels of the key variables. This may yield important clues for prevention and intervention, even within normative samples. In particular, the fostering of valuable coping resources, especially in population groups that may be expected to encounter substantial life stressors, may have valuable payoffs. Of course, the reduction of the life stressors for these high-risk groups through changes to the overall quality of the social environment might be expected to have an even bigger payoff in health outcomes, but these may be harder to achieve.

Patterns of High-Level Competence

One of the commonalities between the study of developmental psychopathology and the study of developmental diversity more generally is that, from both perspectives, diversity is viewed as arising out of developmental histories that represent the confluence of numerous processes. These developmental processes do not yield a single linear scale of competence, either between normalcy and deviance or within normative populations.

This multidimensional perspective on diversity has had relatively little effect on educational practice, which continues more or less within a linear, single-scale paradigm (Keating, 1995b). One of the areas in which this is most clearly evident is in the treatment of high-level competence. Most programs for gifted students, for example, use global IQ as if it were the most informative or useful criterion for determining the need for adaptive instruction (Keating, 1991).

We decided to investigate the nature of competence within a highly restricted sample of highly competent students, to explore whether there were identifiable patterns of competence within this selective group (Matthews & Keating, 1995). We used a theoretical model that we have called habits of mind (Keating, 1996b) that seeks to integrate two notions: (1) domain-general habits of learning, which have social, emotional, and attentional roots in early development and which shape the diverse ways that individuals learn and attend (Miller, 1995; Miller, Keating, MacLean, Marshment, & Keenan, 1996); and (2) domain-specific expertise, which addresses the cumulative progress in acquiring specific skills (Keating, 1990b; Keating & Crane, 1990).

The basic notion is that what we assess as intellectual or cognitive competence in fact reflects a complex, developmental pattern incorporating fundamental habits of how we think and solve problems, how we interact with others, and how we react emotionally to situations of different types. The notion of habits of mind is intended to capture this integrative perspective.

The challenge for this (as for any other) integrative model is to generate empirical evidence that captures these aspects. We addressed this challenge here in several ways. First, we selected a population that was quite homogeneous on the traditional, linear dimension of intelligence in order to show that substantial diversity can be observed within this group. We studied early adolescents who had been identified as gifted on the basis of standard aptitude and achievement indices (in the top 5% of their school board). The homogeneity of the sample tends to work against the hypothesis in its restrictiveness, but provides the opportunity to explore diversity in a sample unaffected by generally low motivation or low cooperation, which can inflate performance correlations.

Second, we selected early adolescence—ages 11–14, in grades six, seven, and eight—as the developmental period of interest because a variety of cognitive competencies are known to emerge at this time (Keating, 1980, 1990a). The increase in interindividual and intraindividual differentiation, as noted above, affords the potential for examining domain specific diversity.

Third, we included two traditional academic domains (mathematical or logical thinking, and linguistic or verbal competence) along with a nonacademic domain, social competence, assessed both cognitively and behaviorally (through peer and self-ratings).

To summarize, we expected two patterns: First, distinct configurations of competence should emerge even among this highly select group, with students achieving exceptional competence in some but not all domains. Second, there should be interpretable patterns of self-report and personality differences associated with these differences in expertise, reflecting the differing socialization experiences that may lead to different patterns of competence. If different patterns of high-level competence reflect meaningful differences in developmental pathways, these different patterns should be associated with aspects of personality functioning such as self-concept and goals.

The sample we used in this study—highly competent early adolescents—is extremely restricted in range and, thereby, affords a robust look at patterns of diversity. The sample included 130 subjects in grades six, seven, and eight, and included all students enrolled in full-time gifted classes in a large public school board in southern Ontario. The sample was of mixed socioeconomic status and race, although predominantly middle class and white. Of the 122 subjects who completed all components of the testing, there were 60 boys and 62 girls.

Gifted students (or in our preferred terminology, developmentally advanced students) are typically viewed as relatively homogeneous because we more often use age-appropriate rather than developmentally appropriate tasks to assess their competence. Due to ceiling effects, they will necessarily appear similar in the former assessments (Keating, 1991). To avoid this, we used achievement tests that were standardized on students two grades ahead of the actual grade placement and included challenging tests of critical thinking for both the logical-mathematical (Arlin, 1984) and linguistic (Astington & Olson, 1990) domains. We assessed social competence in two ways: (1) using two social cognitive tasks, a means-ends problem-solving task adapted from Platt and Spivack (1977) and a life-planning judgment task adapted from Crites (1978); and (2) through peer ratings of social competence and of peer acceptance.

We explored the resulting patterns of performance in several ways analytically, in the hope that they would yield converging evidence: theoretically convergent / discriminant correlational analyses, factor analysis, and distributional overlap.

The correlations conformed rather closely to our expectations. Most of the correlations between measures that were hypothesized to belong to different domains were not significantly different from zero. Except for the social cognitive measures, the correlations between measures considered to be in the same domain were significant. Two other patterns among the correlations were notable. First, reading performance did correlate with many of the measures, which is to be expected due to method covariance. Second, the social cognitive measures did not correlate with each other or with the social competence measures. This distinction between social cognition and social competence perhaps reflects the noninteractive nature of these social cognitive assessments. A hierarchical convergent-discriminant analysis of the confidence intervals around interdomain and intradomain correlations showed that the measures within each of the three domains correlated better with each other than with measures outside their domain. This degree of consistency is striking given the overall restricted range of this sample.

In the factor analysis, a three-factor solution was observed that corresponds to linguistic, logical-mathematical, and social domains, as shown in Table 3. The only exceptions to predicted patterns were the social cognitive measures (as noted above) and Arlin's (1984) Test of Formal Reasoning, which seemed to apportion its variance between the logical-mathematical and the linguistic factors. We interpreted the first factor as a linguistic factor, in that its two significant loadings were the two measures of the linguistic domain: Stanford Reading Achievement (.75) and Astington-Olson (.56). The second factor was interpreted as a social competence factor, with significant loadings on the peer acceptance (.44), self-perceived social acceptance (.51), and social competence ratings (.79). The third appears to be a logical-mathematical factor, loading significantly on Stanford Mathematics (.95). It appears that the social domain is best considered a coherent and separate domain of competence only when measured contextually, that is, when actual social competence—rather than the ability to answer questions on a social cognitive task—is assessed.

Another approach looked at domain specificity on an individual level: What percentage of the most competent students in one traditional academic domain (linguistic or logical-mathematical) were also among the most competent in the other domain? As competence on one of the academic measures increased, the likelihood of competence at the equivalent level on the other measure decreased, from a 78% overlap at the median to a 26% overlap at the top decile. Although statistically predictable, this pattern further supports a domain-specific model of

...elf-concept than their peers

...02). There were no other

...ly in the essentially equiva-

...rceived competence. This

...ived competence supports

...ctors beyond competence

...ent, all of whom had high

...der of their importance, 6

...ademic or career achieve-

...d a significant difference

...social achievers ranking

...competent students in-

...ifferences in patterns of

...ifferences (which were

...illustrated by the fact

...various achievement

...views competence as

...d as habits of mind

...ment and personality

...ighly restricted and

...viduals may pursue

...e over another, in

...opment. One such

...ller proportion of

...choose to pursue

...ung men.

...arch Council of

...roportion of stu-

...universities. We

...tively untapped

...study in small

...pointed to the

...young women's

...ating, 1993).

...e the effective-

...regular public

	acto	
	2	
	-0.02	
	0.07	
	0.11	0.9
	-0.10	0.29
	0.44*	-0.02
	0.51*	0.01
	0.79*	-0.02
	0.24	0.07
5	0.03	-0.22

...ighly capable in the general

...mpetence.

...competence, we investigated

...ce and students' self-reported

...f students: (1) top achievers,

...gical-mathematical, linguistic,

...tom one third in all domains

...he top 5% of their normative

...c achievers, in the top one third

...in the social domain; (4) social

...l competence domain, but below

...t self-rated social competence was

...1, so that we could examine it as a

...in was defined as the two peer-rated

...social cognitive measures.

...n Profile for Children (Harter, 1985)

...elf-perceived competence: Academic,

...lobal. A repeated measures (or profile)

...elf-perception) yielded a highly signifi-

...indicating that there were significant

...ported self-concept. There was also a

...ndicating that the groups had different

...sure.

...achievers reported higher social self-con-

...$(F = 4.68; p < .05)$ and that the academic

chievers reported higher academic
hievement patterns ($F = 6.27$; $p <$
p differences in self-concept, most nota
given by all four groups on global self-p
between objectively assessed and self-perce
n that this diversity includes psychosocial fa
ven within a group that is overall highly compet
of global self-perceived competence.

Ve also had students rank a series of 12 life goals in or
which reflected social affiliation goals and 6 reflected ac
ment goals. A 4 (Group) x 2 (Measure) ANOVA reveale
in the groups' scores ($F = 4.77$; $p < .01$), attributable to the
social goals as more important than the other three groups

These consistent patterns of diversity among highly
cluded life goals and self-concept as well as achievement; d
achievement could not be attributed to general ability d
roughly equated); and sources of self-efficacy were diverse,
that global self-concept was essentially equal across the
groups.

Overall, these findings are consistent with a model that
arising from developmental histories that may be capture
(Keating, 1996b). They suggest that the *integration* of achieve
factors yields coherent patterns of diversity even with a
high-functioning group of young adolescents.

Gender Patterns in Achievement and Aspirations

Another outcome of these patterns of diversity is that indi
developmental pathways that value one domain of competenc
ways that may be detrimental or constraining to their later devel
pattern that has been studied extensively is the relatively sma
mathematically or scientifically talented young women who
advanced study in those areas as compared to equally talented yo

Several years ago, the Social Science and Humanities Rese
Canada cited numerous surveys that pointed to a decline in the p
dents pursuing science, mathematics, and engineering in Canadian
proposed to research why we were losing one major source of rela
talent, young women, who were selecting these fields for furthe
numbers. Evidence from developmental and educational research
adolescent transition as a potentially critical period for this decline in
participation and success in these areas (Eccles & Midgley, 1989; K

As one component of this investigation, we were able to examin
ness of an ongoing intervention program that was designed within a

Table 4. Mean Perceived Self-Concept Scores by Pattern of Competence

Pattern Group	Self-Concept Domain					
	Academic	Social	Sports	Appearance	Behavior	Global
Top achievers (n = 8)	19.2	18.2	18.6	18.2	16.9	20.5
Low achievers (n = 16)	16.9	15.8	16.9	16.4	18.2	19.0
Academic achievers (n = 27)	19.5	15.4	16.1	16.6	18.9	19.4
Social achievers (n = 11)	17.7	19.2	18.2	16.5	18.5	19.0

Source: Matthews and Keating, 1995, Table 3.

high school to enhance such participation during the secondary school years. Initially, our principal focus was on collecting detailed data on young women who had participated in the special program. We collected data on these students' performance in mathematics and science while in the program (grades 9 and 10) and in later grades. We also gathered information on a range of psychological and social factors that the research literature suggested might play a critical role in achievement and persistence in these fields, using student self-reports on a survey questionnaire.

The research evidence from international studies has generally supported the effectiveness of all girls' classes in enhancing performance and persistence, but such evaluations have not addressed whether such interventions can work outside the special circumstances of all girls' private schools or special laboratory conditions. It is crucial to understand the impact of this intervention because it has taken place in the context of a regular, coeducational high school and has functioned in that setting for eight years. This afforded adequate numbers of students for reliable statistical comparisons.

The intervention itself is quite modest. Selected girls are grouped together for mathematics and science classes taught by women instructors (who are regular faculty members of the high school) in grades 9 and 10, after which they reenter the regular integrated high school courses in these subjects. During grades 9 and 10, they take all other classes with the general student population. If this modest intervention has a significant impact on these critical developmental outcomes, the prospects for enhancing participation of young women in mathematics and science become far more feasible.

In order to evaluate the effectiveness of this nearly unique intervention, it was essential to compare the performance of these girls with relevant other groups of students. Because the program used pre–high school mathematics performance as a screening mechanism for inclusion in the all girls' classes, it was necessary to obtain data on comparably talented but nonparticipating girls and boys. Because the existence of the all-girls' classes changes the composition of other grade 9 and 10 mathematics classes, it was important also to have similar comparison groups from another high school in which such a program did not exist. To observe whether the effect of the program versus mathematics ability was more important, and whether ability has a similar effect on outcomes for boys and girls, it was also necessary to secure a sample of the full population of both boys and girls.

Reviewing all these considerations, we decided that it would be impossible to anticipate all the potentially valuable comparisons in advance. We opted to shift to a full population model for the study, rather than to select particular comparison groups in advance. This also has the potential to enhance the statistical reliability of our findings because they are based on a larger sample and because no prior selection of comparison groups is imposed. We collected the detailed questionnaire data on the full population of the two selected high schools, one in which the program was operating and another matched to the target school on parental socioeconomic status and on the proportion of graduating students attending university.

We began our analyses with a look at participation in mathematics and science in the full population (grades 9–13, $N = 1369$). We found that at grade 12, girls experience a substantial drop in their perceptions of the importance of math and science for future career plans and in their enrollment in advanced math courses, compared to boys (MacLean, Keating, Miller, & Shuart, 1994). Although this is somewhat later in the high school years than some other investigators have reported, it appears to be the same developmental phenomenon. Note, too, that in Ontario most advanced-level students have traditionally gone on to grade 13 (equivalent to the first year of university), and thus the proportion of girls entering math or science courses of study in university may not be much greater than in jurisdictions where the decline shows up earlier, even given the later developmental arrival of the gender difference in participation in this sample.

We also found in the full population that the factors predicting mathematics participation were different for boys than girls. In hierarchical regression analyses, we found that boys continue taking math courses when their parents are highly educated, when they see themselves as competent in math, and when they view pursuing one's own interests as an important goal. For girls, in addition to parental education, math participation is predicted by high levels of teacher support and rating the life goal of understanding other people as less important than other girls do. This concordance of internal motivations and achievement for boys who pursue advanced study in these areas, compared with the apparent requirement of external support and gender nonconformity for girls who do so, is a pattern that has been widely discussed (Eccles & Midgely, 1989).

In contrast, girls with experience in the all-girls' program perform better than boys and than girls without such experience. In a multivariate analysis of variance of five outcome variables, the group effect was highly significant ($p < .001$). In follow-up univariate ANOVAs, we found that girls from the special program take

Table 5. Grade and Sex Differences in Four Indicators of Mathematics and Science Achievement in High School With ANOVA Effects

	Grade 9		Grade 10		Grade 11		Grade 12	
	M	F	M	F	M	F	M	F
1. Math for career?	3.7	3.7	3.6	3.7	3.5	3.4	3.6	3.0
N	181	204	200	187	145	158	155	136
2. Science for career?	3.5	3.6	3.4	3.5	3.2	3.3	3.2	2.7
N	183	203	199	186	145	159	153	137
3. Math advanced courses?	83	87	76	82	64	68	52	49
N	693*	696	507	486	306	299	158	138
4. Last math grade	3.3	3.1	2.7	3.0	2.5	2.7	2.3	2.5

ANOVA Summary Indicating p-Values

Variable	Grade	Sex	Grade by Sex
1	.001	NS	.05
2	.001	NS	.05
3*	.001	NS	.05
4	.001	NS	NS

*Ns are for total sample; ANOVA is for Grade 12 students only, using grade as a within-subject repeated measure. Source: MacLean et al., 1994, Table 1.

more math and science courses overall than either boys or girls without such experience, and also earn higher marks in advanced senior level mathematics courses and science courses than do boys, even after matching for math grades on entry to high school (MacLean, Sasse, Keating, Stewart, & Miller, 1995).

Girls who were matched on elementary school mathematics performance, but who did not participate in the special program, were less likely to initiate advanced study in mathematics and science, as defined by a criterion of at least one advanced level course in each area (67% versus 100%). About 82% of the matched boys met this criterion. Among those girls not in the special program who did begin such study, however, performance on most outcomes was generally comparable (except for total number of math and science courses taken). This suggests that the observed program effects may work as an "inoculation," helping a larger percentage of girls to make it into the pool of prospective advanced students, after which normal processes leading to further participation may become engaged. In this ongoing longitudinal study, we hope to determine whether these effects continue to hold through the university experience.

To date, we have been able to demonstrate substantial effects of participation in the all-girls' program. These effects are evident in several crucial outcome indicators following participation in the program. The magnitude of these effects are not

Table 6. Univariate ANOVAs for Five Performance Outcomes

	Boys $N = 63$	Girls From Regular Program $N = 20$	Girls From All-Girls' Program $N = 59$	p-Value	Fisher's Protected LSD at $p < .05$
Total math and science courses taken	8.84	7.65	9.85	< .001	All girls > Boys > Regular Program Girls
Senior advanced math courses taken	2.33	2.30	2.58	< .07	None None
Senior advanced math courses average marks	67.40	74.25	74.98	< .01	All Girls > Boys
Senior advanced science courses taken	2.83	3.20	3.00	NS	None
Senior advanced science courses average marks	67.20	74.80	77.61	< .01	All Girls > Boys

Source: MacLean et al., 1995, Table 1.

only statistically significant but also practically meaningful. Subsequent performance in these advanced courses is enhanced in both quantity and quality.

These results strongly support the effectiveness of this program designed to enhance the participation and performance of girls in mathematics and sciences. This is particularly meaningful for several reasons: the program operated in a regular public coeducational secondary school; in addition to grouping, the only other formal difference was the assignment of women teachers to these classes; the effects of the program reported here are longitudinal, following participation in the special program; and the outcomes are educational performance criteria that are crucial to further study in these fields. The critical transition period of adolescence appears to offer the possibility of shifting developmental pathways through a relatively modest intervention.

Dealing With Diversity

It appears from these examples that the patterns of developmental diversity within nonpathological populations are as complex as those discovered from the perspective of developmental psychopathology. Furthermore, many of the developmental processes that give rise to the diversity are similar. Efforts toward developmental integration that make coordinated use of experimental, correlational, and historical analyses appear to be fruitful in both pursuits.

Beyond the interesting intellectual analogues, what are the applied implications of this integrative perspective on developmental diversity? Consider what goals are served by a better understanding of developmental diversity. As already noted, the increasing diversity of modern populations is likely to continue or even accelerate at the same time that we are finding it difficult to meet the demands for greater inclusiveness and participation by all groups in social and economic life. Clearly, we need to learn to deal more effectively with this diversity in our social institutions and practices (Keating, 1995a; 1996b). We may state as a reasonable goal the enhancement of population health, coping, and competence, although to some that may sound utopian rather than reasonable.

Asking how much we can increase the health, well-being, and competence of children and adolescents across the full range of diversity begs a related question: How much can we change the physical and social environment that supports optimal development? It is on such policy questions that a developmental approach becomes even more crucial because the degree of leverage for change is a function of the developmental period at which the intervention (or prevention program) is aimed and a function of our ability to understand the key processes and contexts that are operating at each point in development.

The prospects for success in enhancing the developmental outcomes for diverse populations require that we apply a productive conceptual framework for developmental diversity when we consider social and institutional changes. The work we have summarized here supports the notion that an integrative perspective on diversity is fruitful for both normal and deviant populations and that adolescence is a key transition in the development of diversity that demands our close attention.

REFERENCES

Arlin, P. K. (1984). *Arlin Test of Formal Reasoning*. East Aurora, NY: Slosson Educational Publications.

Astington, J. W., & Olson, D. R. (1990). Metacognitive and metalinguistic language: Learning to talk about thought. *Applied Psychology, 39*(1), 77–87.

Bradburn, N. M. (1969). *The structure of psychological well-being*. Chicago: Aldine.

Bronfenbrenner, U. (1975). Nature with nurture: A re-interpretation of the evidence. In A. Montagu (Ed.), *Race and IQ* (pp. 114–144). New York: Oxford University Press.

Brooks-Gunn, J., & Chase-Lansdale, P. L. (Eds.). (1991). Secondary data analyses in developmental psychology [Special section]. *Developmental Psychology, 27*(6), 899–951.

Campbell, D. T., & Stanley, J. C. (1963). Experimental and quasi-experimental designs for research on teaching. In N. L. Gage (Ed.), *Handbook of research on teaching* (pp. 171–246). Chicago: Rand McNally.

Cicchetti, D. (1993). Developmental psychopathology: Reactions, reflections, projections. *Developmental Review, 13*, 471–502.

Cole, M., & Means, B. (1981). *Comparative studies of how people think.* Cambridge, MA: Harvard University Press.

Collins, L. M., & Horn, J. L. (Eds.). (1991). *Best methods for the analysis of change.* Washington, DC: American Psychological Association.

Compas, B. E. (1987a). Coping with stress during childhood and adolescence. *Psychological Bulletin, 101*, 393–403.

Compas, B. E. (1987b). Stress and life events during childhood and adolescence. *Clinical Psychological Review, 7*, 225–302.

Compas, B. E., Banez, G. A., Malcarne, V., & Worsham, N. (1991). Perceived control and coping with stress: A developmental perspective. *Journal of Social Issues, 47*, 23–34.

Cook, T. D., & Campbell, D. T. (1979). *Quasi-experimentation.* Boston: Houghton Mifflin.

Costello, C. G. (1992). Conceptual problems in current research on cognitive vulnerability to psychopathology. *Cognitive Therapy and Research, 16*, 379–390.

Crites, J. O. (1978). *CMI administration and use manual* (2nd ed.). Monterey, CA: CTB / McGraw-Hill.

Cronbach, L. J. (1957). The two disciplines of scientific psychology. *American Psychologist, 12*, 671–684.

Csikszentmihalyi, M. (1990). *Flow: The psychology of optimal experience.* New York: Harper & Row.

Daedalus. (1994). Health and wealth [Special issue]. *Journal of the American Academy of Arts and Sciences, 123*, 1–216.

Derogatis, L., Lipman, R., Rickels, K., Uhlenroth, E., & Cole, L. (1974). The Hopkins Symptom Checklist: A self report inventory. *Behavioral Science, 19*, 1–15.

DuBois, D. L., Felner, R. D., Brand, S., Adam, A. M., & Evans, E. G. (1992). A prospective study of life stress, social support and adaptation in early adolescence. *Child Development, 63*, 542–557.

Eccles, J., & Midgley, C. (1989). Stage–environment fit: Developmentally appropriate classrooms for young adolescents. In C. Ames & R. E. Ames (Eds.), *Research on motivation in education: Vol. 3. Goals and cognitions* (pp. 139–186). San Diego: Academic Press.

Farne, M., Sebellico, A., Gnugnolli, D., & Corallo, A. (1992). Personality variables as moderators between hassles and subjective indications of distress. *Stress Medicine, 8*, 161–165.

Feyerabend, P. (1975). *Against method.* London: New Left Books.

Glymour, C., Scheines, R., Spirtes, P., & Kelly, K. (1987). *Discovering causal structure.* Orlando, FL: Academic Press.

Gould, S. J. (1986). Evolution and the triumph of homology, or why history matters. *American Scientist, 74*, 60–69.

Green, B. F. (1992). Exposé or smear? The Burt affair. *Psychological Science, 3*, 328–331.

Harter, S. (1985). *Manual for the Self-Perception Profile for Children.* Denver: University of Denver.

Harter, S. (1990). Self and identity development. In S. S. Feldman & G. R. Elliott (Eds.), *At the threshold: The developing adolescent* (pp. 352–387). Cambridge, MA: Harvard University Press.

Holmes, T. H., & Rahe, R. H. (1967). The social readjustment rating scale. *Journal of Psychosomatic Research, 11*, 213–218.

Keating, D. P. (1980). Thinking processes in adolescence. In J. Adelson (Ed.), *Handbook of adolescent psychology* (pp. 211–246. New York: Wiley.

Keating, D. P. (1990a). Adolescent thinking. In S. Feldman & G. Elliott (Eds.), *At the threshold: The developing adolescent* (pp. 54–89). Cambridge, MA: Harvard University Press.

Keating, D. P. (1990b). Charting pathways to the development of expertise. *Educational Psychologist, 25,* 243–267.

Keating, D. P. (1991). Curriculum options for the developmentally advanced: A developmental alternative for gifted education. *Education Exceptionality Canada, 1,* 53–83.

Keating, D. P. (1993). Developmental diversity in mathematical and scientific competence. In L. Penner, G. M. Batsche, H. M. Knoff, & D. L. Nelson (Eds.), *The challenge in mathematics and science education: Psychology's response* (pp. 315–339). Washington, DC: American Psychological Association.

Keating, D. P. (1995a). The learning society in the information age. In S. A. Rosell (Ed.), *Changing maps: Governing in a world of rapid change* (pp. 205–229). Ottawa: Carleton University Press.

Keating, D. P. (1995b). The transformation of schooling: Dealing with developmental diversity. In J. Lupart, A. McKeough, & C. Yewchuck (Eds.), *Schools in transition: Rethinking regular and special education* (pp. 119–139). Toronto: Nelson.

Keating, D. P. (1996a). Habits of mind: Developmental diversity in competence and coping. In D. K. Detterman (Ed.), *Current topics in human intelligence: Vol. 5. The environment* (pp. 31–44). Norwood, NJ: Ablex.

Keating, D. P. (1996b). Habits of mind for a learning society: Educating for human development. In D. R. Olson & N. Torrance (Eds.), *Handbook of education and human development: New models of learning, teaching and schooling* (pp. 461–481). Oxford: Blackwell.

Keating, D. P., & Crane, L. L. (1990). Domain-general and domain-specific processes in proportional reasoning. *Merrill-Palmer Quarterly, 36,* 411–424.

Keating, D. P., Menna, R., & Matthews, S. (1992). *Adolescent cognitive development in everyday life.* Paper presented at the biennial meeting of the Society for Research on Adolescence, Washington, DC.

Keating, D. P., & Rosen, H. (Eds.). (1990). *Constructivist perspectives on developmental psychopathology and atypical development.* Hillsdale, NJ: Erlbaum.

Keating, D. P., & Sasse, D. K. (1996). Cognitive socialization in adolescence: Critical period for a critical habit of mind. In G. Adams, R. Montemayor, & T. Gullotta (Eds.), *Psychosocial development during adolescence* (pp. 232–258). Thousand Oaks, CA: Sage.

Kline, M. (1980). *Mathematics: The loss of certainty.* New York: Oxford University Press.

MacLean, D. J., Keating, D. P., Miller, F. K., & Shuart, V. E. (1994, February). *Adolescents' decisions to pursue math and science: Social and psychological factors.* Paper presented at the biennial meeting of the Society for Research on Adolescence, San Diego, CA.

MacLean, D. J., Sasse, D. K., Keating, D. P., Stewart, B. E., & Miller, F. K. (1995, April). *All girls' mathematics and science instruction in early adolescence: Longitudinal effects.* Paper presented at the annual meeting of the American Educational Research Association, San Francisco.

Matthews, D .J., & Keating, D. P. (1995). Domain specificity and habits of mind: An investigation of patterns of high-level development. *Journal of Early Adolescence, 15,* 319–343.

Menna, R., & Keating, D. P. (1992, June). *Coping with emotions in adolescence and early adulthood: Developmental changes.* Paper presented at the annual meeting of the Canadian Psychological Association, Quebec City.

Menna, R., & Keating, D. P. (1993, August). *Coping with worries in adolescence and early adulthood: Developmental changes.* Paper presented at the biennial meeting of the American Psychological Association, Toronto.

Menna, R., Keating, D. P., & Oatley, K. (1994, February). *Cognitive and emotional factors affecting stress reactivity of adolescents and young adults.* Paper presented at the meeting of the Society for Research on Adolescence, San Diego, CA.

Miller, F. (1995). *Attention regulation, emotion regulation, and cognitive competence from infancy to school entry.* Unpublished master's thesis, Brock University, St. Catharines, Ontario.

Miller, F. K., Keating, D. P., MacLean, D. J., Marshment, R. P., & Keenan, T. R. (1996, August). *The prediction of preschool cognitive competence from infant attentional, emotional, and social regulation.* Poster presentation at the biennial meeting of the International Society for the Study of Behavioural Development, Quebec City.

Morrison, F. M., Lord, C. J., & Keating, D. P. (Eds.). (1984–1989). *Applied developmental psychology* (Vols. 1–3). New York: Academic Press.

Nock, S. L., & Rossi, P. H. (1979). Household types and social standing. *Social Forces, 57,* 1325–1345.

Pearlin, L., & Schooler, C. (1978). The structure of coping. *Journal of Health and Social Behavior, 19,* 1–21.

Platt, J., & Spivack, G. (1977). *Measures of interpersonal cognitive problem-solving for adults and adolescents.* Unpublished manuscript. (Available from Community Mental Health and Mental Retardation Center, Department of Mental Health Sciences, Hahnemann Medical College and Hospital, Philadelphia, PA 19102)

Procidano, M. E., & Heller, K. (1983). Measures of perceived social support from friends and from family: Three validation studies. *American Journal of Community Psychology, 11,* 1–24.

Rosenberg, M. (1965). *Society and the adolescent self-image.* Princeton, NJ: Princeton University Press.

Sarason, I., Johnson, J., & Siegel, J. M. (1978). Assessing the impact of life changes: Development of the Life Experiences Survey. *Journal of Consulting and Clinical Psychology, 46,* 932–946.

Seligman, M. E. P. (1990). *Learned optimism.* New York: Pocket Books.

Sroufe, L. A., & Rutter, M. (1984). The domain of developmental psychopathology. *Child Development, 55,* 17–29.

Thelen, E., & Ulrich, B. D. (1991). Hidden skills. *Monographs of the Society for Research in Child Development, 56,* (1, Serial No. 223).

Von Eye, A. (Ed.) (1990). *Statistical methods in longitudinal research* (Vols. 1–2). San Diego, CA: Academic Press.

Werner, E., & Smith, R. (1992). *Overcoming the odds: High risk children from birth to adulthood.* New York: Cornell University Press.

Windle, M. (1992). A longitudinal study of stress buffering for adolescent problem behaviors. *Developmental Psychology, 28,* 522–530.

Author Index

Subject Index